MINERALOGY

MINERALOGY

Editor

Wilhelmina Faerber

scitus
academics

Mineralogy

Edited by **Wilhelmina Faerber**

This Book is Printed on Acid-Free Paper. ♾

ISBN: 9781681179001

Library of Congress Control Number: 2017948928

SCITUS Academics LLC,

616, Corporate Way, Suite 2, 4766,

Valley Cottage, NY 10989

www.scitusacademics.com

Printed in United States of America.

Preface

The science of mineralogy is concerned with studying minerals and their physical and chemical properties. The learning of mineralogy not only upsurges our knowledge about Earth's history, it also delivers important evidence about Earth materials which are economically noteworthy and so are important to any contemporary industrial society. Although the importance of geological mineralogy has been emphasized for some time, it appears from the present texts that the descriptive approach is still being followed in the teaching of mineralogy. Within mineralogy there are also those who study how minerals are formed, where they are geographically located, as well as their potential uses. Like many sciences, mineralogy has its origins in several ancient civilizations, and it has been concerned primarily with the various methods of classification of minerals for most of its history. During the last 200 years, the latitude of mineralogy has transformed considerably from a simply descriptive discipline to a highly ground-breaking field of science by incorporating ideas, techniques, and aspects of many contiguous areas such as physics, chemistry, material science, biology, archeology, and medicine. Modern mineralogy pursues to understand the links among mineral structure, properties, and processes in a variety of potentially technologically useful solids. The constant requirement of human societies for new materials--ranging from building materials to pharmaceuticals--acted as a driving force for the diversity of mineralogical research and finally led to the innovations of the mineralogical sciences permeating many aspects of our daily life. The physical properties of a mineral are determined by its chemical composition and its crystalline structure, within the limits of the permissible variation in chemical composition. Modern-day mineralogy has been expanded by advances in other sciences, such as biology and chemistry, to shed even more light on the nature of the materials that form the earth we live on.

This volume 'Mineralogy' describes the basic principles of mineral chemistry and structure. The book chapters have been classified according to their characteristics in topics and applications. Therefore, it is dedicated to the characterization and utilization of minerals in deposits. This book will be of valuable for a wide variety of readers, whether they wish to use it in their work, or simply to extend their knowledge. The book is envisioned to chemists, physicists, engineers, and the students of geology, geophysics, and soil science, but it will also be helpful to the more advanced students of mineralogy who are looking for a future research.

Table of Contents

CHAPTER 1

Mineralogy of Iron Ores of Different Alumina Levels from Singhbhum Belt and Their Implication on Sintering Process

Moni Sinha[1], Sri Harsha Nistala[1], Sanjay Chandra[1], Tilak Raj Mankhand[2]

[1]*Research and Development, Tata Steel, Jamshedpur, India*

[2]*Department of Metallurgical Engineering, Indian Institute of Technology (Banaras Hindu University), Varanasi, India*

ABSTRACT

To increase the life of the mines, ores of lower quality are blended with the higher quality ores. In general, the blending of ores at mines site is controlled by the chemical composition of the ores by controlling the Al_2O_3 and SiO_2 levels on the lowest side possible. However, if there is a change in mineralogy, like an increase in hydrous phase the composition may still be the same, but this difference may affect the sinterability and sinter quality. The mineralogical investigation of iron ores and their chemistry are required to avoid the increase in deleterious elements during sintering. By the know-how of the behavior of individual iron ores in terms of their sinterability, in advance, the quality of the iron ores blend can be optimized to produce quality sinter and productivity. Characterisation studies conducted on iron ores of Singhbhum Orissa belt revealed that hematite, goethite, clay, gibbsite and quartz occur in all the iron ore samples but in variable proportions. Traces of alumina, present as solid solution in the iron oxide minerals, has also contributed Al_2O_3 to the ores.

Keywords: Singhbum, Iron Ore, Sintering, Alumina

1. INTRODUCTION

Iron is found in many mineralogical forms, but only a few are used as commercial sources, the most important of which are oxides (Hematite Fe_2O_3, Magnetite Fe_3O_4, and Limonite $Fe_2O_3 \cdot 2H_2O$ & Goethite $Fe_2O_3 \cdot H_2O$). Other less important sources are Carbonates ($FeCO_3$), Sulphides (Pyrite FeS_2) and Silicates like Chamosite, Grunerite etc. with complex chemical formulae. The main impurities in iron ore are SiO_2, Al_2O_3, S, P, TiO_2, and Na_2O & K_2O. Each steel company has its own blast furnace practice and the impact of these impurities varies accordingly. Iron ore is being used for iron and steel production. It is being used in two forms, lump and fines. The lump is used directly in the blast furnace or for direct reduction in alternative ironmaking processes. The fines are used as agglomerates (sinter/pellet) after some more preparation.

The performance of a sinter depends both on the characteristics of the raw materials and the operational parameters. Among raw materials due to its major quantity, the iron ore plays the most significant role in sintering and hence the iron ore should have better metallurgical properties like high reducibility >70%, low degradation index, high softening and melt-down characteristics. Sintering process is extremely complex due to very short time. The material has complex melting behavior due to different constituents. The physico-chemical nature of raw materials governs the reactions of the mix. The degree of complexity further increases with increasing number of ore fines in the mixture. Quite different behaviors of different iron ore fines are observed at higher temperatures. The complexity of the reaction poses difficulties to optimize the process and hence the quality of the sinter. In the sintering process, the iron ore corresponds to the mixture of hematite, goethite and other gangue mineral typologies. However the goethite and gangue contents are relatively lower than the hematite, but the goethite + gangue to hematite ratio influences the reactivity during sintering. The minerals association in iron ores and their chemistries affect the formation and composition of sinter phases during sintering. Goethite is important as it carries deleterious impurities for steel making. The effect of texture is equally important as it affects the reactivity and/or assimilation of iron ore minerals and subsequently the formation of pores during sintering.

Thus increasing the knowledge on the characteristics of different ore types and its influence on sinter quality will also enable controlling the physicochemical properties of primary sinter melts and will be beneficial to control sintering process to produce quality sinters.

2. INFLUENCE OF IRON ORE MINERALOGY ON SINTER STRENGTH

The iron bearing minerals have different characteristics. The metallurgical properties of an ore depend on its specific mineralogy and textural characteristics, including the nature of the pore spaces. The iron-bearing minerals themselves contain impurities and the gangue minerals contain some iron phases. These may either be chemically bonded or so finely intergrown that they are difficult to remove by physical beneficiation. The nature of the gangue, particularly its amount and distribution of the alumina-bearing minerals is important apart from their fusibility and sinterability [1] [2].

2.1. Mineralogy of Iron Ore

The mineralogy of iron ore varies from quite simple hematite and quartzite with little of aluminium mineral to hydrated hematite along with complex gangue minerals like hydrated aluminium silicate or kaolinite $Al_4(SiO_{10})(OH_8)$ and trihydratedaluminium oxide or gibbsite $Al(OH)_3$. The hematite can be partially hydrated (goethite, FeOOH) or deoxidized (magnetite Fe_3O_4). In some ore bodies the quartzite is the predominant gangue mineral and significantly low alumina bearing mineral. Some ore bodies, on the other hand, may contain mainly magnetite and only a very little hematite. The gangue may vary from simple to complex type (Figure 1) as mentioned above. The extent of hematite or magnetite and total iron content and any of the impurities varies from source to source. All these types may have different behavior during sintering.

2.2. Effect of Goethite

Goethite in general comprises a low amount in the ores; however, it can be relevant in the reactivity during sintering. Goethite often carries aluminium in solid solution as well as minor, but significant, amounts of adsorbed phosphorus [3]. It has a variable microtexture and internal porosity.

Goethitic ores can vary greatly in hardness ranging from finely crystalline, soft and clayey (limonitic) to hard and vitreous in appearance. These variations results in an associated variation in metallurgical behaviour of the ores.

Although goethite is completely consumed during the sintering processes, its control in the sintering burden is very important due to the impurities present in its composition, which can affect in different ways the distribution of elements in the sinter phases [4] . The compositional aspects of sinters, which are inherited from the ore minerals, are partially related to the reactivity behavior, which can be completely distinct if the hematite/goethite ratio varies largely. The variation of this ratio can indirectly affect the performance of sinters like mechanical strength and reducibility [3]. Goethitic ores reduces the heat available during the meling period and results in the formation of more viscous high Fe_2O_3 melts, thus more coke is required to maintain the level of melt superheat and thus prevent deterioration in sinter strength [1]. Due to presence of goethite there is deterioration in granulation efficiency, for the same bed permeability, the bed containing goethitic ore have a faster flame front speed than the bed with denser ores. Goethitic ores assist in the production of more porous sinters, which are more reducible as these ores/phases are porous nature and have high combined water [1].

2.3. Effect of Al Bearing Minerals: Gibbsite and Kaolinite

It is generally believed that too much alumina impairs the sinter quality, particularly the low temperature reduction degradation characteristics [5] [6]. For every 1% increase in alumina coke breeze consumption increase by 5 - 10 kg/t for an increase in temp by 30°C. High alumina percentage in ores results in high sintering temperature, longer sintering time, increase in fuel rate and lower productivity. These can be attributed to poor reactivity of the phase and high viscosity of the primary melt [2].

Iron ores can have two major types of alumina bearing phases, alumino silicates like kaolinite, while the other phase is gibbsite (alumina hydroxide). However, it was observed that kaolinite containing ore generates melilite as a slag forming mineral whereas gibbsite type ores forms calcium ferrite [2]. These in turn have different effect on the cold strength of the sinter [7].

2.4. Effect of Ore Texture

Ore micro-pelletising efficiency and reactivity in sintering and melting reactions are controlled, to a first approximation by crystal size. These properties are enhanced as the crystal size decreases. Crystal morphology is also important because of its influence on particle surface roughness. Low ore porosity and the presence of liberated and fine quartz particles, are beneficial to sinter physical and mechanical properties.

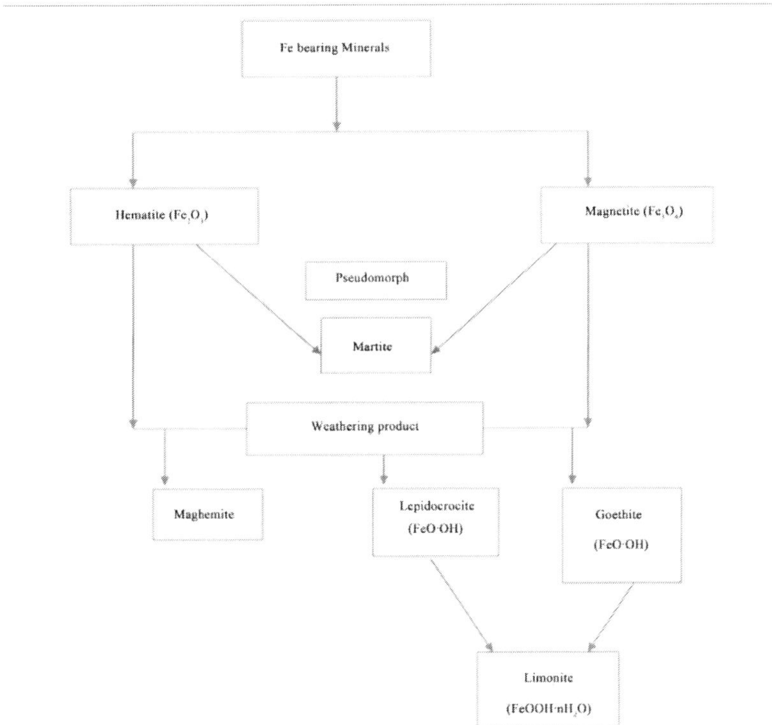

Figure 1. Flow chart showing the relation between different iron minerals.

As the ore contains less alumina less of it will go to sinter's molten part to form multi-component hematite or SFCA (Silico Ferrite of Calcium and Alumina) and the RDI (Reduction Degradation Index) will be remarkably improved. The smaller crystal size and higher open porosity of ores benefits the reducibility of the sinters produced with low energy input as these crystals remain in the relict or unmoltenore particles found in the sinter. To summarise Iron ores with fine crystal size coupled with low degree of hydration favours high productivity and low coke rate in sintering. Iron ores with medium crystal size and very low degree of hydration favours sinter tumbling strength and RDI. In another study carried out by Okazaki et al. [8] it was observed that in Indian ore there were originally many pores in a part of goethite, which coalesced during heating; furthermore porosity in martite grains seemed to reduce. Surface of the Indian Ore turned granular during sintering upto 1200°C. The change implied that martite had re-crystallised, or the goethite decomposed to grow such granules. In an assimilation of this ore with limestone powder SFCA developed well between relict ores different mineral phases have variations in crystal size, shape and orientation that reflect differences in their mode of formation (tectonism, degree of metamorphism, supergene changes, etc.). Significantly, they express varying degrees of microporosity and grain density, which are relevant for determining hardness, friability and metallurgical properties such as reducibility and reactivity with fluxes. There is, again, an associated variation in metallurgical behaviour [9]. Some effect of ore texture on sinter microstructure is shown in Figure 2 below.

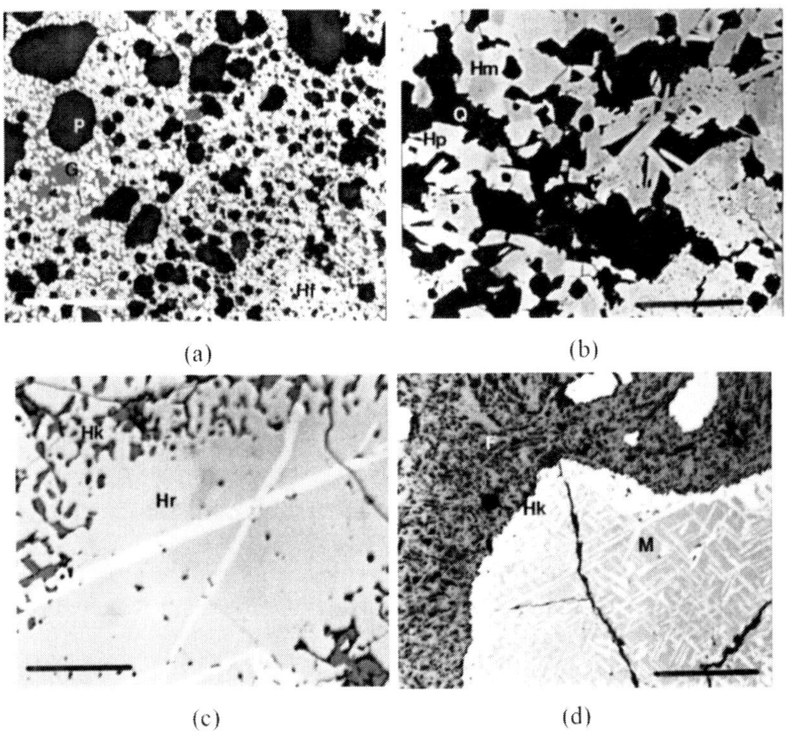

Figure 2. Reflected-light photomicrographs illustrating ore textures when encapsulated in sinter: (a) Finely granular hematite set in a glassy matrix from a type 4 (fluviatile) ore; (b) Relict martitic and microplaty hematite from a siliceous type 4/2 (medium-grade hematite) ore; (c) Coarsely crystalline hematite from concentrates made from a type 2 (BIF) and showing the typical reduction/oxidation reaction marginal to the sinter

melt; (d) Partly oxidized magnetite from a type 7 ore showing the reaction and oxidation rim against the calcium ferrite-rich sinter melt. F, SFCA; G, glass; H, hematite; f, finely crystalline; k, recrystallised; m, martitic; p, microplaty; r, relict; M, magnetite; P, pore space; Q, quartz. Length of scale bar, 0.1 mm.

3. GEOLOGY OF THE IRON ORE DEPOSITS

The Singhbhum Orissa Iron Ore Craton (SOIOC) contains a large number of iron ore deposits associated with the Iron Ore Group (IOG) (Figure 3). On the basis of tectonic analysis, the regional structure is a low NNE plunging synclinorium overturned towards SE. The iron ores found in these areas are a Banded Iron Formation (BIF; mainly present as BHJ, banded hematite jasper) hosted deposits. These ores are considered to be of secondary in nature resulting from the supergene enrichment of the host rocks and by gradual removal of silica from BIFs and locally from the ferruginous shales of the Iron ore group.

Mineralogically the iron ore deposits consist mainly of hematite together with variable amount of iron hydroxides (goethite, lepidocrocite, limonite). Magnetite occurs in small amount in some of the deposits, particularly in the deposits of the eastern basin; martite grains (pseudomorph of magnetite) are fairly common. Specularite is noted in some ores. The gangue minerals include quartz, chert, gibbsite and clay minerals (chiefly kaolinite).

4. METHODOLOGY

Characteristaion studies of three different types of iron ore from Joda mines were carried out. Each of the ores is classified as hard (HO), friable (FO) or soft ore (SO). The ores are further subdivided based on their alumina levels (2%, 4%, and 6%). The samples were crushed to get a size suitable for sintering (*i.e.*, −10 mm). The crushed samples were characterised size wise for their chemical analysis. XRD, EPMA and Image analysis were done for characterisation of gangue and ore minerals. The aim of the mineralogical investigation is to describe the mineral associations and relationships to ascertain the character of the Al-bearing minerals present in the samples. The chemical analysis was done by ICP AES instrument.

The samples of all the three iron ores and their size fractions mentioned earlier, were pulverised and subjected to qualitative X-ray diffraction (XRD) analysis, using Cu $K\alpha$ radiation, a 2θ scan range of 10° - 90, a step size of

0.02° 2θ, and a counting time of 1 s per step. Only crystalline phases in amounts sufficient to diffract (usually ≥3 - 4 mass%) under the conditions employed, were detected. The −1 mm fraction was subjected to electron microprobe examination.

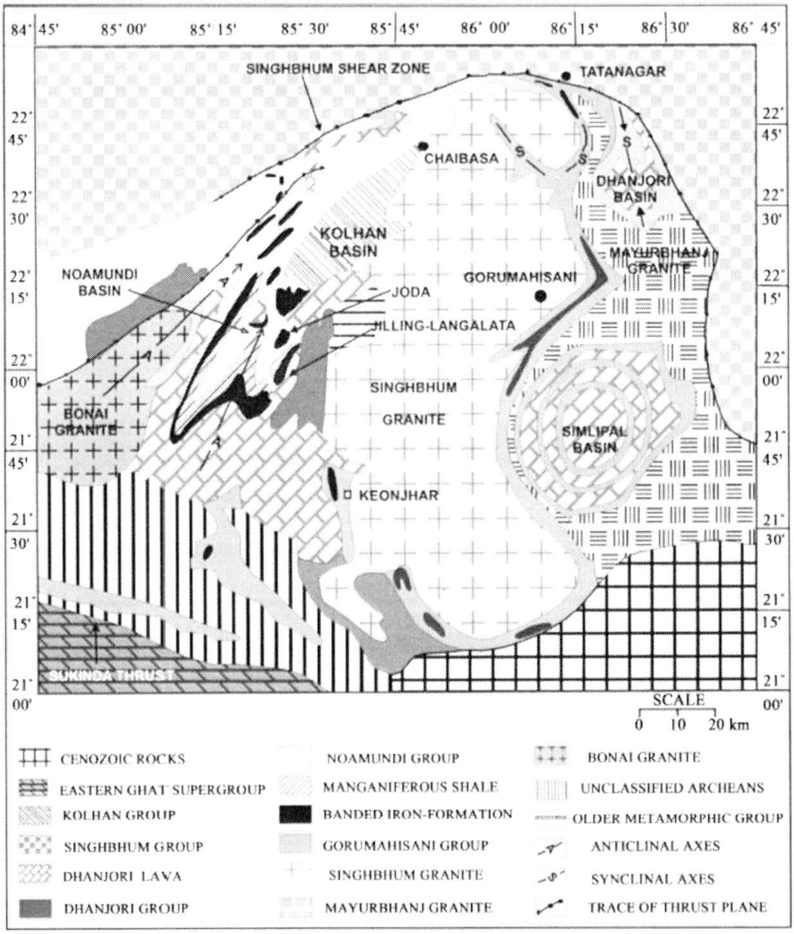

Figure 3. Generalised geological map of Singhbhum Orissa iron ore craton.

For EPMA and Image analysis same samples were used. EPMA was done by JEOL JXA-8230. For the optical microscopy each size fraction was crushed to −3 + 1 mm and was mounted and polished. Carl Zeiss Axiovision 4 reflected light microscope integrated with an image analyser was used for the optical phase identification and quantification.

5. RESULTS AND DISCUSSION

Characterisation studies of nine different types of iron ore from Joda mines were carried out. The size wise analysis of the ores (HO, FO, SO) are as presented in Table 1.

5.1. Microscopic and EPMA Studies

The mineral chemistry and association of different iron and gangue bearing phases was observed to be similar for different size fraction for individual type of ore. The amount of phases present and their elemental composition in different ore types are as given below in Table 2 and Table 3 respectively. It can be observed from Table 2 that as the alumina % of ore types increases the amount of gangue minerals especially that of clay increases. The Hematite percentage decreases correspondingly.

The Table 3 gives a picture of the elemental composition of phases present in different types of ores. It is evident that the gangue content (Alumina and Silica) increases with increase in weathering intensity of ores (HO → FO → SO). Hematite is also present as re-precipitated hematite or secondary hematite and contains alumina as an impurity. This suggests that this type of hematite is re-precipitated from goethite. Goethite contains alumina and silica as a major impurity. Clay is ferruginous in nature.

A detailed description of individual types of ore texture is discussed below.

5.1.1. Hard Ore

Microscopic studies of hard ores indicate massive (Figure 4(a)) and banded structures of hematite (Figure 4(b)). Specularite and fine grained hematite grains are present as aggregates. Goethite exhibits a replacement texture with hematite (Figure 4(c)).

The EPMA studies show presence of alumina alongwith the iron minerals in voids (Figure 5, point 1 & 3). The alumina is also present in goethite as layered texture (Figure 5, point 4). The hematite grains shows presence of insignificant amount of alumina may be present in adsorbed state (Figure 5, point 2 & 5).

5.1.2. Flaky Ores

Ore microscopic studies of flaky-ores indicate that the clay minerals generally occur in the form of layering along with the hematite grains, as inter-granular cavity fillings and as veins (Figure 6(a)). Clay minerals are present as intercavity fillers (Figure 6(b)). Goethite is also found to be present in this ore type.

This type of ore shows an extensive amount of replacement as it has undergone higher degree of weathering. The goethite occurs as layered structure with alumina and clay minerals as alternate band (Figure 7). However, the hematite grains shows presence of insignificant amount of alumina in adsorbed state similar to the HO type ore as mentioned earlier.

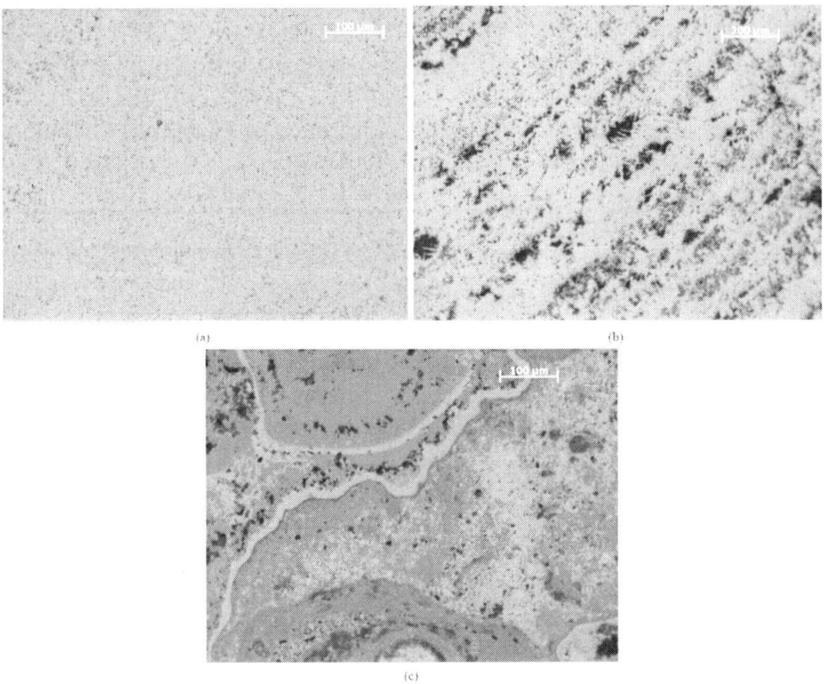

Figure 4. (a) Massive ore; (b) Banded structure; (c) Goethite.

5.1.3. Soft Ore

As the name suggests this ore is very soft in nature and has undergone cycles of weathering. This resembles the hard laminated ore in texture/appearance. This ore also seems to be hydrated and has undergone weathering, as a result goethiticminerals is in abundance. The replacement texture is visible. The ore mineral has been leached out, and the voids thus created is filled in by the clay and goethitic minerals (Figure 8(a) and Figure 8(b)).

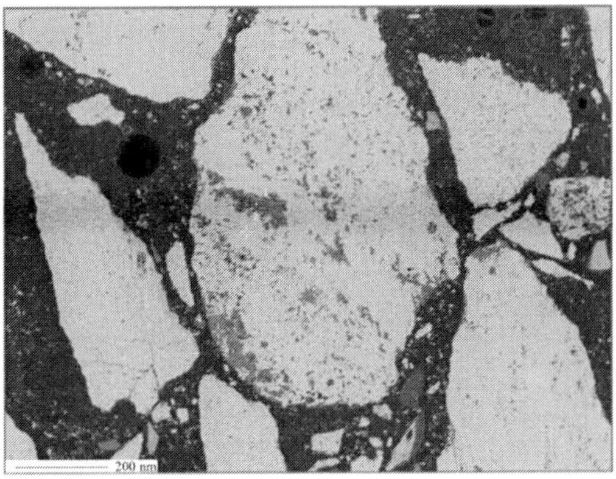

Point	Element mass%		
HO	Al_2O_3	SiO_2	FeO
1	33.96	1.09	64.95
2	0.01	0.58	99.41
3	31.30	1.21	67.49
4	7.28	0.85	91.87
5	0.12	0.00	99.88

Figure 5. EPMA analysis of HO type ore. Figure on top is a BSE image from EPMA.

Apart from hematite, goethite dominates the soft ore mineralogy. It is mainly present as replacement texture and occurs in bands like previous ores (Figure 9). The alumina content in goethite and other gangue minerals are significantly higher than the other ore types.

5.2. XRD Analysis

The XRD pattern of the iron ores indicates presence of hematite and goethite, while clay minerals and gibbsite are not traceable. It can be observed clearly from the pattern that as the alumina level as well as weathering intensity increases (HO → FO → SO), the goethite peak intensity is higher (Figure 10). The peak shape however indicates a poorly crystalline goethite. This indicates the presence of alumina in goethite. This is confirmed by the EPMA studies of the iron ores (Figure 5, Figure 7 and Figure 9).

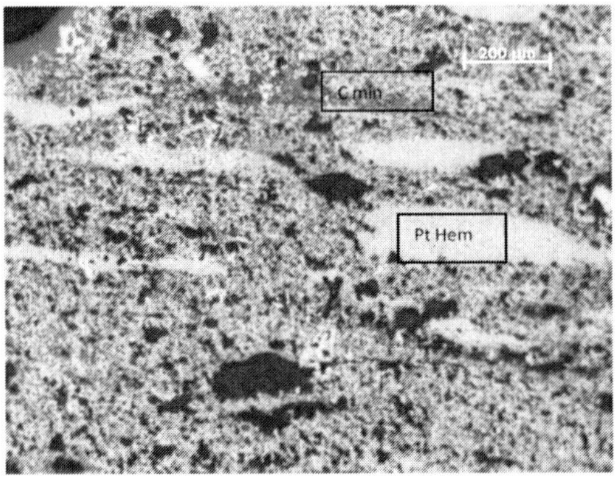

(a)

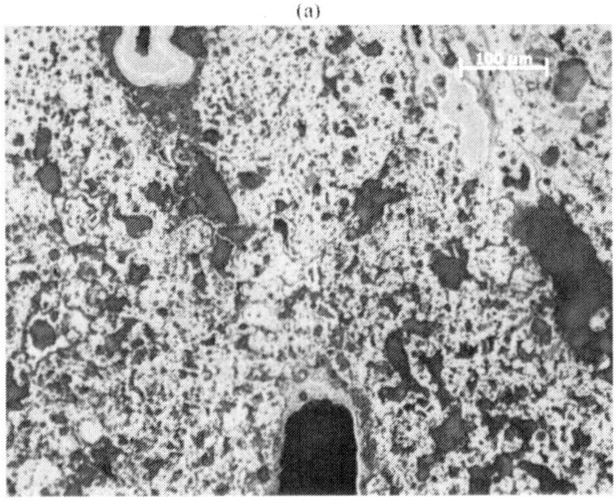

(b)

Figure 6. (a) Microplaty hematite (Pt hem) and Clay minerals (C min) in FO type ore; (b) Goethite with clay minerals as intercavity fillers.

Table 1. Size wise chemical analysis of ores.

Sample ID	Fe(T)	SiO_2	Al_2O_3	LOI	Sample ID	Fe (T)	SiO_2	Al_2O_3	LOI
HO 2% BULK	66.92	0.32	1.16	1.89	**FO 2% BULK**	**65.65**	**2.18**	**1.63**	**1.59**
HO 2% (+8 mm)	67.59	0.59	0.56	2.00	FO 2% (+8 mm)	67.4	0.24	0.9	1.68
HO 2% (+6 mm)	67.47	0.86	0.90	1.00	FO 2% (+6 mm)	67.36	0.17	1.02	1.75
HO 2% (+3 mm)	67.30	0.76	1.15	1.09	FO 2% (+3 mm)	67.58	0.28	0.84	1.50
HO 2% (+1 mm)	65.92	1.02	1.40	1.81	FO 2% (+1 mm)	65.16	1.96	1.36	2.16
HO 2% (−1 mm)	64.28	1.50	2.28	2.50	FO 2% (−1 mm)	63.08	3.69	2.1	2.51
HO 4% BULK	62.00	2.35	3.19	4.32	**FO 4% BULK**	**61.04**	**4.43**	**4**	**3.80**
HO 4% (+8 mm)	61.86	2.13	3.43	5.06	FO 4% (+8 mm)	64.85	2.27	2.03	2.68
HO 4% (+6 mm)	64.90	1.59	2.19	2.84	FO 4% (+6 mm)	64.82	2.35	2.11	2.60
HO 4% (+3 mm)	63.28	1.99	2.92	3.92	FO 4% (+3 mm)	64	2.9	2.44	2.92
HO 4% (+1 mm)	54.92	3.96	5.36	7.98	FO 4% (+1 mm)	61.22	3.7	3.9	3.73

HO 4% (−1 mm)	65.6	1.39	2.04	2.30	FO 4% (−1 mm)	56.2	7	6.42	5.42
HO 6% BULK	**53.65**	**4.61**	**7.78**	**9.70**	**FO 6% BULK**	57.6	6.08	5.65	5.10
HO 6% (+8 mm)	60.58	2.68	3.98	5.96	FO 6% (+8 mm)	60.04	4.86	4.6	4.10
HO 6% (+6 mm)	55.28	3.87	5.73	10.21	FO 6% (+6 mm)	59.61	5.15	5.05	4.20
HO 6% (+3 mm)	58.64	3.08	4.72	7.66	FO 6% (+3 mm)	59.45	4.79	4.86	4.77
HO 6% (+1 mm)	56.34	3.52	5.52	9.40	FO 6% (+1 mm)	57.35	6.08	5.85	5.52
HO 6% (−1 mm)	55.41	3.76	5.68	10.22	FO 6% (−1 mm)	51.2	10.18	8.88	7.05
Sample ID	**Fe (T)**	**SiO_2**	**Al_2O_3**	**LOI**					
SO 2% BULK	66.34	1.01	1.24	1.73					
SO 2% (+8 mm)	65.71	1.91	1.21	1.31					
SO 2% (+6 mm)	66.39	1.17	1.02	1.22					
SO 2% (+3 mm)	64.33	0.81	0.86	4.86					
SO 2% (+1 mm)	66.23	1.20	1.14	1.27					
SO 2% (−1 mm)	64.82	1.83	2.05	2.09					
SO 4% BULK	60.67	2.34	3.82	6.14					
SO 4% (+8 mm)	64.64	1.84	2.05	2.92					
SO 4% (+6 mm)	63.55	1.81	2.92	3.94					
SO 4% (+3 mm)	63.23	1.66	2.94	4.48					
SO 4% (+1 mm)	61.72	1.8	3.33	5.93					
SO 4% (−1 mm)	57.64	3.77	4.8	7.62					
SO 6% BULK	56.33	4.13	6.04	8.12					
SO 6% (+8 mm)	60.78	2.47	4.82	4.98					
SO 6% (+6 mm)	59.59	2.61	5.73	5.78					
SO 6% (+3 mm)	56.76	3.12	7.13	7.68					
SO 6% (+1 mm)	55.60	3.52	8.09	7.95					
SO 6% (−1 mm)	53.88	4.21	8.66	9.04					

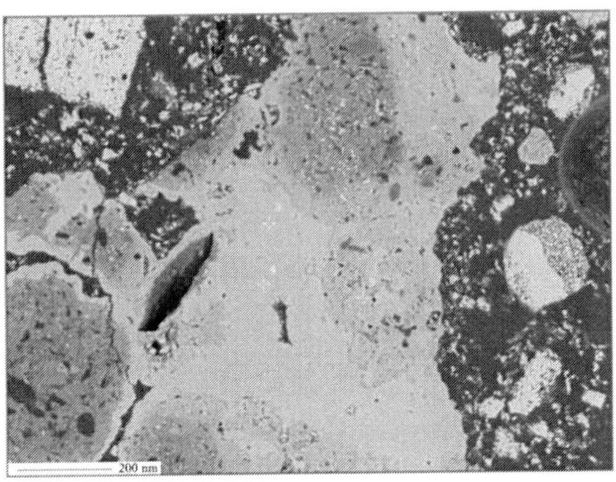

Point	Element mass %		
FO	Al_2O_3	SiO_2	FeO
1	16.36	0.41	83.23
2	16.15	0.85	83.00
3	0.00	0.68	99.32
4	0.14	0.61	99.26
5	9.48	0.59	89.93
6	9.90	0.50	89.61

Figure 7. EPMA analysis of FO type ore. Figure on top is a BSE image from EPMA.

Table 2. Phase % in different ore types.

Sample ID	Hematite	Goethite	Gibbsite	Clay
HO 2%	91.92	3.13	2.19	2.76
HO 4%	84.37	3.19	4.01	8.43
HO 6%	68.98	5.17	4.41	21.44
FO 2%	89.10	3.17	3.08	4.66
FO 4%	81.10	4.12	3.78	11.01
FO 6%	74.06	5.55	3.20	17.19
SO 2%	90.03	3.20	2.34	4.43
SO 4%	80.60	4.10	3.61	11.69
SO 6%	74.42	5.43	3.42	16.72

Table 3. Elemental composition of different phases by EPMA.

Phases	Ore types	FeO	Al_2O_3	SiO_2
	HO	99.5	0.1	0.4
Hematite	FO	98.7	0.6	0.6
	SO	98.2	1.0	0.9
	HO	91.9	7.1	1.0
H (Secondary)	FO	92.2	6.9	0.9
	SO	94.6	4.2	1.2
	HO	89.1	8.1	2.8
Goethite	FO	84.6	11.1	4.3
	SO	91.4	6.8	1.8
	HO	33.4	32.7	33.9
Clay (Ferruginous)	FO	14.3	38.2	47.6
	SO	19.1	38.0	42.9
	HO	1.7	98.2	0.1
Gibbsite	FO	4.1	95.2	0.7
	SO	0.4	99.6	0.0

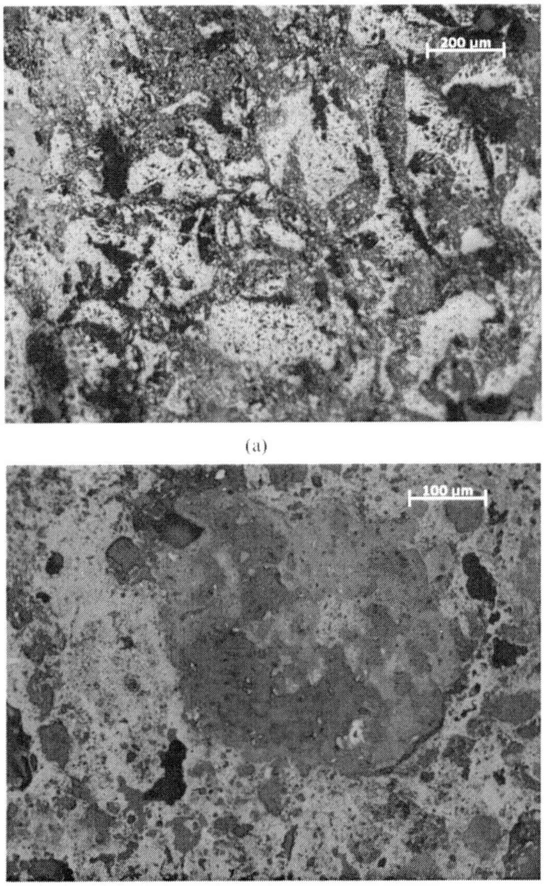

Figure 8. (a) Replacement texture; (b) Goethite.

6. SUMMARY

1) In general, hematite and goethite are major iron bearing minerals. The Al-bearing minerals in the assemblage are gibbsite, clay and goethite. Silica is mainly in the form of quartz and occur in all the iron ore samples. Occasionally accessory rutile and/or zircon were also observed.

2) Hematite is present as microplaty, specular or banded texture. EPMA studies show presence of alumina adsorbed on the surface of the hematite grains. The concentration is very low (~0.14 - 0.3) as detected by EPMA. The above is true for all the ore types.

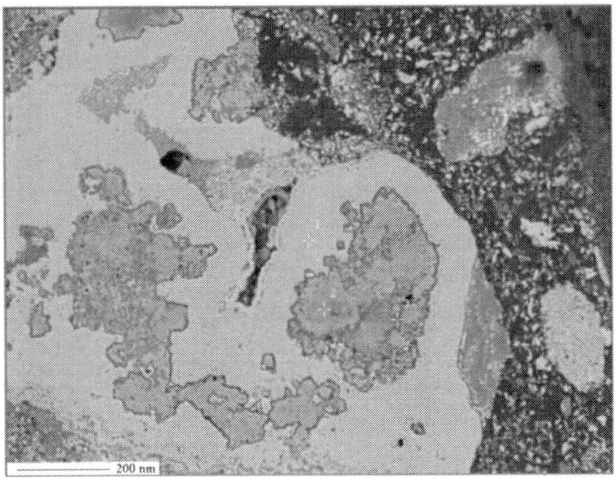

Point	Element mass %		
SO	Al_2O_3	SiO_2	FeO
1	0.49	0.77	98.75
2	13.82	0.53	85.65
3	9.82	2.00	88.17
4	19.04	12.27	68.68
5	13.95	5.52	80.53

Figure 9. EPMA analysis of SO type ore. Figure on top is a BSE image from EPMA.

3) Hematite is also present as re-precipitated hematite or secondary hematite and contains alumina as an impurity. This suggests that this type of hematite is re-precipitated from goethite.

4) As the alumina % of ore types increases the amount of gangue minerals especially that of clay increases. The Hematite percentage decreases correspondingly.

5) It is evident that the gangue content (Alumina and Silica) increases with increase in weathering intensity of ores (HO → FO → SO).

6) The ores also contains significant amount of secondary iron minerals (goethite) and clay. The goethite commonly found in this iron ores is colloform banded goethite and massive goethite. Goethite exhibits replacement relationship with hematite.

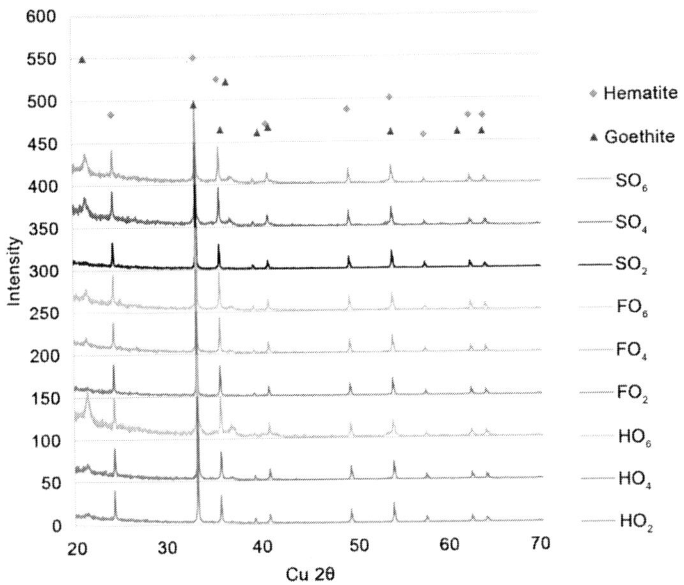

Figure 10. XRD pattern of bulk ore.

7) Hematite dominates the bulk samples, whereas clay is a significant component of the −1 mm size fraction.

8) The associated gangue minerals esp. clay is ferruginous in nature. The ultrafine coating particles consist of hematite, poorly crystalline goethite and clay minerals.

9) Alumina and clay minerals tend to get deposited in the cavities and thus are present as fillers in the cavity, cracks and pores of the mineral grains.

10) As expected, the higher the alumina percentage, higher the weight loss was observed in the mentioned ore types.

7. CONCLUSIONS

XRD studies revealed hematite as the major minerals with subordinate amounts of goethite, confirming the microscopic findings. Qualitative mapping and quantitative EPMA studies on these ores indicated the presence of gibbsite as the only alumina bearing phase. Clay is one of the major sources of alumina. Traces of alumina, present as solid solution in the hematite phase, have also contributed to the ores. Electron microscopic studies revealed that gibbsite grains are in the size range of 10 - 50 microns and are intimately and intricately present along with the iron oxide phases.

The studied ore types have a significant amount of goethite, and especially the 4% and 6% alumina level ores are very high in goethite and clay minerals, and thus the solid fuel rate requirement is expected to be higher than the other hematite ores. However the reducibility is expected to improve. Due to high gangue content, the amount of amorphous phases, as well, in sinter can be high.

The sinter feed is a blend of different types of iron ores. Thus, the information about the ratio of mixing of different ores can give an insight about the quality expected. Also, mineralogy of a given iron ore is one of the most relevant aspects and should be considered in preparation of sinter blends. By increasing the knowledge on the characteristics of different ore types, its influence on sinter properties will help define measures which can neutralize, dilute/localize the effects of alumina on the sinter quality.

REFERENCES

1. Loo, C.E. (2005) A Perspective of Geothitic Ore Sintering Fundamentals. ISIJ, 45, 436-448. http://dx.doi.org/10.2355/isijinternational.45.436
2. Lu, L., Holmes, R.J. and Manuel, J.R. (2007) Effects of Alumina on Sintering Performance of Hematite Iron Ores. ISIJ, 47, 349-358. http://dx.doi.org/10.2355/isijinternational.47.349
3. Pecchio, et al., Eds. (2004) Applied Mineralogy. ICAM-BR, S&ã Paulo.
4. Yang., L.X. (2005) Sintering Fundamentals of Magnetite Alone and Blended with Hematite and Hematite/Goethite Ores. ISIJ International, 45, 469.
5. Panigrahy, S.C., Rigaud, M.A.J. and Dilewijns, J. (1985) Effect of Dolomite Addition on the Properties of Sintered Ore Produced from the High Iron Ore of Alumina. Steel Research, 56, 35.
6. Pimenta, H.P. and Seshadri, V. (2002) Influence of Al2O3 and TiO2 Degradation Behaviour of Sinter and Hematite at Low Temperatures on Reduction. Ironmaking Steelmaking, 29, 175-179. http://dx.doi.org/10.1179/030192302225001992
7. Caparoli, L. (1998) ICSTI, Iron Ore Micro Structure Properties—Performance Relation in Sintering Process. Ironmaking Steelmaking Conference Proceedings (B), 3, 123-135.
8. Okazaki, et al. (2001) Advanced Approach to Intelligent Iron Making Processes. Effect of Mineralogical Properties of Iron Ore on Pore Formation of Sinter. Tetsu-to-Hagané, 87, 298-304.
9. Goldring, D.C. (2003) Iron Ore Categorisation for the Iron and Steel Industry. Applied Earth Science, 112, 5-17. http://dx.doi.org/10.1179/0371745032501162

CHAPTER 2

Mineralogy and Beneficiation of Lamujärvi Syenites

Xiaosheng Yang[1], Jukka Laukkanen[1], Akseli Torppa[2], Neea Heino[1]

[1]*Geological Survey of Finland, GTK Mintec, Outokumpu, Finland*

[2]*Geological Survey of Finland, Kuopio, Finland*

ABSTRACT

The Lamujärvi syenites intrusions occur within Svecofennian (ca. 1.92 Ga) gneisses, ca. 70 km west of the Otanmaki alkaline area in central Finland. Previous studies have revealed zones in the syenites that display strong enrichment in Zr (1587 ppm), Nb (up to 685 ppm), Ta (up to 82 ppm), and REE (up to 5350 ppm). The major REE-bearing minerals in the Lamujarvi syenites are allanite and monazite. A sample of the Lamujärvi syenites with the REE content of 4700 ppm was studied on the mineralogy and beneficiation at GTK Mintec. The mineralogical analysis was conducted by MLA. The REE-bearing minerals allanite, parisite, zircon, and apatite were identified at significant concentrations and high contents of titanite (15.4%), biotite (30.9%) and magnetite (8.2%) were displayed in the sample. The beneficiation experiments by flotation, wet low and high gradient magnetic separations and gravity separation with a shaking table were carried out. A composite REE-Ti concentrate with REE 23,000 ppm, Zr 5000 ppm, Nb 3400 ppm and TiO_2 25.5% at REE and Ti recoveries of 57% and 54% was obtained. In addition, biotite and magnetite concentrates were gained as by-products.

Keywords: The Lamujärvi Syenites, Mineralogy, Beneficiation, Allanite, Titanite

1. INTRODUCTION

No REE are currently mined in Europe, but potential resources are known to be widespread, and many are being explored [1] . Finland has high potential for exploration of REE and the known examples of REE mineralisation occur in carbonatitic and alkaline intrusives, granites, hydrothermal rocks and kaolinitic saprolites (Figure 1). There are currently no economic REE deposits in Finland, although REE were extracted between 1963 and 1972 from apatite concentrate mined as a by-product in the Korsnäs lead (Pb)-REE mine in western Finland [2]. REE could potentially be extracted in the future as a by-product of mining for commodities such as phosphorus (P) and gold (Au).

As shown in Figure 1, Finland is situated in the central part of the Fennoscandian shield, which is the largest outcropping Precambrian domain in Europe. The shield hosts a few sub-economic to economic REE deposits mostly with the intrusions of the Devonian Kola alkaline province in Russia [3] [4]. The Sokli carbonatite complex (total area 20 km^2) and the Iivaara ijolite complex in Finland are the westernmost intrusions of the Devonian Kola alkaline province. Other potentially economic REE deposits in Fennoscandia include the Norra Kärr in Sweden. Currently, the most promising REE deposit is associated with the Sokli carbonatite complex in north eastern Finland. The Sokli carbonatite (approximately 360 - 380 Ma) hosts an unexploited deeply weathered phosphate deposit enriched in niobium (Nb), tantalum (Ta), zirconium (Zr), REE and uranium (U). The carbonatite intrusion consists of a magmatic phoscorite-carbonatite core, surrounded by metacarbonatite and a wide fenite aureole, altogether about 9 km in diameter. Late-stage carbonatite veins in the magmatic core and in the fenite zone have high potential for REE mineralisation [5] [6].

The Lamujärvi plutons are located ca. 70 km west of the Otanmäki area, which hosts Nb-REE mineralisations in alkaline gneisses, but despite the spatial proximity, in light of the 200 Ma age difference suggested by recent U-Pb (zircon) dating of the Lamujärvi syenite at ca. 1.85 Ga and the Otanmäki granite at ca. 2.05 Ma, a genetic relationship between the two occurrences is improbable. The Lamujärvi syenites in central Finland display strong enrichment in Zr (1587 ppm), Nb (up to 685 ppm), Ta (up to 82 ppm) and REE (up to 5350 ppm). The major REE-bearing minerals in the enriched rocks are allanite and monazite [7]. Meanwhile, other valuable minerals such as titanite, biotite and magnetite are at high concentrations which could be economically recovered. In this paper, the mineralogy and the beneficiation of the Lamujärvi syenites are studied.

2. EXPERIMENTAL AND REAGENTS

A sample of the Lamujarvi syenites over 100 kg collected from Lamujärvi area was received at GTK Mintec in 2012. The sample was crushed by a jaw and a roll crusher to −1.5 mm and split into 1.5 kg amount of subsamples which were stored for beneficiation experiments. One of the subsamples was further split into 100 g amount of samples for chemical and mineralogical analyses.

Chemical analyses of the sample were conducted by XRF, ICP-OES and ICP-MS at Labtium Oy in Finland. A Mineral Liberation Analyzer (MLA) was used for mineralogical analysis at GTK Mintec which consists of the standard modern SEM (FEI Quanta 600) with the energy dispersive X-ray analyzer (EDAX Genesis with two detectors) and the software package developed originally by JKTech (Australia) [8] [9] . The modal mineralogy of the samples was analyzed by the XMOD-STD and XBSE-STD methods, which were point counting methods, and the grain size was analyzed by the XBSE method.

The samples of 1.5 kg with the size of −1.5 mm were ground at a laboratory rod mill as the head samples for beneficiation experiments. Flotation experiments were carried out using an Outotec laboratory flotation machine. Flotation cells with the volumes of 4 and 1.5 liters were used in rougher and cleaner flotation, respectively. Tap water was used for wet grinding and flotation. The room temperature (20°C - 23°C) was applied for flotation conditioning. A laboratory shaking table was used for gravity concentration at the conditions of inclination 3° - 4°, feed rate 0.3 kg/h, stroke 350/min, slurry density 25% and rate of rinse water 6 l/min. Wet low intensity magnetic separation (WLIMS) experiments were performed by using a Sala laboratory roller type magnetic separator with the magnetic field of 0.07 Tesla. A laboratory high gradient magnetic separator (Sala HGMS 10-15- 20 SCR) was used for the high gradient magnetic separation (HGMS) experiments. The matrix type of 3.5 XRO with the aperture size of 750 μm was used. The magnetic intensities of 1.5 and 2 Tesla were applied for roughing process and 0.2 Tesla for cleaning process.

To avoid negative effects of slimes on beneficiation desliming was conducted by a hydraulic sedimentation method to remove the fine fraction of −20 μm after material grinding. Acid leaching testwork by using hydrochloric acid (HCl) was performed for the treatment of slimes. The 37% HCl was used and diluted into required concentration for leaching. The volume of the leaching tank was 1000 ml, the ratio of solid to acid solution 75 g/500 ml and the agitation speed 180 rpm. The acid concentration was 6 mol/l and leaching temperature 60°C which was controlled by circulation of heated water. The leaching times tested were 60 min, 120 min and 180 min.

After leaching the residues were filtered, dried, weighted and assayed by XRF.

In biotite flotation, two cationic collectors tallow amine and Aero 3030C supplied by Cytec were used and in REE-Ti flotation hydroxamate type of reagent Aero6494 and sulfosuccinamate type of reagent Aero 845 from Cytec, FS-2 an anonic collector, and petroleum sulfonate were applied.

3. RESULTS AND DISCUSSION

3.1. Chemical Compositions

The REE contents determined by ICP-OES and ICP-MS are shown in Table 1. Total sixteen rare earth elements were found but seven of them including La, Ce, Pr, Nd, Sm, Gd and Y with significant contents over 100 ppm. The REE with the highest content is Ce that is 2060 ppm. The total REE (TREE) is 4729 ppm or 0.47% and the ratio of the light (LREE) and heavy REE (HREE) is 91%/9%. The contents of the major elements determined by XRF are shown in Table 2. The elements Nb and Zr were found to be with significant contents, i.e., ZrO 0.18% or 1800 ppm, Nb 0.066% or 660 ppm. The contents of Th and U are ignorable. The relatively high concentrations of TiO_2, FeO and MgO in the sample are explained by abundant titanite, biotite and magnetite which could be recovered as co-product and by-products.

3.2. Mineralogical Studies

Modal mineralogy of the sample with the chemical formulas and physical properties of the contained minerals are shown in Table 3. A total of 18 minerals were identified by MLA (over 52000 points were measured) in which 89.5% are silicates, and other remained are Fe-oxides 8.6%, phosphates 1.24% and carbonates 0.48%. The main silicate minerals are biotite 31.6%, plagioclase 30.9%, titanite 15.4%, garnet 2.4% and allanite 1.9%. The Fe-oxides are magnetite 8.2%, goethite 0.3% and ilmenite 0.1%. The phosphate is apatite 1.24% and the carbonate calcite 0.48%. The identified REE minerals are allanite and parisite. Most the REE are contained in allanite because of very low concentration of parisite. Meanwhile, REE may be also contained in apatite and zircon in trace contents.

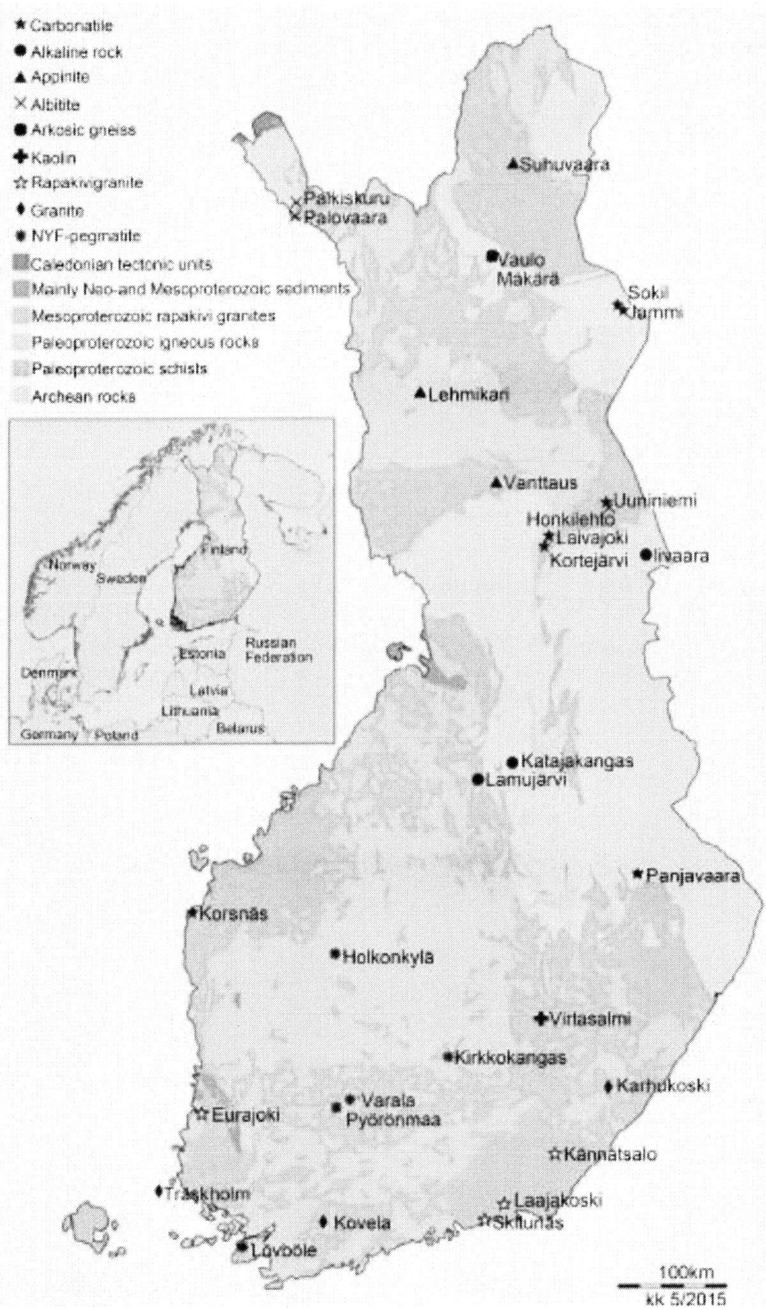

Figure 1. Simplified geological map of Finland showing the distribution of REE occurrences (the occurrences are classified by rock types) [7].

The SEM images of REE minerals allanite, parisite and zircon are shown in Figure 2. The allanite particles are clearly found in liberated and coarse grain sizes. The particles of zircon also appear in smaller grain sizes and some are liberated. But the parisite particles are found in very fine grain size and tightly locked in other minerals like silicates, titanite, calcite and allanite etc. The grain sizes of REE minerals allanite, parisite and zircon are presented in Figure 3. Allanite has a coarse grain size, *i.e.*, P80 = 400 µm, however, zircon and parisite are much finely grained, *i.e.*, P80 = 70 µm for zircon and P80 = 50 µm for parisite.

3.3. Beneficiation Studies

3.3.1. Design of Flowsheet

As shown in Table 3, the sample contains high contents of magnetite and biotite. Magnetite is a ferromagnetic mineral and easily separated by low intensity magnetic separation. With a good floatability by using amine collector biotite can be recovered by flotation effectively. As the most important REE mineral in the sample allanite is a paramagnetic mineral and also a heavier mineral comparing to other major gangue minerals in the sample (plagioclase, K_feldspar and quartz) and could be concentrated by gravity and high gradient or intensity magnetic separations, and flotation. Meanwhile, titanite is also a paramagnetic and heavier mineral and could be recovered with allanite as co-product.

Table 1. The REE contents in the sample.

La ppm	Ce ppm	Pr ppm	Nd ppm	Sm ppm	Eu ppm	Gd ppm	Tb ppm	Dy ppm
859	2060	261	990	140	7	110	12	52

Ho ppm	Er ppm	Tm ppm	Yb ppm	Lu ppm	Y ppm	TREE ppm	TREE wt%	LR/HR[a]
8	22	2	12	2	193	4729	0.47	91%/9%

[a]Light REE to heavy REE ratio.

Rough experiments of gravity separation, high gradient magnetic separation and flotation were performed and a combination flowsheet was determined as illustrated in Figure 4. After grinding the slimes of −20 µm is removed by desliming. Then magnetite is recovered by WLIMS followed by biotite flotation. Finally, allanite and titanite are concentrated by gravity concentration with shaking tabling, flotation and HGMS as bulk REE-Ti concentrates.

Table 2. The major element contents in the sample (wt%).

SiO_2	TiO_2	Al_2O_3	FeO	CaO	MgO	Na_2O
44.8	5.35	14.2	14.6	5.91	3.18	3.92
K_2O	ZrO	P_2O_5	Nb	Th	U	C
4.06	0.18	0.56	0.066	0.004	0.00	0.119

Table 3. Modal mineralogy of the sample.

Mineral	Formula	Wt%	S.G	Magnetic property
Quartz	SiO_2	0.84	2.59 - 2.63	
Plagioclase	$NaAlSi_3O_8$-$CaAl_2Si_2O_8$	30.87	2.62 - 2.76	Diamagnetic
K_feldspar	$KAlSi_3O_8$	6.04	2.55 - 2.76	Diamagnetic
Biotite	$K(Mg, Fe)_3(AlSi_3O_{10})(F, OH)_2$	31.63	2.7 - 3.3	Paramagnetic
Chlorite	$NaClO_2$	0.03		Diamagnetic
Fe_tschermakite	$Ca_2Fe^{2+}{}_3AlFe^{3+}(Si_6Al_2O_{22})(OH)_2$	0.09	3.41	
Garnet	$(SiO_4)_3$	2.41	3.1 - 4.3	
Titanite	$CaTiSiO_5$	15.43	3.48 to 3.60	Paramagnetic
Calcite	$CaCO_3$	0.48	2.71	
Apatite	$Ca_5(PO_4)_3(F, Cl, OH)$	1.24	3.16 - 3.22	Diamagnetic
Zircon	$ZrSiO_4$	0.37	4.6 - 4.7	Diamagnetic
Allanite	$(Ce, Ca, Y, La)_2(Al, Fe^{+3})_3$	1.87	3.5 - 4.2	Paramagnetic
Parisite	$Ca(Ce, La)_2(CO_3)_3F_2$	0.01	4.34	
Goethite	$FeO(OH)$	0.29	3.3 - 4.3	
Magnetite	Fe_3O_4	8.21	5.17 - 5.18	Ferromagnetic
Ilmenite	$FeTiO_3$	0.12	4.70 - 4.79	Paramagnetic
Pseudorutile	$Fe_2Ti_3O_9$	0.01		
Unclassified		0.05		
Total		100.00		

3.3.2. Parametric Experiments and Results

The amount of the feed sample for each experiment was 3 kg and the grinding size was P80 = 85 μm which was determined based on overall analysis of mineral associations and liberations. Through roughing and cleaning process of WLIMS the magnetite concentrate with the grade of FeO

87% was obtained with low losses of REE and TiO_2 (<1.0%). The WLIMS tailings (rougher and cleaner tailings combination) was reported to the biotite flotation in which the slurry was conditioned for 5 min at a high density (>60%) with addition of amine collector and kerosene at pH 3.0 by using H_2SO_4 and then the biotite flotation was performed at pH 3.0. Two kinds of amine collectors tallow amine and Aero3030C were tested with kerosene as the assistant collector. The biotite flotation results indicated by MgO grade and recovery using the two collectors are shown in Figure 5. It is demonstrated that higher recovery of MgO (biotite) was obtained by using Aero 3030C. The losses of REE (indicated by Ce) and TiO_2 during the biotite flotation for both collectors are compared in Figure 6. Generally, the loss of REE in the biotite concentrate was quite low (<20%) for both collectors but slightly lower recovery of REE was obtained while using tallow amine. Comparing to Aero 3030C, the loss of TiO_2 in the biotite concentrate was significantly lower while using tallow amine.

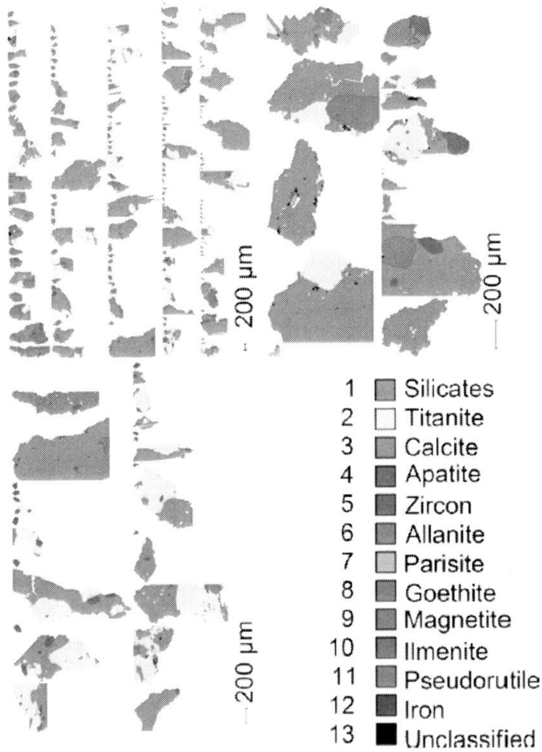

Figure 2. SEM images of REE minerals, allanite, parisite and zircon.

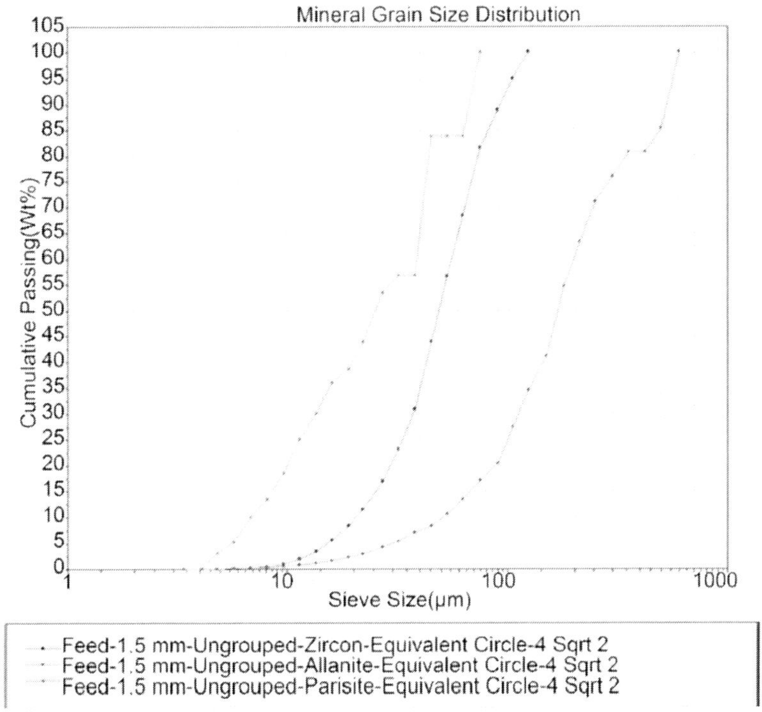

Figure 3. The grain sizes of REE minerals allanite, parisite and zircon.

After the biotite flotation gravity concentration by shaking tabling, flotation and HGMS were performed for recovery of REE-bearing minerals (mainly allanite) and titanite. The operation conditions of shaking tabling were inclination 3° - 4°, feed rate 0.3 kg/h, stroke 350/min, slurry density 25% and rate of rinse water 6 l/min which were determined by observing the stratification and distribution status of the material on the table. During the tabling most coarse particles of allanite and titanite were concentrated as the Gra REE-Ti concentrate and the tailings from the shaking table mainly containing fine allanite and titanite particles was reported to the REE-Ti flotation circuit where the slurry was conditioned for 5 min with additions of chemicals at high density (>60%) prior to flotation. The pH values were adjusted to the optimal depending on collectors which were in the range of pH 6 to 9. In the roughing stage three collectors Aero 6494, FS-2 and P Sulfonate (petroleum sulfonate) combined with Aero 845 were tested. Experimental results of the rougher flotation with different collectors displayed by the relationships between Ce and TiO_2 recoveries and the weight recoveries (yields) of concentrate are shown in Figure 7. Comparing to Aero 6494 and FS-2, the combination of P Sulfonate and Aero 845 as the collector shows significantly more selective for flotation of REE (Ce) with higher Ce recoveries at the same yields. For flotation of TiO_2 both collector FS-2 and the combination of P Sulfonate and Aero 845 have a very good selectivity although FS-2 performs slightly more efficient for TiO_2 flotation.

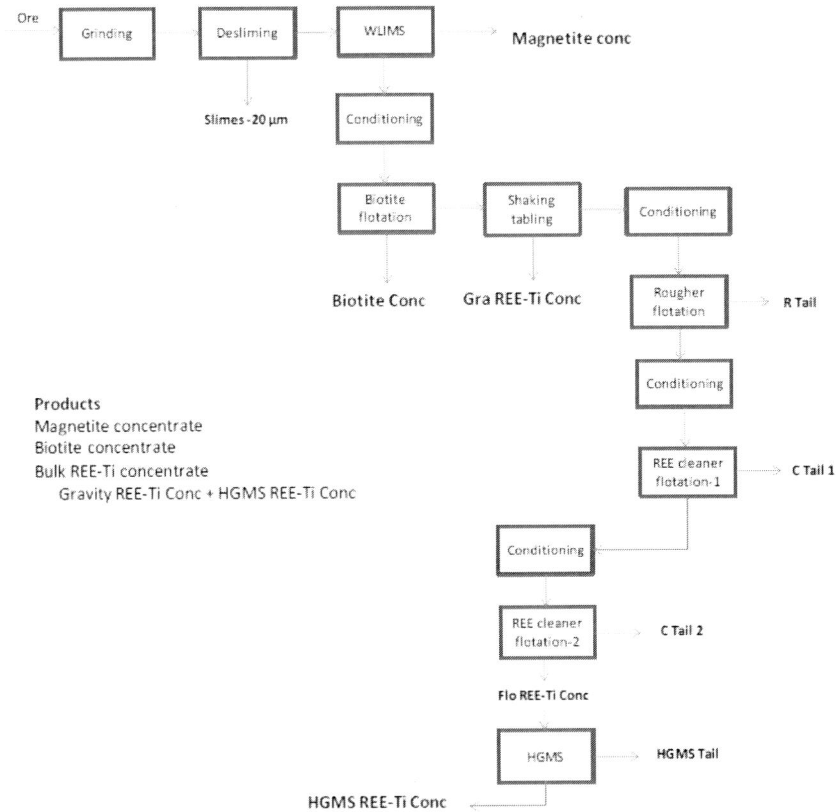

Figure 4. Beneficiation flowsheet of the Lamujärvi syenites.

Being more selective for flotation of REE the combination of P Sulfonate and Aero 845 was selected as the collector for the cleaner flotation and the experimental results are shown in Figure 8. It is seen that the curves of Ce and TiO_2 grade-recovery display high efficiencies of both REE and TiO_2flotation.

The suitable magnetic intensities of the high gradient separation were defined by roughly testing, that is, 1.5 and 2 Tesla were applied for roughing process and 0.2 Tesla for cleaning process.

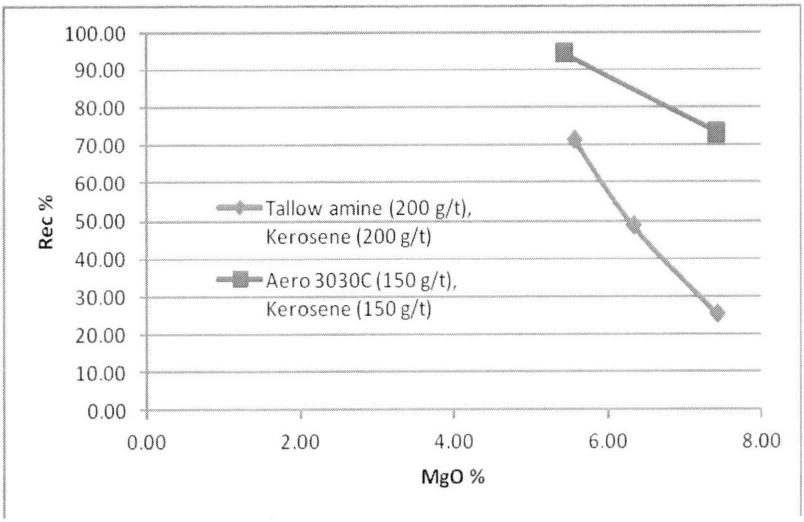

Figure 5. MaO grade and recovery of biotite flotation concentrate.

3.3.3. Overall Beneficiation Results

The overall beneficiation results of the Lamujärvi syenites are presented in Table 4 and Table 5. The REE-Ti concentrate which is the composite of gravity and HGMS REE-Ti concentrates has the grade of REE (La, Ce, Y) 15,235 ppm (TREE around 23,000 ppm) at the recovery of 57.2% and the grade of TiO_2 25.5% at the recovery of 54.0%. In addition, as the by-products, the magnetic concentrate with the grade of FeO 86.6% at the recovery of 45.6% and the biotite concentrate with the grade of MgO 6.6% at the recovery of 51.0% were obtained.

For analyzing the experimental errors the element contents of the feed by calculation from the experimental results (Calc'd Feed (X_1)) and by the analysis (Analysis Feed (X_2)) are listed in Table 4 and Table 5. The standard deviation (SD) between the calculated feed (X_1) and the analysis feed (X_2) was determined by following Equation.

$$SD = \sqrt{\frac{1}{2-1}\sum\left(X_i - X_m\right)^2}$$

where X_i is the content of element i and X_m is the mean value of X_1 and X_2.

The values of SD for all the elements listed in Table 4 and Table 5 present that the experimental errors are very small.

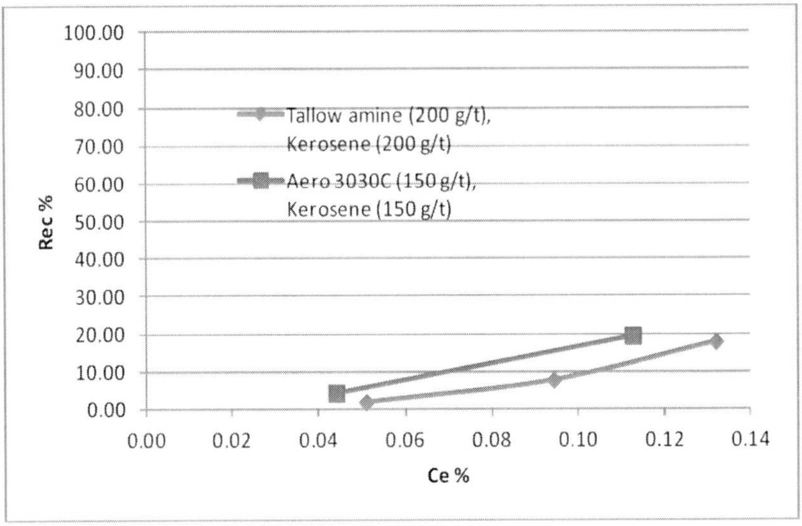

(a)

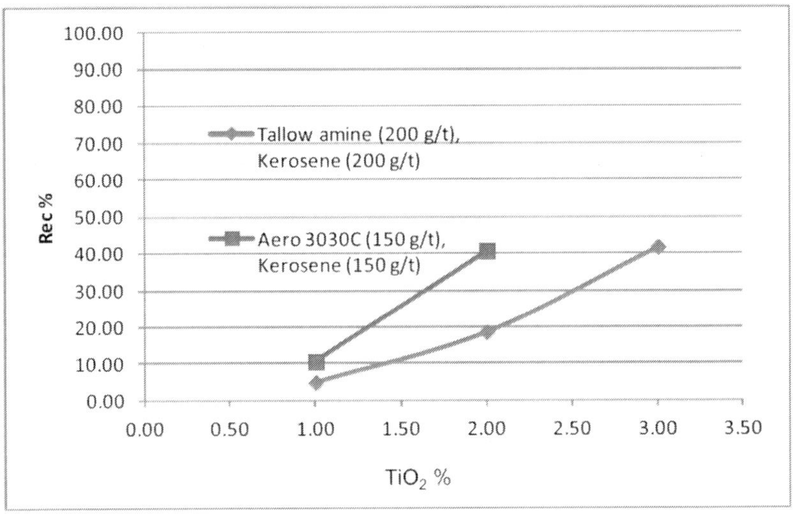

(b)

Figure 6. Recoveries and grades of Ce and TiO2 in biotite concentrate.

3.3.4. Treatment of Slimes by HCl Leaching

It is noticed in Table 4 and Table 5 that 23.4% of REE and 20.8% of TiO_2 were deposited in the slimes fraction which is hardly enriched by physical methods (flotation, gravity and magnetic separations) due to very fine particles. Acid leaching testwork using hydrochloric acid (HCl) was performed to assess the leachabilty of REE for the slimes. The extraction rates of elements in the solution are shown in Table 6. It is revealed that the recoveries of La and Ce in the solution are high. The 89.3% La and 79.8% Ce were obtained at the leaching time 180 min. The extraction of P_2O_5 was over 91%. That is, allanite as the major REE-containing mineral in the slimes and also apatite were mostly decomposed. Meanwhile, the extraction of TiO_2 was low (<18.0%) and SiO_2 as am impurity was remained in the solid in almost 100%.

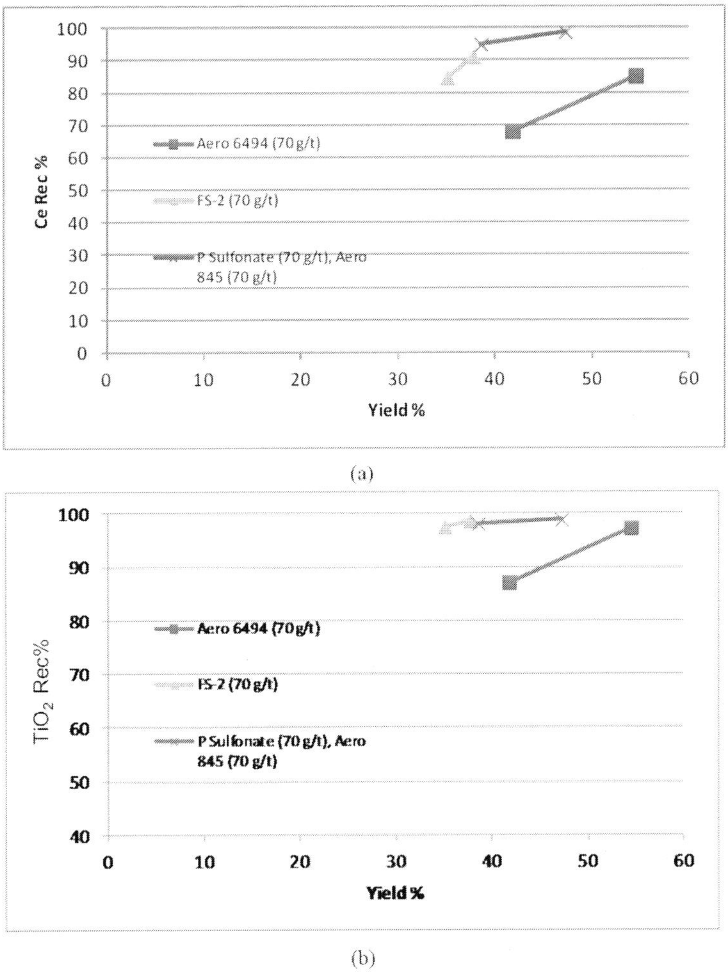

Figure 7. Experimental results of the rougher flotation with different collectors.

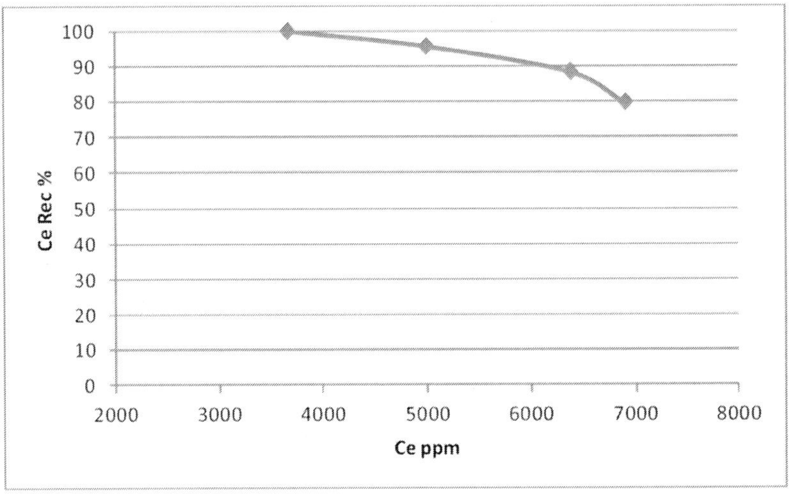

(a)

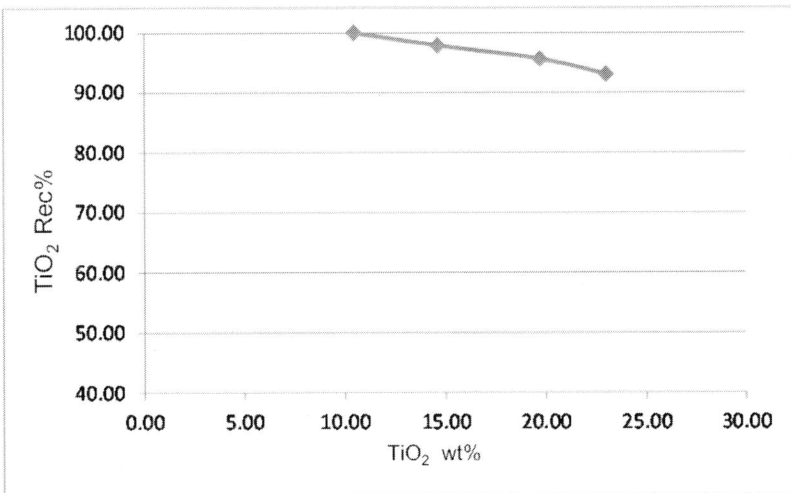

(b)

Figure 8. Experimental results of the cleaner flotation (P sulfonate 25 g/t and Aero 845 50 g/t; conditioning for 5 min; pH 6).

3.3.5. Process Mineralogy

Modal mineralogy of the flotation REE-Ti concentrate (Flo REE-Ti Conc) was determined by MLA analysis. The modal mineralogies of Flo REE-Ti Conc and the Feed are compared in Table 7. It is seen that target minerals allanite, apatite and titanite were significantly enriched, that is, allanite was enriched from 1.87% to 3.85%, apatite from 1.24% to 5.62%, and titanite from to 15.43% to 78.57%. Major silicate minerals biotite, plagioclase and K_felspar but garnet was removed in the flotation concentrate. Meanwhile, the contents of Fe-oxides are low in the Flo REE-Ti Conc. Garnet was concentrated into Flo REE-Ti Conc which could be due to its similarity to allanite, apatite and titanite on crystal properties and floatability [10] .

Table 4. Beneficiation results of the Lamujärvi syenites (assayed by XRF).

Product	Yield	La		Ce		Y		REE (La, Ce, Y)	
	%	ppm	Rec %	ppm	Rec %	ppm	Rec %	ppm	Rec %
Slimes (−20 μm)	29.1	800	24.9	1690	23.0	120	18.1	2610	23.4
Magnetite Conc	8.3	120	1.1	180	0.7	11	0.5	311	0.8
Biotite Conc	23.4	270	6.8	742	8.1	127	15.4	1140	8.6
REE-Ti Conc	12.4	4215	55.4	10,089	57.8	931	58.7	**15,235**	**57.2**
Tail	26.8	416	11.8	839	10.4	54	7.3	1310	10.7
Calc'd Feed (X_1)	100.0	940	100.0	2157	100.0	196	100.0	3293	100.0
Analysis Feed (X_2)		859		2060		193		3113	
Standard deviation (SD)		57.3		68.4		1.6		127.3	

Table 5. Beneficiation results of the Lamujärvi syenites (assayed by XRF).

Product	Yield	FeO		SiO_2		MgO		TiO_2	
	%	%	Rec %	%	Rec %	%	Rec %	%	Rec %
Slimes (−20 μm)	29.1	12.0	22.2	47.7	31.4	4.5	43.6	4.1	20.8
Magnetite Conc	8.3	**86.6**	**45.6**	1.89	0.4	0.1	0.2	0.5	0.7
Biotite Conc	23.4	17.6	26.1	39.9	21.1	**6.6**	**51.0**	4.3	17.3
REE-Ti Conc	12.4	4.8	3.8	31.6	9.2	0.2	1.0	25.5	54.0
Tail	26.8	1.4	2.4	62.8	37.9	0.5	4.1	1.6	7.1
Calc'd Feed (X_1)	100.0	15.8	100.0	44.1	100.0	3.0	100.0	5.8	100.0
Analysis Feed (X_2)		14.6		44.8		3.2		5.4	
Standard Deviation (SD)		0.8		0.5		0.1		0.3	

The contents of biotite, allanite and titanite in the gravity (Gra), HGMS and composite (COMP) REE-Ti concentrates and the biotite concentrate were calculated and the results are shown in Table 8.

Biotite was concentrated from 31.6% to 66.3% in the biotite concentrate. The contents of biotite in the REE-Ti concentrates were reasonably low, that is, 1.52%, 0.91% and 2.0% in the COMP, Gra and HGMS REE-Ti Concs, respectively.

Table 6. Testwork results of acid leaching using HCl for the slimes.

Leaching time, min	Extraction rate %					
	La	Ce	Y	TiO_2	SiO_2	P_2O_5
60	81.8	73.0	19.9	17.6	3.3	96.5
120	89.3	77.8	19.9	17.1	0.0	93.7
180	89.3	79.8	25.2	18.0	0.0	91.4

Allanite was concentrated in the REE-Ti Concs, that is, from 1.9% in the feed to 5.48% in COMP REE-Ti Conc, 5.9% in Gra REE-Ti Conc and 5.2% in HGMS REE-Ti Conc. Meanwhile, titanite was significantly enriched in the REE-Ti Concs, that is, from 15.4% in the feed to 71.19% in COMP REE-Ti Conc, 66.1% in Gra REE-Ti Conc and 75.2% in HGMS REE-Ti Conc. The contents of allanite and titanite in the biotite concentrate are comparably low at 1.2% and 10.9%.

Modal mineralogy of the acid leaching residue of the slimes by MLA is shown in Table 9. The content of REE mineral allanite is dropped from 1.87% in feed to 0.06% in leach residue or it was almost dissolved into the solution, but titanite is still remained in the solid at a high content of 17.8%. Most remained minerals (>99%) in the residue are silicates including plagioclase 49.5%, titanite 17.8%, quartz 11.8%, K_felspar 10.9%, biotite 7.1% and garnet 1.9%.

3.3.6. REE-Ti Separation

The analyses of process mineralogy (Table 8) present that in bulk REE-Ti concentrates allanite and titanite consist of 70% to 80%. The further investigations in the continuation of this project will focus on the separation of allanite and titanite in the bulk REE-Ti concentrates.

According to the mineralogical studies of leaching residue of the slimes (Table 9) acid leaching with HCl is proved an effective approach to separate allanite and titanite by dissolving allanite into the solution and remaining titanite in the residue. Due to their similar floatability flotation would be challengeable to separate allanite and titanite. But testwork will be carried on to study the possibility such as by using selective collectors (cationic or anionic) and adjusting pH values etc.

4. CONCLUSIONS

Mineralogy and beneficiation of the Lamujärvi syenites were studied. In the sample, seven rare earth elements including La, Ce, Pr, Nd, Sm, Gd and Y were found with significant contents over 100 ppm and the total REE (TREE) is 4729 ppm. Meanwhile, the sample contains a high content of TiO_2.

A total of 18 minerals were identified by MLA analysis in which 89.5% silicates, 8.6% Fe-oxides, 1.24% phosphates and 0.48% carbonates are contained. Allanite with content of 1.9% is the major REE-containing mineral and has a coarse grain size. Parisite was identified as another REE-containing mineral but at very low concentration and fine grain size. In addition, the sample has high contents of titanite (15.4%), biotite (31.6%) and magnetite (8.2%). Because allanite and titanite have similar physical (gravitational and magnetic) and surface properties, they could be recovered as co-products by gravity and magnetic separations and flotation. Biotite and magnetite could be recovered as by-products by cationic flotation and low intensity magnetic separation methods, respectively.

Beneficiation flowsheet was developed containing sequent grinding, desliming, WLIMS, shaking tabling, flotation and HGMS. The composite REE-Ti concentrate with the grade of REE (La, Ce, Y) 15235 ppm (TREE 23000 ppm) at the recovery of 57.2% and the grade of TiO_2 25.5% at the recovery of 54.0% was obtained. In addition, as the by-products, the magnetic concentrate with the grade of FeO 86.6% at the recovery of 45.6% and the biotite concentrate with the grade of MgO 6.6% at the recovery of 51.0% were achieved. The REE in the slimes fraction were efficiently leached (recoveries of 89.3% La and 79.8% Ce) with hydrochloric acid (HCl) at concentration 6 mol/l and temperature 60°C.

Table 7. Modal mineralogy comparison of feed and REE-Ti flotation concentrate.

Mineral	Wt%		Mineral	Wt%	
	Feed	Flo REE-Ti Conc		Feed	Flo REE-Ti Conc
Quartz	0.84	0.29	Titanite	15.43	78.57
Plagioclase	30.87	1.31	Calcite	0.48	0.00
K_feldspar	6.04	0.65	Apatite	1.24	5.62
Biotite	31.63	2.23	Magnetite	8.21	0.08
Garnet	2.41	6.15	Ilmenite	0.12	0.21
Allanite	1.87	3.85	Goethite	0.29	0.01
Zircon	0.37	0.41			
Parisite	0.01	0.02	Total	99.81	99.41

Table 8. Contents of biotite allanite and titanite in REE-Ti concentrates and biotite concentrate.

Mineral	Wt%				
	Feed	COMP REE-Ti Conc	Gra REE-Ti Conc	HGMS REE-Ti Conc	Biotite Conc
Biotite	31.63	1.52	0.91	2.03	66.30
Allanite	1.87	5.48	5.87	5.17	1.23
Titanite	15.43	71.19	66.14	75.16	10.92

Table 9. Modal mineralogy of the acid leaching residue of the slimes.

Mineral	wt%		Mineral	wt%	
	Feed	Leach Residue	Feed	Feed	Leach Residue
Quartz	0.84	11.82	Zircon	0.37	0.16
Plagioclase	30.87	49.47	Allanite	1.87	0.06
K_feldspar	6.04	10.86	Apatite	1.24	0.01
Biotite	31.63	7.11	Calcite	0.48	0.01
Garnet	2.41	1.90	Ilmenite	0.12	0.18
Titanite	15.43	17.77	Total	91.30	99.30

Studies of process mineralogy present that allanite and titanite are successfully concentrated and that major silicate minerals biotite, plagioclase and K_felspar are significantly removed by gravity separation, high gradient magnetic separation and flotation processes for REE-Ti. Meanwhile, acid leaching with HCl could be an effective approach to separate allanite and titanite for further treatment of the REE-Ti concentrate.

ACKNOWLEDGEMENTS

The authors are grateful to the Ministry for Foreign Affairs of Finland for the financial support.

REFERENCES

1. Al Ani, T. and Sarapää, O. (2013) Geochemistry and Mineral Phases of REE in Jammi Carbonatite Veins and Fenites, Southern End of the Sokli Complex, NE Finland. Geochemistry: Exploration, Environment, Analysis, 13, 217-224. http://dx.doi.org/10.1144/geochem2011-088
2. Vartiainen, H. (1980) The Petrography, Mineralogy and Petrochemistry of the Sokli Carbonatite Massif, Northern Finland. Geological Survey of Finland, Bulletin 313, 126.

3. Sarapää, O., et al. (2013) Rare Earth Exploration Potential in Finland. Journal of Geochemical Exploration, 133, 25-41. http://dx.doi.org/10.1016/j.gexplo.2013.05.003
4. Fandrich, R., et al. (2007) Modern SEM-Based Mineral Liberation Analysis. International Journal of Mineral Processing, 84, 310-320. http://dx.doi.org/10.1016/ j.minpro. 2006.07.018
5. Gu, Y. (2003) Automated Scanning Electron Microscope Based Mineral Liberation Analysis An Introduction to JKMRC/FEI Mineral Liberation Analyser. Journal of Minerals & Materials Characterization & Engineering, 2, 33-41. http://dx.doi.org/10.4236/ jmmce.2003.21003
6. Zeh, A. (2004) Crystal Size Distribution (CSD) and Textural Evolution of Accessory Apatite, Titanite and Allanite during Four Stages of Metamorphism: An Example from the Moine Supergroup. Scotland. Journal of Petrology, 45, 2101-2132. http://dx.doi.org/10. 1093/petrology/egh049
7. Zaitsev, A.N., et al. (2014) Rare-Earth Elements Minerals in Carbonatites of the Kola Alkaline Province (Northern Fennoscandia). 1st European Rare Earth Resources Conference, Milos Island, 4-7 September 2014, 343-347.
8. Arzamastsev, A., et al. (2008) The Khibina and Lovozero Alkaline Massifs: Geology and Unique Mineralization. 33 IGC Excursion No. 47, Apatity, 22 July-2 August 2008, 1-58.
9. Sarapää, O. and Sarala, P. (2013) Rare Earth Element and Gold Exploration in Glaciated Terrain—Example from the Mäkärä Area, Northern Finland. Geochemistry: Exploration, Environment, Analysis, 13, 131-143. http://dx.doi.org/10.1144/geochem2012-136
10. Goodenough, K.M., et al. (2016) Europe's Rare Earth Element Resource Potential: An Overview of REE Metallogenetic Provinces and Their Geodynamic Setting. Ore Geology Reviews, 72, 838-856. http://dx.doi.org/10.1016/j.oregeorev.2015.09.019

CHAPTER 3

Synergistic Adsorption and Flotation of New Mixed Cationic/Nonionic Collectors on Muscovite

Hao Jiang *, Wanying Ji, Qinhong Yang, Longhua Xu, Chen Zhao and Yuehua Hu

School of Minerals Processing and Bioengineering, Central South University, Changsha 410083, China

ABSTRACT

The mixed cationic collector cetyltrimethylammonium chloride (CTAC) and nonionic collector octanol (OCT) was found to exhibit a synergistic effect on the flotation and adsorption of muscovite. To understand the underlying synergistic mechanism, flotation, contact angle, surface tension, and adsorption measurements were carried out. The results obtained from flotation measurements indicated that the mixed CTAC/OCT exhibits a better collecting ability than CTAC or OCT. The recovery of muscovite with CTAC only rapidly decreased from 97.25% at pH 2.64 to 75.26% at pH 5.82, followed by a flat horizontal at a pH is higher than 6. In contrast, a high recovery of greater than 85% muscovite was observed using mixed CTAC/OCT at α_{CTAC} = 0.67 (the mole ratio of CTAC:OCT = 2:1) over the investigated pH range. From the surface activity parameters (CMC, γ_{CMC}, T_{max}, A_{min}) estimated from surface measurements and interaction parameters (β^m, β^σ), in addition to the micellar and interfacial compositions ($x_1{}^m$, $x_1{}^\sigma$) obtained from the theory of regular solutions, a synergistic effect is evident in the mixed micelle and at the water/air interface. Moreover, the mixed CTAC/OCT at α_{CTAC} = 0.67 exhibited the maximum synergistic interaction. The results obtained from surface tension measurements indicated that the mixed CTAC/OCT exhibits considerably higher surface activities compared to single CTAC or OCT. The contact angle results confirmed that the mixed CTAC/OCT is a better collector than the individual CTAC or OCT for the flotation of muscovite. According to the results obtained from adsorption experiments, compared with that of individual CTAC or OCT, the amounts of CTAC and OCT adsorbed on the muscovite surface are considerably increase in the mixed systems because of co-adsorption. Based on these results, the mixed CTAC/OCT exhibits a remarkable synergistic effect during the flotation and adsorption of muscovite.

Keywords: muscovite; mixed collectors; synergism; flotation; adsorption

1. INTRODUCTION

Muscovite has been commonly used in a wide range of industrial applications because of its distinctive properties, e.g., dielectric properties, heat resistance, and chemical resistance [1]. It is a typical two-layer-structured aluminosilicate, with a negatively charged tetrahedral layer owing to the isomorphic substitution of approximately 1/4 Si^{4+} ions by Al^{3+} ions. These permanent charges are balanced by the interlayer K^+ ions, which serve as a bridge between the sheets [2,3]. The layer-to-layer cleavage of muscovite occurs easily, and the cleavage surface exposes K^+, Na^+, and $[SiO_4]^{4-}$ ions, which can easily undergo hydrolysis and break away from the surface. Owing to the crystal structure of muscovite, the cleavage surface of muscovite is extremely hydrophilic and negatively charged. Hence, the cationic collectors are typically used for the flotation of muscovite. Unfortunately, several silicates accompanying muscovite are negatively charged under neutral or alkaline conditions and are doped into the flotation foam. In other words, it is apparently impossible to separate muscovite from these minerals using only cationic collectors [4].

In recent years, instead of pure components, mixed surfactants have been widely used as collectors for flotation because they typically exhibit better physical and chemical properties and synergistic effects compared to their corresponding single surfactants [5,6]. Several studies have reported the use of various mixed surfactant systems (nonionic/nonionic, ionic/ionic, and ionic/nonionic) for flotation. A collector mixture of dodecyl amine hydrochloride and sodium oleate exhibit enhanced activity for the flotation of muscovite [7]. The use of the mixed collectors of alkyl diamine/sulfonate has been used for the flotation of feldspar from quartz under acid conditions [8,9]. At an optimal molar ratio of 9:1 for sodium oleate and dodecyl ammonium chloride, high selectivity has been reported for the flotation of spodumene from feldspar [10,11]. Among numerous composite surfactant systems, the mixed cationic/nonionic surfactants are notable because they form spontaneous vesicles in the water phase and exert strong synergistic effects. Such synergistic effects can reduce the total amount of reagents used in industry, thereby reducing environmental pollution and economic costs [12,13,14,15].

The adsorption of surfactants at the solid–liquid interface is crucial in numerous application, such as emulsification, foaming control, detergents, surfactant-based separation, surface wetting modification, and flotation [16,17]. Although the adsorption mechanism of a single surfactant at the solid–liquid interface has been thoroughly investigated, few studies have reported the adsorption of mixed surfactant solutions. Furthermore, the use of mixed collectors for enhanced flotation and selectivity has only recently been recognized. Therefore, several studies focusing on mixed cationic/nonionic surfactant systems have recently been carried out [18,19,20,21]. In general, these mixed surfactant systems typically exhibit considerably higher surface activities than their individual components, leading to better wetting, solubilizing, and foaming behavior. In addition, the synergistic effects of mixed surfactants at the air/water interface and their co-adsorption on mineral surfaces have been reported [22,23,24,25,26,27]. Although comprehensive studies on mixed surfactants in aqueous systems have been conducted, several limitations exist. Hence, considerable efforts should be focused on understanding the underlying synergistic adsorption and flotation mechanism of mixed cationic/nonionic surfactants on minerals.

The objective of this work is to understand the synergistic flotation and adsorption of binary cationic/nonionic surfactant mixtures on muscovite. A binary surfactant mixture of cetyltrimethylammonium chloride (CTAC) and octanol (OCT) was used as collectors. The flotation performance of muscovite using the mixed CTAC/OCT was investigated by flotation measurements. In addition, surface tension, contact angle, and adsorption measurements were carried out to examine the adsorption characteristics and possible synergic effects at the gas-liquid or solid-liquid interfaces. Based on these measurements, the role of mixed cationic/nonionic collectors in the flotation and adsorption of muscovite is delineated.

2. MATERIALS AND METHODS

2.1. Materials and Reagents

Muscovite samples used in this work were obtained from Hebei, China. These mineral samples were crushed to a size of less than 2 mm and added into a porcelain ball mill. Next, the fine particles were screened to obtain a size fraction of −149 + 74 μm. Table 1 and Figure 1 show the compositional analysis of muscovite and its X-ray diffraction pattern, respectively. The results indicated that the muscovite sample was highly pure. The BET surface area of muscovite was measured as 2.431 m2/g. The rock mineral samples provided by Guangzhou Richone Electronic Co., Ltd. (Guangzhou, China) were used for contact angle measurements. Analytically pure samples of CTAC and OCT purchased from Nanjing Robiot Co., Ltd. Analytical-grade hydrochloric acid (HCl) and sodium hydroxide (NaOH) were used as pH modifiers. Distilled water (resistivity = 18.2 MΩ cm) was obtained from an ultra-pure water machine.

Table 1. Chemical composition of the mineral sample (mass fraction, %).

Composition	Al_2O_3	SiO_2	Fe_2O_3	TiO_2	CaO	MgO	K_2O	Na_2O	H_2O
Muscovite	28.96	46.55	5.30	0.74	0.19	0.97	10.53	3.24	3.23

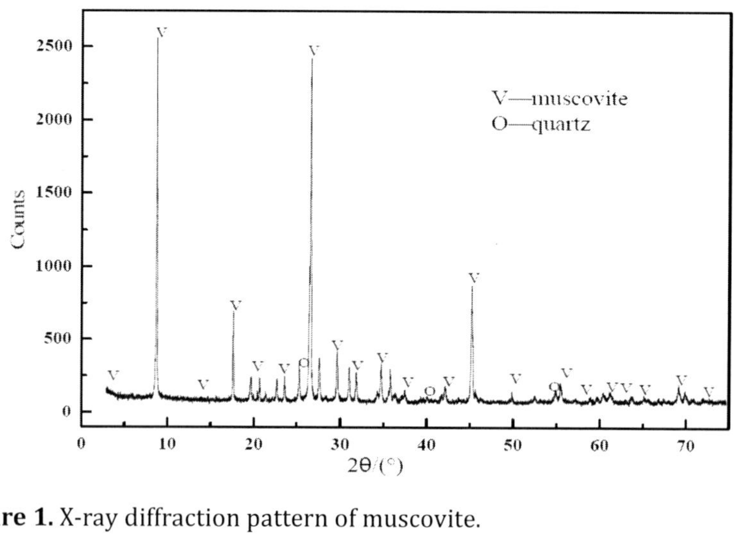

Figure 1. X-ray diffraction pattern of muscovite.

2.2. Flotation Experiments

Flotation tests with muscovite were carried out in a 40 mL flotation cell, and the speed of fixed impeller was 1600 r/min. The air flow rate was 0.15 m^3/h. In each flotation test, 2 g of a muscovite sample and 35 mL of deionized water were used. Reagent grade HCl and NaOH were used to adjust the pH. After adding the required reagents, the pulp was conditioned for 3 min, followed by a flotation collection period of 3 min. The froth products and tails were separately weighed after filtration and drying, the recovery rate was calculated. Each measurement was performed at least five times, and the average was reported as the final value.

2.3. Surface Tension Measurements

Surface tension was measured on a BZY-1 automatic tensiometer (Shanghai Hengping Instrument and Meter Factory, Shanghai, China) with a platinum plate at an environmental temperature of 25 °C. The tensiometer was calibrated using distilled water. The platinum plate was cleaned using deionized water and heated by an alcohol flame to ensure the complete removal of the organic matter and other impurities. Before the measurements, each sample was stabilized for 10 min until reaching equilibrium. Notably, each measurement was performed at least five times.

2.4. Contact Angle Measurements

The MiniLab ILMS (GBX, Rémy, France) instrument was used to measure the contact angles of muscovite. A sticky tape was used to freshly cleave the mineral sample surface, which permitted the observation of the perfect cleavage of muscovite along its basal planes. Reagent grade HCl and NaOH were used to adjust the solution pH. First, the sample was conditioned at a certain concentration solution for 30 min at 25 °C, followed by washing with washed with deionized water, drying under ultra-pure grade nitrogen, and performing the contact angle measurements. Notably, the sessile liquid drop technique was used to measure the contact angles of the samples. A single liquid drop (volume ca. 5 μL) was deposited onto the solid surface, and the static contact angle was measured. Each experiment was repeated at least five times at different locations of the dry solid muscovite sample and the final value of the contact angle was calculated as an average.

2.5. Adsorption Measurements

A TOC-LCPH analyzer (Shimadzu, Kyoto, Japan) was used to measure the amount of adsorbed reagents on the mineral surface. First, mineral particles

(2.0 g) were placed in a Plexiglas cell (40 mL) filled with 35 mL of deionized water. After adding a certain amount of collectors, the suspension was stirred for 3 min in the micro-flotation cell. Then, the conditioned pulp suspension obtained from the previous procedure was added to a 50 mL vial. After shaking the vials for 24 h at 25 °C, the supernatants were measured using the TOC-LCPH analyzer. The amount of the adsorbed collectors was calculated as follow:

$$\Gamma = \frac{V(C_0 - C)}{mA} \quad (1)$$

Here, C_0 and C are the initial and supernatant concentrations of quaternary ammonium salts, respectively. V is the solution volume, m is the sample mass, and A is the particle specific surface area.

The total organic carbon (TOC) and total organic nitrogen (TON) in mixed CTAC/OCT were estimated using the TOC-LCPH analyzer. Nevertheless, to calculate the amount of the adsorbed OCT using a mixed CTAC/OCT system, the amount of the adsorbed CTAC must be calculated. Through the investigation, the concentrations of organic carbon (OC) and organic nitrogen (ON) in CTAC exhibited a linear relationship (Figure 2).

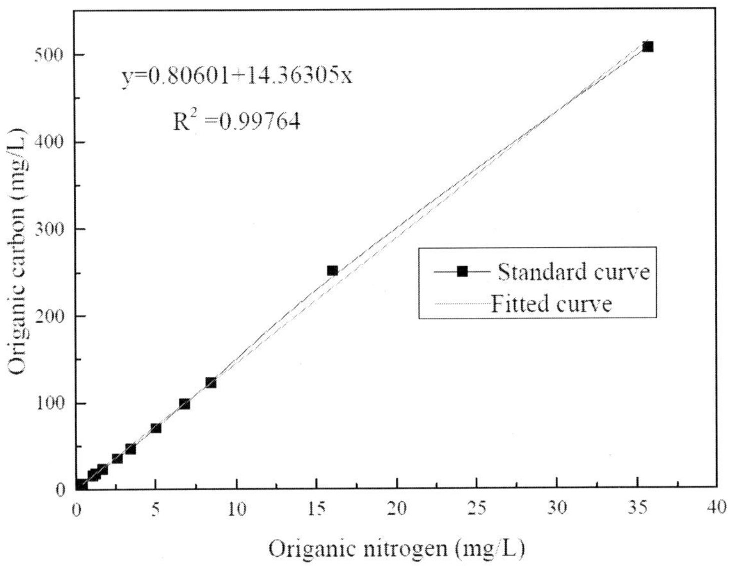

Figure 2. Standard curves of the linear relationship between organic carbon (OC) and organic nitrogen (ON) in cetyltrimethylammonium chloride (CTAC).

So, according to Figure 2, the OC of CTAC can be obtained by the amount of ON and OC of OCT can be obtained by subtracting the OC of CTAC from the TOC of CTAC/OCT. The environmental temperature was maintained at 25 °C and each measurement was performed at least three times.

3. RESULTS AND DISCUSSION

3.1. Flotation Results

Figure 3a shows the effect of the collector concentration of CTAC, OCT, and the mixed CTAC/OCT surfactants on the flotation of muscovite at pH 7. The recovery of muscovite increased with increasing concentrations of CTAC and the mixed CTAC/OCT. However, the recovery of muscovite with OCT rarely changed with increasing concentration. In case of CTAC/OCT, the recoveries of muscovite rapidly increased at low collector concentrations, particularly at α_{CTAC} = 0.67 (the mole ratio of CTAC:OCT = 2:1), which exceeded that of CTAC alone. For example, at a collector concentration of 8 × 10^{-4} mol/L, the flotation recovery of muscovite in the presence of the mixed surfactant was approximately 73.25%, whereas it was only approximately 45.9% for CTAC alone. The mixed CTAC/OCT collector is clearly beneficial for the flotation of muscovite. Therefore, a CTAC/OCT mole ratio of 2/1 (α_{CTAC} = 0.67) is selected for the subsequent studies, related to the significant improvement in the recovery of muscovite.

Figure 3b shows the recovery of muscovite as a function of pH at a concentration of 1 × 10^{-4} mol/L for single or the mixed surfactants. The results indicate that the flotation of muscovite is almost not possible using OCT in the presence of OCT within the studied pH range. The recovery of muscovite using only CTAC rapidly decreased from 97.25% at pH 2.64 to 75.26% at pH 5.82, followed by a flat horizontal at a pH greater than 6. However, the increase in the pH of the pulp exerted less impact on the recovery of muscovite using a mixed surfactant: the recovery of greater than 85% was maintained for muscovite using mixed surfactants in the investigated of pH range.

3.2. Surface Properties of Mixed Surfactants (CTAC/OCT)

Figure 4 shows the surface tension (γ) vs. log concentration (C) plots at different molar ratios of mixed surfactants (CTAC/OCT) at pH 7. Table 2 lists the critical micelle concentration (CMC) values and surface tension at the CMC (γ_{CMC}) for the corresponding surfactant mixture. For all mixed systems, the CMC and γ_{CMC} values were less than those obtained for the pure systems, indicating that mixed surfactants exhibit a stronger ability to reduce the surface tension of a solution compared to individual surfactants. Therefore, micro bubbles are more easily generated in mixed systems, which are more advantageous as they collide with particles and capture them.

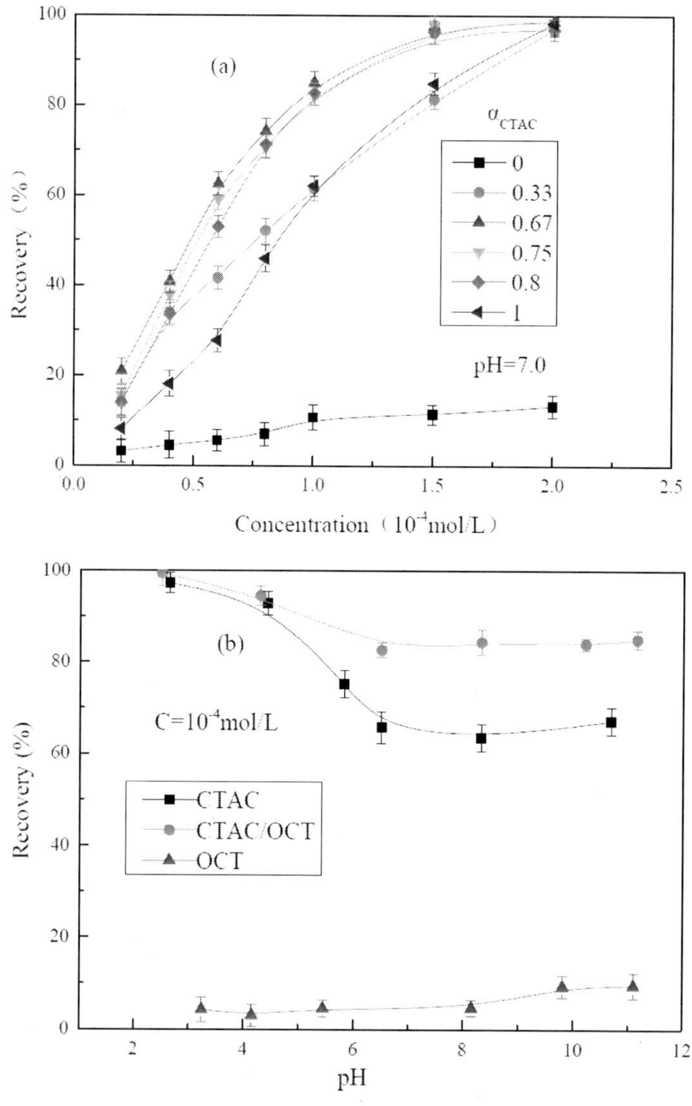

Figure 3. Flotation recoveries of muscovite with CTAC, OCT and mixed CTAC/OCT as a function of (a) concentration and (b) pH.

Moreover, the molar ratio of mixed CTAC/OCT clearly affected the value of CMC. The CMC values gradually decreased with increasing CTAC mole fraction and reached a minimum value α_{CTAC} = 0.67, followed by the increase in the CMC values with increasing CTAC mole fraction. The decrease in the CMC values at the beginning is possibly related to the neutralization of hydrophilic groups by their electrostatic interactions. On the other hand, when α_{CTAC} was greater 0.67, the individual surfactant exhibited self-interactions, thereby increasing the CMC.

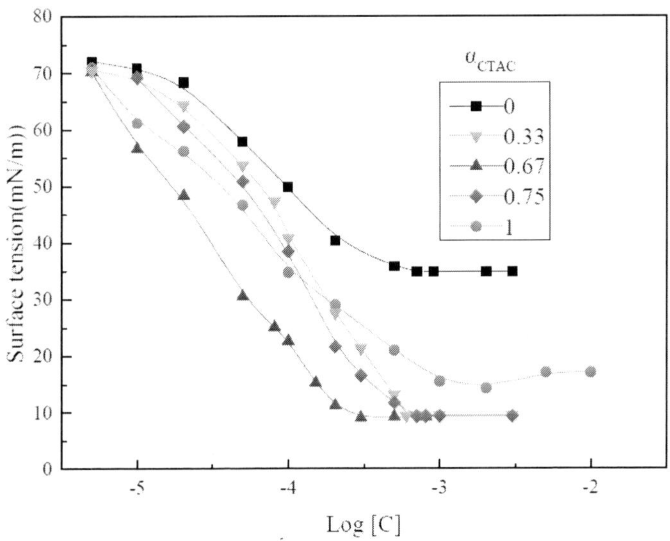

Figure 4. Surface tension vs. log concentration plots for CTAC, nonionic collector octanol (OCT), and mixed collectors CTAC/OCT at various molar ratios.

Furthermore, the minimum area per molecule (A_{min}) and the saturation adsorption value (Γ_{max}) are factors for confirming effective adsorption at the air/water interface [26]. The Γ_{max} and A_{min} values were calculated by the Gibbs adsorption Equations (2) and (3), respectively:

$$\Gamma_{max} = -\frac{1}{2.303nRT}\left(\frac{\partial \gamma}{\partial \log C}\right)_T \quad (2)$$

$$A_{min} = \frac{1}{\Gamma N_A} \quad (3)$$

Here, $(\partial\gamma/\partial\log C)$ is the slope from the plot of surface tension vs. the logarithm of the total surfactant concentration at 25 °C, T is the absolute temperature (298.15 K), R is the universal gas constant (8.314 J mol^{-1} K^{-1}), and NA is the Avogadro number (6.023 × 10^{23} mol^{-1}). The units of Γ_{max} is μmol/m^2, and A_{min} is in $Å^2$. As has been reported previously [27], n = 1 and n = 2 for mixed CTAC/OCT and a single system, respectively.

Table 2 summarizes the Γ_{max} and A_{min} values for some of the mixtures investigated. In the investigated molar ratio range, the highest Γ_{max} and the lowest A_{min} values for the mixed collectors were observed at α_{CTAC} = 0.67. Thus, the molecular arrangement in mixed CTAC/OCT is considerably denser compared to the individual surfactant, related to the fact that the strong in all of the mixed systems dominates other weaker interactions.

Table 2. Surface properties of CTAC/OCT, α_{CTAC} = 0.0.33, 0.67, 0.75, and 1.00 in aqueous solutions at 25 °C.

α_{CTAC}	CMC (10^{-4} mol/L)	γ_{CMC}	Γ_{max}	A_{min}
0	5.49	34.81	1.78	93.27
0.33	4.51	9.50	5.33	31.15
0.67	2.18	9.24	6.36	26.11
0.75	4.28	9.36	6.08	27.31
1	10.08	15.54	2.21	75.13

3.3. Synergism of the Mixed Collectors (CTAC/OCT)

Based on the regular solution theory [28,29], the nature and strength of the interaction between two components can be ascertained by calculating β parameters (including β^m, the interaction parameter for mixed micelle formation in an aqueous solution, and β^σ, the interaction parameter for mixed adsorption film formation at the air/water interface). Notably, the β value only reflects the molecular interactions between the head groups of two surfactants at the interface or in micelles. However, it does not consider the interaction between the hydrophobic alkyl chains of the surfactants [27]. In general, β parameters are easily obtained from the surface tension or from CMC data.

According to the regular solution theory and Rubingh's theory of nonideal mixing, the mole fraction of CTAC in the mixed micelles (x_1^m) and the micellar interaction parameter (β^m) arecalculated using Equations (4) and (5), respectively:

$$\frac{(x_1^m)^2 \ln(\alpha_1 c_{12}^m / x_1^m c_1^m)}{(1 - x_1^m)^2 \ln[(1 - \alpha_1) c_{12}^m / (1 - x_1^m) c_2^m]} = 1 \tag{4}$$

$$\beta^m = \frac{\ln(\alpha_1 c_{12}^m / x_1^m c_1^m)}{(1 - x_1^m)^2} \tag{5}$$

Here, x_1^m denotes the mixed micelles, which are related to α_1 (CTAC mole fraction), c_1^m, c_2^m, and c_{12}^m are CMC values for CTAC, OCT, and their mixture, respectively.

Moreover, using Rosen's approach, the mole fraction of CTAC in a mixed monolayer (x_1^σ) and the interfacial interaction parameter (β^σ) are calculated from Equations (6) and (7), respectively:

$$\frac{(x_1^\sigma)^2 \ln(\alpha_1 c_{12}^s / x_1^\sigma c_1^s)}{(1 - x_1^\sigma)^2 \ln[(1 - \alpha_1) c_{12}^s / (1 - x_1^\sigma) c_2^s]} = 1 \tag{6}$$

$$\beta^\sigma = \frac{1 \ln(\alpha_1 c_{12}^s / x_1^\sigma c_1^s)}{(1 - x_1^\sigma)^2} \tag{7}$$

Here, c_1^s, c_2^s, and c_{12}^s are the molar concentrations of CTAC, OCT, and their mixture at a mole fraction of CTAC α_1, required to produce a given surface tension reduction (corresponding to $\gamma = 35$ mN/m in this study).

As can be observed from Table 3, β^m and β^σ values of all mixed systems were negative, indicating that the mixtures exhibit a stronger attractive interaction in the mixed micelles and at the solution/air interface compared to the two components before mixing. With increasing CTAC mole fraction, the β^m and β^σ values became more negative and reached the minimum at $\alpha_{CTAC} = 0.67$, indicating that the mixture at $\alpha_{CTAC} = 0.67$ exhibits a stronger attractive interaction compared to the other two mixed systems in the micelle. Notably, the β^σ values were more negative than β^m, suggesting that the interactions at the solution/air interface are stronger compared to mixed micelles.

Furthermore, synergism in the mixed micelle formation exists when the CMC of a mixture was less than that of the individual surfactants. Such a synergistic effect is observed s under two conditions, β^m must be negative and $|\beta^m|$ must be greater than $|\ln(c_1^m/c_2^m)|$, respectively. All the condition were already satisfied (Table 2). The $|\beta^m|$ values were greater than those of $|\ln(c_1^m/c_2^m)|$ in all of the mixed collector systems, and all β^m values were negative. This result further supported that mixed CTAC/OCT exerts a synergistic effect.

Table 3. Molecular interaction and synergistic parameters for mixed collectors CTAC/OCT at α_{CTAC} = 0.33, 0.67 and 0.75, in 0.1 mol/L NaCl aqueous solutions.

α_{CTAC}	x_1^m	β^m	$\|\ln(c_1^m/c_2^m)\|$	x_1^σ	β^σ
0.33	0.40	−2.08	0.608	0.68	−2.25
0.67	0.51	−6.66	-	0.71	−10.07
0.75	0.32	−1.94	-	0.59	−2.04

3.4. Wettability of Muscovite with Mixed Collectors (CTAC/OCT)

Contact angle measurements are crucial for investigating the wetting behavior of reagents on the muscovite surface. Before the sample is conditioned in the corresponding collector solution, we have measured the contact angle of muscovite with fresh cleavage, as muscovite surface is very hydrophilic, contact angle is only a few degrees. Figure 5a shows the effect of concentration on the contact angles of muscovite. The contact angle of muscovite in the presence of CTAC/OCT was significantly greater than that in the presence of single CTAC at low surfactant concentration. On the other hand, at high surfactant concentration, the contact angles of muscovite using single CTAC and the mixed CTAC/OCT were maintained constant (approximately 86°). However, in the presence of single OCT, a relatively small contact angle was observed. Although the contact angle of the mica in the presence of OCT is larger than that in the water, OCT is a nonionic surfactant, the surface activity of OCT is relatively low, which cannot produce stable foam to capture muscovite particles. According to the above results, OCT itself cannot allow for good flotation unless CTAC is present, implying that CTAC plays a leading role in the overall synergistic adsorption and OCT only plays a supporting role.

Figure 5b shows the effect of pH on the contact angle of muscovite surface by using 10^{-4} mol/L of corresponding reagents. At pH greater than 7, the contact angle of muscovite gradually increased with increasing pH and was maintained at approximately 80° using CTAC. In addition, the contact angle using OCT alone was still low. In contrast, with the use of mixed CTAC/OCT, pH marginally affected the contact angle of muscovite over the investigated pH range. Hence, the hydrophobicity of muscovite is gradually enhanced using the mixed CTAC/OCT. Notably, considerably different results between contact angle and flotation measurements were observed. This discrepancy results from the different particle size of the samples. A fine size fraction (−149 + 74 μm) was used in flotation tests, while the plane was utilized for the contact angle tests. The contact angle test only reflects the properties of the cleaved muscovite surface (basal plane), but does not consider the edge surface, which also affects the flotation behavior.

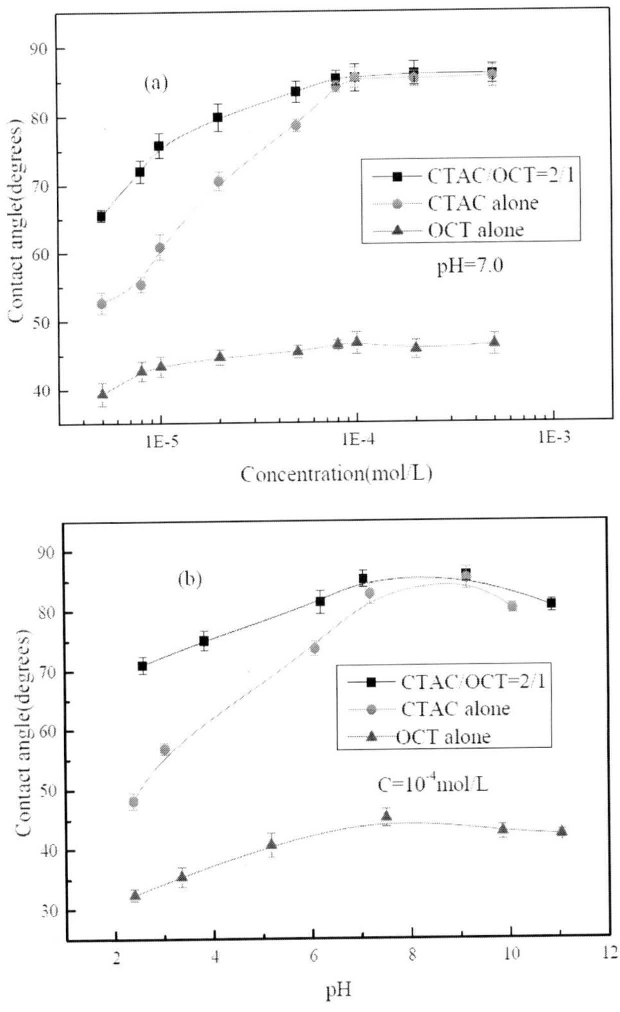

Figure 5. Contact angles of muscovite as a function of (a) concentration and (b) pH.

3.5. Adsorption Results

The amount of the adsorbed collectors on the mineral surface directly affects the flotation behaviors of the minerals. Figure 6 shows the measured adsorption isotherms for the collectors adsorbed on muscovite as a function of the equilibrium concentration. The solution pH was controlled at approximately 7. The changing trends of the curves clearly coincided with the first two regions of the four-region model. In the first stage, the electrostatic adsorption between surfactant ions and the negatively charged (001) basal planes is the dominant driving force for the considerable increase in the adsorbed amounts greatly, in addition to the cation exchange adsorption between the surfactant and the potassium ions released on the muscovite surface. In the second stage, the nonpolar carbon chains of collectors were highly intertwined with one another and formed a hydrophobic layer on the muscovite surfaces, affording a hydrophobic state [16].

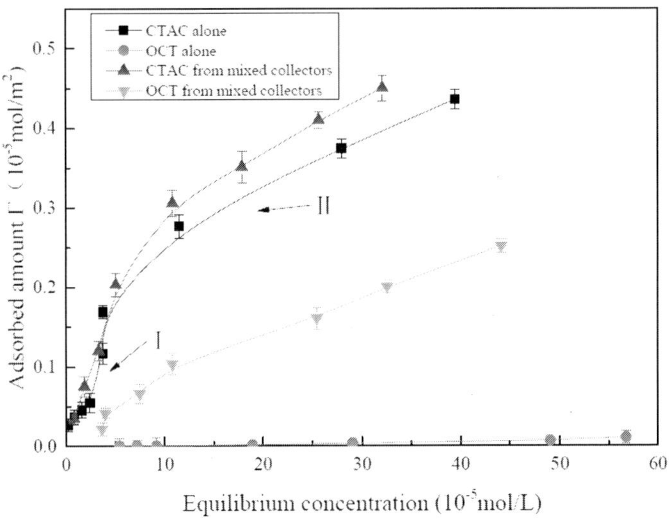

Figure 6. Adsorption isotherms of CTAC and OCT on muscovite from mixed CTAC/OCT solutions at a feed molar ratio of 2:1 CTAC/OCT (α_{CTAC} = 0.67) and pH 7.0.

In addition, a small amount of OCT was adsorbed on the muscovite surface in the single OCT solution. However, in the presence of mixed CTAC/OCT, the adsorbed amount of OCT on the muscovite surface significantly increased. This increase can be explained by the co-adsorption OCT adsorbed between the molecules of CTAC through the hydrogen bonding and hydrophobic interactions, which can decrease the electrostatic repulsions between CTAC molecules and considerably enhance the adsorption of CTAC and OCT on the mineral surface. Because of the co-adsorption of CTAC and OCT on the muscovite surface in the mixed collectors, the amounts of adsorbed CTAC and OCT are distinctly increased. Such a synergistic effect between CTAC and OCT led to considerably stronger hydrophobicity on the muscovite surface and a better surface activity compared to single CTAC or OCT. Thus, the mixed CTAC/OCT can facilitate the flotation of muscovite.

4. CONCLUSIONS

In this study, a cationic collector cetyltrimethylammonium chloride (CTAC) and nonionic collector octanol (OCT) are found to exhibit a synergistic effect on the flotation and adsorption of muscovite. The following conclusions are therefore drawn: the mixed CTAC/OCT can facilitate the flotation of muscovite, indicating that that the mixed surfactants exhibit a stronger collecting ability than the individual CTAC or OCT, particularly at $\alpha_{CTAC} = 0.67$. The results obtained from surface tension tests and contact angle tests further confirm the synergistic effects of the mixed CTAC/OCT on the gas-liquid and solid-liquid interface. The maximum synergistic interaction is observed at $\alpha_{CTAC} = 0.67$. According to the adsorption experiments, the adsorbed amounts of CTAC and OCT are distinctly enhanced due to the co-adsorption of CTAC and OCT on the muscovite surface in the presence of the mixture.

ACKNOWLEDGMENTS

The authors would like to thank the National Natural Science Foundation of China (Grant Nos. 51674207, 50974134 and 51304162) for their support.

AUTHOR CONTRIBUTIONS

Hao Jiang, conceived of and designed the experiments. Wangying Ji, Qinghong Yang, and Chen Zhao performed the experiments and analyzed the data. Longhua Xu, Yuehua Hu, and Hao Jiang contributed reagents and materials. Wangying Ji and Qinghong Yang wrote the paper.

CONFLICTS OF INTEREST

The authors declare no conflict of interest.

REFERENCES

1. Xu, L.; Wu, H.; Dong, F.; Wang, L.; Wang, Z.; Xiao, J. Flotation and adsorption of mixed cationic/anionic collectors on muscovite mica. Miner. Eng. 2013, 41, 41–45.
2. Yan, L.; Masliyah, J.H.; Xu, Z. Interaction of divalent cations with basal planes and edge surfaces of phyllosilicate minerals: Muscovite and talc. J. Colloid Interface Sci. 2013, 404, 183–191.
3. Xu, L.; Hu, Y.; Dong, F.; Gao, Z.; Wu, H.; Wang, Z. Anisotropic adsorption of oleate on diaspore and kaolinite crystals: Implications for their flotation separation. Appl. Surf. Sci. 2014, 321, 331–338.
4. Nakazawa, H.; Ogasawara, S.; Sato, H. Separation of sericite from quartz by amine flotation. Nippon Kogyo Kaishi 1988, 104, 803–807.
5. Rao, K.H.; Forssberg, K.S.E. Mixed collector systems in flotation. Int. J. Miner. Process. 1997, 51, 67–79.
6. Von Rybinski, W.; Schwuger, M.J. Adsorption of Surfactant Mixtures in Froth Flotation. Langmuir 1986, 2, 639–643.
7. Xu, L.; Hu, Y.; Tian, J.; Wu, H.; Wang, L.; Yang, Y.; Wang, Z. Synergistic effect of mixed cationic/anionic collectors on flotation and adsorption of muscovite. Colloids Surf. A 2016, 492, 181–189.
8. Poorkani, M.; Banisi, S. Industrial use of nitrogen in flotation of molybdenite at the Sarcheshmeh copper complex. Miner. Eng. 2005, 18, 735–738.
9. Ejtemaei, M.; Gharabaghi, M.; Irannajad, M. A review of zinc oxide mineral beneficiation using flotation method. Adv. Colloid Interface Sci. 2014, 206, 68–78.
10. Tian, J.; Xu, L.; Deng, W.; Jiang, H.; Gao, Z.; Hu, Y. Adsorption mechanism of new mixed anionic/cationic collectors in a spodumene-feldspar flotation system. Chem. Eng. Sci. 2017, 164, 99–107.
11. Xu, L.; Hu, Y.; Tian, J.; Wu, H.; Yang, Y.; Zeng, X.; Wang, Z.; Wang, J. Selective flotation separation of spodumene from feldspar using new mixed anionic/cationic collectors. Miner. Eng. 2016, 89, 84–92.
12. Desai, T.R.; Dixit, S.G. Interaction and Viscous Properties of Aqueous Solutions of Mixed Cationic and Nonionic Surfactants. J. Colloid Interface Sci. 1996, 177, 471–477.
13. Penfold, J.; Staples, E.J.; Tucker, I.; Thompson, L.J. Adsorption of Mixed Cationic and Nonionic Surfactants at the Hydrophilic Silicon Surface from Aqueous Solution:Studied by Specular Neutron Reflection. Langmuir 1997, 13, 6638–6643.
14. Dixit, S.G.; Vanjara, A.K.; Nagarkar, J.; Nikoorazm, M.; Desai, T. Co-adsorption of quaternary ammonium compounds—Nonionic surfactants on soild-liquid interface. Colloids Surf. A 2002, 205, 39–46.
15. Desai, T.R.; Dixit, S.G. Coadsorption of Cationic-NonionicSurfactant Mixtures on Polytetra Fluoroethylene (PTFE) Surfave. J. Colloid Interface Sci. 1996, 179, 544–551.
16. Somasundaran, P.; Krishnakumar, S. Adsorption of surfactants and polymers at the solid-liquid interface. Colloids Surf. A 1997, 123, 491–513.
17. Blanco, E.; Rodriguez-Abreu, C.; Schulz, P.; Ruso, J.M. Effect of alkyl chain asymmetry on catanionic mixtures of hydrogenated and fluorinated surfactants. J. Colloid Interface Sci. 2010, 341, 261–266.
18. Wang, L.; Hu, Y.; Liu, J.; Sun, Y.; Sun, W. Flotation and adsorption of muscovite using mixed cationic-nonionic surfactants as collector. Powder Technol. 2015, 276, 26–33.

19. Fang, X.; Zhao, S.; Mao, S.; Yu, J.; Du, Y. Mixed micelles of cationic-nonionic surfactants: NMR self-diffusion studies of Triton X-100 and cetyltrimethylammonium bromide in aqueous solution. Colloid Polym. Sci. 2003, 281, 455–460.
20. Jones, T.; Rubingh, D.N. Mechanism of Detergency in Systems Containing Cationic and Nonionic Surfactants. Ind. Eng. Chem. Prod. Res. Dev. 1982, 21, 176–182.
21. Wang, L.; Hu, Y.; Sun, W.; Sun, Y. Molecular dynamics simulation study of the interaction of mixed cationic/anionic surfactants with muscovite. Appl. Surf. Sci. 2015, 327, 364–370.
22. Vora, S.; George, A.; Desai, H.; Bahadur, P. Mixed Micelles of Some Anionic-Anionic, Cationic-Cationic, and Ionic-Nonionic Surfactants in Aqueous Media. J. Surfactants Deterg. 1999, 2, 213–221.
23. Hao, L.S.; Deng, Y.T.; Zhou, L.S.; Ye, H.; Nan, Y.Q.; Hu, P. Mixed micellization and the dissociated Margules model for cationic/anionic surfactant systems. J. Phys. Chem. B 2012, 116, 5213–5225.
24. Rojas, Y.V.; Phan, C.M.; Lou, X. Dynamic surface tension studies on poly(N-vinylcaprolactam/N-vinylpyrrolidone/N,N-dimethylaminoethyl methacrylate) at the air-liquid interface. Colloids Surf. A 2010, 355, 99–103.
25. Maiti, K.; Bhattacharya, S.C.; Moulik, S.P.; Panda, A.K. Physicochemistry of the binary interacting mixtures of cetylpyridinium chloride (CPC) and sodium dodecylsulfate (SDS) with special reference to the catanionic ion-pair (coacervate) behavior. Colloids Surf. A 2010, 355, 88–98.
26. Zhao, S.; Zhu, H.; Li, X.; Hu, Z.; Cao, D. Interaction of novel anionic gemini surfactants with cetyltrimethylammonium bromide. J. Colloid Interface Sci. 2010, 350, 480–485.
27. Wang, R.; Li, Y.; Li, Y. Interaction Between Cationic and Anionic Surfactants: Detergency and Foaming Properties of Mixed Systems. J. Surfactants Deterg. 2014, 17, 881–888.
28. Rosen, M.J.; Zhou, Q. Molecular Interactions of Surfactants in Mixed Monolayers at the Air/Aqueous Solution Interface and in Mixed Micelles in Aqueous Media:The Regular Solution Approach. Langmuir 2003, 19, 4555–4562.
29. Khan, I.A.; Mohammad, R.; Alam, M.S.; Kabir-ud-Din. Surface Properties and Mixed Micellization of Cationic Gemini Surfactants with Ethyleneamines. J. Chem. Eng. Data 2010, 55, 370–380.

CHAPTER 4

Mineralogical and Chemical Characteristics of Clays Consumed in the District of Abidjan (Côte D'Ivoire)

Vamoussa Coulibaly[1*], Joseph Sei[1], Léon Konan Koffi[1], Samuel Oyetola[1], El-Aïd Jdid[2,3], Fabien Thomas[4,5]

[1]*Laboratoire de Chimie des Matériaux Inorganiques, UFR SSMT, Université Félix Houphouët-Boigny de Cocody-Abidjan, Abidjan, Côte d'Ivoire*

[2]*Université de Lorraine, Laboratoire Géoressources—UMR 7359, Rue du Doyen Roubault, Vandœuvre, France*

[3]*CNRS, Laboratoire Géoressources—UMR 7359, Rue du Doyen Roubault, Vandoeuvre, France*

[4]*Université de Lorraine, Laboratoire Interdisciplinaire des Environnements Continentaux—UMR 7360, 15 rue du Charmois, Vandœuvre, France*

[5]*CNRS, Laboratoire Interdisciplinaire des Environnements Continentaux—UMR 7360, 15 rue du Charmois, Vandœuvre, France*

ABSTRACT

Clay materials of the quarry of Bingerville (district of Abidjan) have been characterized by chemical analysis, X-rays diffraction, infrared spectroscopy and thermal analysis. These materials contain kaolinite (41% - 84%), illite (4% - 10%), quartz (14% - 27%), goethite (2% - 5%) and small amount of rutile (1%) that are not toxic minerals. They also contain some heavy metals which are beneficial to human biological activity (cobalt, copper, molybdenum, zinc) in tiny quantities. Some other heavy metals (lead, cadmium) which are considered as poisonous for human are present in very low content. The samples characterized by relatively fine grains are moderately crystallized. Their specific surface area varied from 26 to 43 $m^2 \cdot g^{-1}$. The mineralogical and physicochemical characteristics of these samples are like ones known for their healing properties. The consumption by internal way of studied materials, although it may be beneficial, requires a sifting to remove coarse grains (ϕ > 2 μm) and a previous microbiological control.

Keywords: Mineralogy, Physicochemical Analysis, Kaolinite, Illite, Edibility

1. INTRODUCTION

Clay materials are natural products widely used in various domains including construction, ceramics [1] , paper [2] , breeding [3] , pharmaceuticals [4] [5] , etc. These multiple uses of clays are due to their properties which are different according to their structure, mineralogical composition and origin.

Particularly, in public health, the clay used as healing material must have fine grain size, high specific surface area, high specific heat, and high adsorption and absorption capacities, while its cooling rate or heat diffusiveness should be low. Clay should also be easy to handle, and give a pleasant sensation when applied directly onto the skin [6] - [8].

In terms of typology, three main groups of natural healing clays can be classified: swelling clay (smectite), illite and kaolinite clays and finally fibrous clays (palygorskite and sepiolite).

Recent investigations showed the antibacterial and bactericidal properties of some clay, based on experiments carried out by [9] - [13], showing the ability to kill a broad spectrum of pathogenic bacteria. One of these bacteria is Mycobacterium ulcerans which causes skin chronic disorder named Buruli ulcer, mostly endemic to central and western Africa. Treated with special clay of bentonite type, the skin lesions or wounds became gradually less acute due to skin tissue regeneration.

In Côte d'Ivoire, clays consumed (especially by pregnant women) for their healing properties, are traditionally prepared and marketed without prior mineralogical and microbiological tests. They are directly ingested or used externally through skin applications.

While this practice provides the consumers a certain well-being, it has drawbacks because these materials, which are generally heterogeneous, may present potential risks for human health, due to toxic minerals or soluble chemical species [14]. Therefore, a better knowledge of the mineralogy and the properties of these materials are critical to prevent risks related to this practice. The aim of the present work is to obtain a comprehensive picture of the mineralogical and chemical composition of samples that are commonly extracted and marketed in the surroundings of Abidjan (Côte d'Ivoire), in order to evaluate their possible negative impact on health.

2. MATERIALS AND METHODS

2.1. Materials

The studied clay materials were sampled from the exploited quarry of Bingerville (district of Abidjan, Côte d'Ivoire), located in the Southeast of the sedimentary basin containing clay and sand formations [15] . These materials were taken according to their color, and subjected to the same heat treatment than the marketed materials commonly called "Lokpo". This treatment involves heating and roaster of the samples for 3 to 5 days in an oven powered by firewood. This treatment confers the clay material an appetizing smell and taste.

For the present study, seven samples used were labeled as follows:

LBF: white Lokpo; LJPF: yellow Lokpo; LRF: red Lokpo; LVF: purple Lokpo; LMF: brown Lokpo; LJFF: dark-yellow Lokpo; LNF: black Lokpo.

The materials labeled LBF, LJPF, LRF, LVP and LMF are marketed as "edible". Conversely, materials LJFF and LNF, probably because of their aspects (color, flavor) are not consumed. The latter were included in this study so as to know if there mineralogy and physicochemical properties are different from the first. For the different analyses the collected samples were dried at 40°C during 24 hours in a stove, and ground in a planetary grinder during 1minute.

2.2. Analytical Methods

Chemical analysis was performed using a Jobin-Yvon 70P quantometer equipped with a plasma ICP-AES (Inductively Coupled Plasma—Atomic Emission Spectroscopy) for major elements and a Perkin Elmer Elan 5000 equipped with a plasma ICP-MS (Inductively Coupled Plasma-Mass Spectroscopy) for the trace elements. Prior to analysis, samples were mixed with lithium metaborate ($LiBO_2$) melted and then dissolved in nitric acid (HNO_3).

X-Ray diffraction patterns on powders and on oriented preparations were recorded using a D8 Advance Bruker diffractometer equipped with Co Kα radiation (λ = 1.7890 Å), operating at 35 kV and 45 mA. The diffraction patterns were obtained in a 2θ angle ranging from 3° to 64° at a scanning rate of 1° min^{-1}. Oriented samples were prepared by extracting the fraction of less than 2 μm of the samples by sedimentation, centrifugation of the supernatant and spreading of the centrifugation pellet on a glass slide.

Infrared spectra in the range of wave numbers from 600 to 4000 cm^{-1} were obtained in the diffuse reflectance mode, using a Infrared Fourrier transformer spectrometer Bruker IFS 55, equipped with a He-Ne laser and a MCT type detector (Mercury telluride and cadmium) broad band (6000 - 600 cm^{-1}) and high sensitivity. For analysis, the samples (70 mg) were diluted in KBr (370 mg) and manually ground.

Differential thermal analyses and thermogravimetric analyses were performed using a Setsy SETARAM 24 DTA/TGA equipment, according to a 5°C·min^{-1} temperature increase from room temperature to 1200°C, under an argon flow.

The particle size distribution was determined using the laser diffraction analyzer Malvern Mastersizer MS 20 analyzing particle sizes between 0.1 and 600 μm with an accuracy of 1% of the median diameter.

The specific surface area was determined by applying the BET method to nitrogen gas adsorption isotherms at 77 K. The adsorption isotherms were recorded on an automatic home-made equipment along with a turbomolecular vacuum group ensuring a dynamic vacuum of 10^{-6} mbar, and a 0 - 1000 mbar Edwards pressure gauge. The gas used is nitrogen higher grade quality (purity > 99.995%) supplied by Alphagaz.

3. RESULTS AND DISCUSSION

3.1. Mineralogical Composition of the Clay Samples

X-ray powder diagrams (Figure 1(a)) showed that samples were mainly composed of kaolinite (7.15 Å and 3.58 Å), illite (9.95 Å and 4.97 Å), goethite (4.15 Å) and quartz (4.25 Å and 3.34 Å). Except sample LJFF, which contained a strong proportion of goethite, the samples showed quite similar mineral composition, the variability of which will be quantified below. Figure 1(b) additionally showed that the fine (<2 μm) fraction is devoid of quartz and contains mainly kaolinite, and secondly illite and/or goethite.

In the 600 - 1200 cm^{-1} range of the infrared spectra (Figure 2(a)), the bands observed at 754 and 792 cm^{-1} are specific to clays from the kaolinite family and are assigned to OH bending toward the surface of the clay sheet [16] [17] . In all samples, additional bands less marked are observed at 831 and 876 cm^{-1}. They are respectively characteristic of the deformation vibration of Al-OH-Fe and Fe-OH-Fe [18] following the substitution of aluminium by iron (structural iron) in the kaolinite structure. The bands observed at 914 and 935 cm^{-1} correspond to the libration of OH bonds in Al-OH groups. The first vibration is related to internal hydroxyls and the second one corresponds to hydroxyls on the surface. Finally, the bands observed at 1010, 1035 and 1110 cm^{-1} are assigned to the stretching vibrations of Si-O bonds.

In the 3000 - 3800 cm^{-1} domain (Figure 2(b)), the infrared spectra of the studied samples showed the four distinct bands characteristic for kaolinite, corresponding to the stretching vibrations of Al_2OH in kaolinite. The band at 3620 cm^{-1} which is very intensive in all samples, corresponds to the stretching vibration of hydroxyls of the median plane of the layer (also called internal hydroxyls) found in both 1:1 and 2:1 minerals. The three other bands respectively at 3653, 3666 and 3695 cm^{-1} correspond to the stretching vibrations of the hydroxyls located in the interlayer space (also called external hydroxyls). These ones appear only in 1:1 minerals and two forms can be distinguished [19] [20] . The band at 3653 cm^{-1} is due to an external hydroxyl forming an angle of 76° with the crystallographic axis c. Those at 3666 and 3695 cm^{-1} are due to the coupling of two external hydroxyls, respectively in phase and out of phase.

In addition, the infrared spectra display a band around 3596 cm^{-1}, assigned to Al-OH-Fe stretching according to Delineau et al. [21].

Thermal analyses (Figure 3) confirmed the mineralogical composition of the studied samples. On the thermograms, the endothermic peak of low intensity was observed between 50°C and 90°C, along with a small mass depletion (0.5% to 1%), which correspond to the loss of water molecules adsorbed on mineral surface (hygroscopic water).

A significant endothermic phenomenon between 250°C and 300°C involving a small mass loss took place for samples LJFF and LJPF. This corresponds to the transformation of goethite to hematite by dehydroxylation [22].

In the range in of 500°C to 525°C, in correlation with temperatures reported by other authors [23] - [25] , the thermograms showed an endothermic peak and a strong mass loss, attributed to the loss of structural hydroxyls from the octahedral sheet of the kaolinite [26] [27] upon formation of metakaolinite [28] .

The slight modulation observed between 560°C and 575°C without loss of mass is characteristic of the ($\alpha \leftrightarrow \beta$) quartz transition [29] [30]. This is confirmed by the presence of an exothermic peak on DTA cooling plot where the reverse action is observed with an exothermic peak at this temperature.

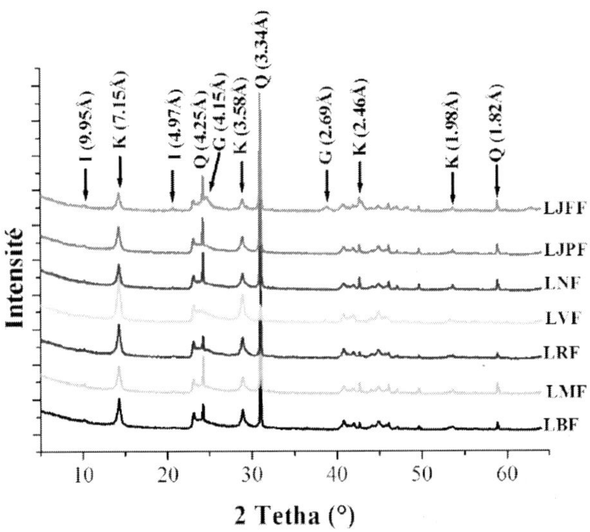

(a)

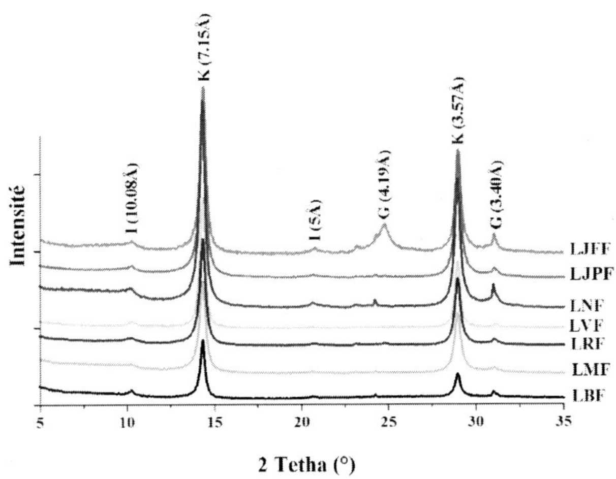

(b)

Figure 1. X-ray patterns of samples (l = 1.7889 Å) (G = Goethite, I = Illite, K = Kaolinite, Q = Quartz); (a) On powders disorientated for the total rock; (b) By oriented deposits for the fine fraction.

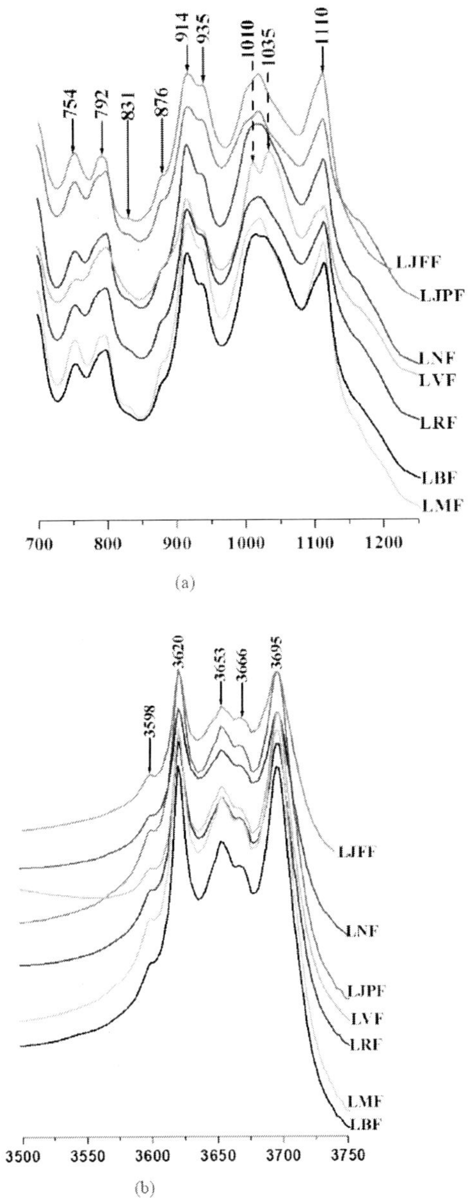

Figure 2. Infrared spectra of samples.

Finally, the exothermic peak toward 955°C indicated the transformation of metakaolinite into spinel [31] - [33].

3.2. Chemical Composition of the Samples

Chemical analysis reveals that, except LJFF, all samples (Table 1) have relatively similar chemical composition. The SiO_2 content range from 45.73% to 57.05%, those of Al_2O_3, from 25.22% to 34.41%, and trace amount of Fe_2O_3, TiO_2, Na_2O, K_2O and MgO are detected. The sample LJFF shows a high concentration of iron oxide (26.02%) and lower concentrations of silica (42.71%) and alumina (18.33%).

The elements identified as trace (Table 2) represent the majority of heavy metals, lanthanides and actinides. Usually these trace metals are trapped into oxy hydroxides of Fe (III) or in the clay particles. They can be retained by sorption or coprecipitation. Samples contain some other heavy metals (cobalt, copper, molybdenum, zinc).

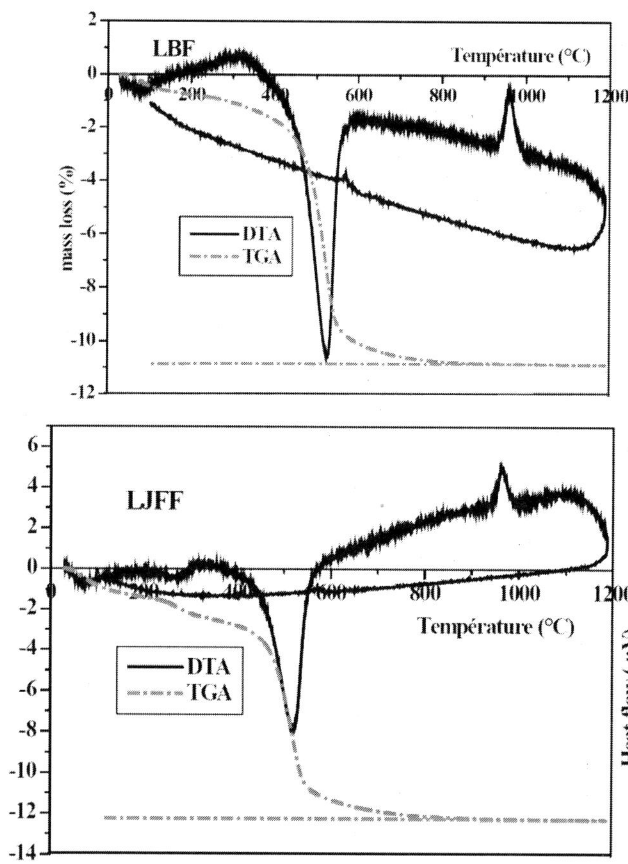

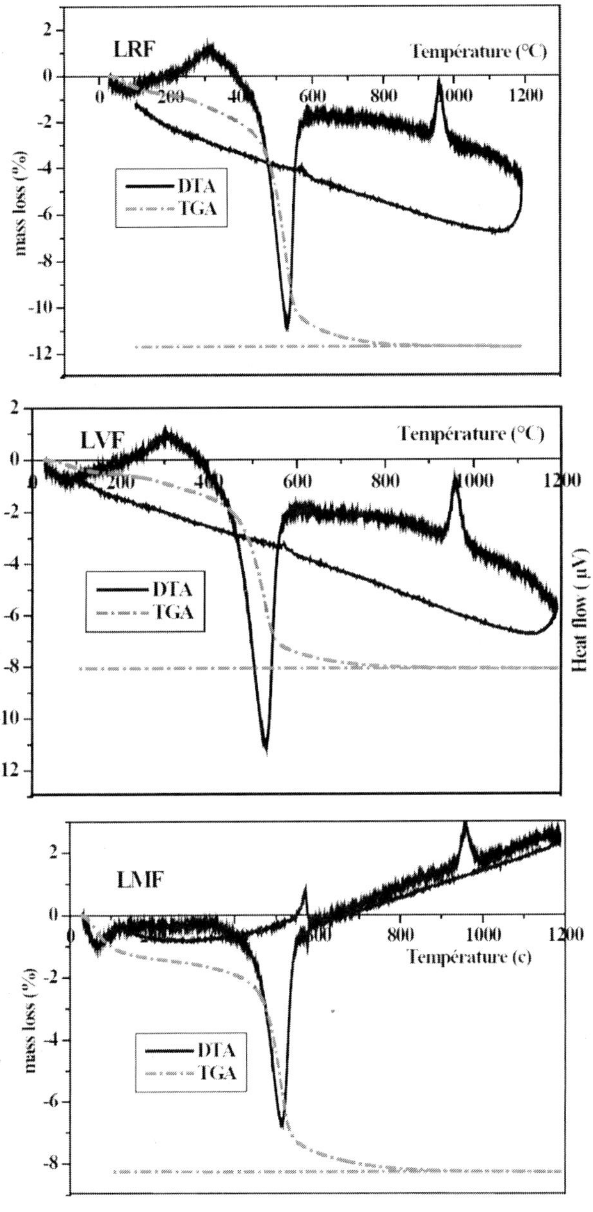
LRF
Température (°C)
0
200
400
600
800
1000
1200
mass loss (%)
2
0
-2
-4
-6
-8
-10
-12
DTA
TGA
LVF
Température (°C)
0
200
400
600
800
1000
1200
2
0
-2
-4
-6
-8
-10
-12
DTA
TGA
Heat flow (µV)
LMF
0
800
1000
1200
Température (c)
mass loss (%)
2
0
-2
-4
-6
-8
DTA
TGA

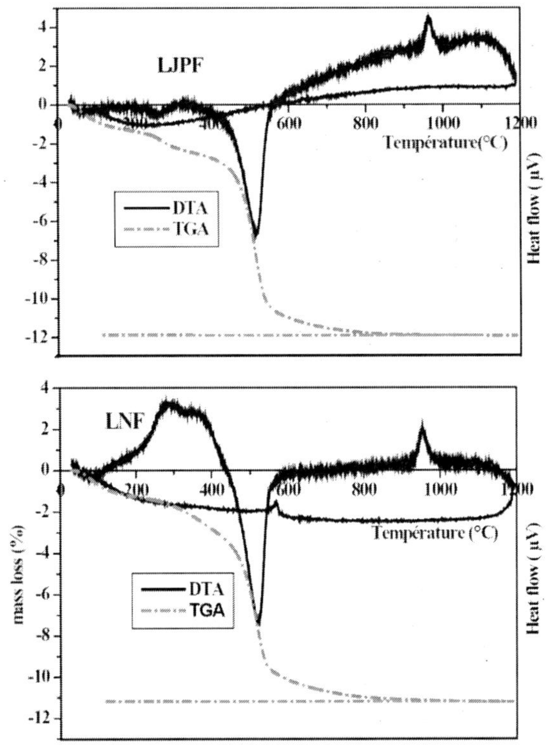

Figure 3. Thermograms of the studied clay samples.

3.3. Quantitative Mineral Composition

Determination of the minerals in a complex sample (Table 3) can be estimated from qualitative mineralogical composition revealed by X-ray diffraction (Figure 1) and infrared spectroscopy (Figure 2), assuming an ideal composition of each of these minerals. For this purpose, the method developed by Njopwouo [34] and Yvon et al. [35] was carried out. This method consists in writing for every element (a) the following relation:

$$Ta = \sum_{i=1}^{n} Mi \times Pi\ (a)$$

Ta = Ta% content in element "*a*" in the material;

Mi: mineral content "*i*" in the material;

Pi(a): proportion of the element "*a*" in the mineral "*i*" (evaluated from the ideal formula).

Table 1. Chemical content in major elements in weight percent (n.d. = not determined).

Samples	Colour	SiO_2	Al_2O_3	Fe_2O_3	MnO	MgO	CaO	Na_2O	K_2O	TiO_2	P_2O_5	H_2O
LBF	White	52.83	29.74	2.34	n.d	0.29	n.d	0.11	1.17	1.06	0.08	12.03
LRF	Red	51.06	30.18	4.48	0.00	0.22	n.d	0.11	1.16	1.13	0.06	11.97
LMF	Brown	56.55	25.44	4.25	0.01	0.36	n.d	0.08	1.01	1.10	0.08	11.46
LVF	Purple	45.73	34.41	4.92	n.d	0.07	n.d	0.06	0.57	1.12	0.09	13.12
LJPF	Yellow	55.33	27.81	3.97	0.00	0.24	n.d	0.06	0.77	0.94	0.21	11.40
LNF	Black	57.05	25.22	2.23	0.01	0.37	n.d	0.07	0.97	1.03	0.10	13.02
LJFF	Dark-yellow	42.71	18.33	26.02	n.d	0.14	n.d	0.05	0.63	0.87	0.12	11.69

Table 2. Chemical composition of trace elements in ppm (n.d. = not determined).

Samples	As	Ba	Be	Bi	Cd	Ce	Co	Cr	Cs	Cu	Dy
LBF	7.77	313.80	1.28	0.35	0.13	120.40	3.53	163.10	7.33	19.34	4.79
LRF	12.26	299.30	1.24	0.40	0.15	53.07	3.17	160.00	5.76	12.97	2.49
LMF	4.55	21.10	0.98	n.d.	0.30	30.42	29.50	94.58	0.51	83.51	7.58
LVF	18.27	159.40	1.13	0.48	0.14	90.75	3.33	173.70	2.05	7.45	2.16
LJPF	14.67	283.90	1.42	0.28	0.15	418.10	2.38	160.70	5.10	11.90	8.09
LNF	12.53	291.70	1.78	0.33	0.20	132.50	21.24	149.60	6.39	19.24	8.09
LJFF	10.75	146.50	1.22	0.25	0.12	29.97	3.09	159.20	3.86	34.63	2.33

Samples	Er	Eu	Ga	Gd	Ge	Hf	Ho	In	La	Lu	Mo	Nb
LBF	2.24	2.16	34.59	6.29	2.04	5.49	0.83	n.d	50.75	0.36	1.073	12.53
LRF	1.50	0.77	37.13	2.61	1.89	5.74	0.49	n.d	31.88	0.29	1.21	13.32
LMF	4.41	2.01	20.18	7.05	1.76	5.06	1.57	n.d	11.93	0.69	0.38	10.85
LVF	1.14	0.96	40.35	2.73	1.57	5.39	0.38	n.d	48.98	0.20	2.64	15.79
LJPF	2.84	5.08	26.87	13.31	2.32	6.30	1.16	n.d	185.60	0.36	0.58	11.20
LNF	3.75	3.16	28.66	10.29	2.17	6.70	1.43	n.d	53.53	0.54	1.03	13.45
LJFF	1.43	0.63	22.28	2.25	1.79	4.64	0.48	n.d	17.35	0.26	0.55	10.78

Samples	Nd	Ni	Pb	Pr	Rb	Sb	Sm	Sn	Sr	Ta	Tb
LBF	49.81	22.52	17.98	13.46	62.08	0.59	9.29	2.79	88.88	1.14	0.91
LRF	18.80	23.19	18.91	5.53	56.74	0.62	3.24	2.90	71.36	1.19	0.41
LMF	21.34	52.18	4.65	4.49	13.05	0.21	6.11	2.19	274.1	0.89	1.20
LVF	29.11	25.47	29.98	9.25	24.59	0.71	4.43	3.04	87.10	1.58	0.40
LJPF	140.50	16.08	100.84	41.91	46.72	0.47	23.56	2.12	131	1.00	1.76
LNF	64.59	96.75	19.21	16.03	63.61	0.64	12.95	2.71	92.55	1.17	1.50
LJFF	12.51	14.34	8.66	3.44	36.81	0.426	2.47	2.02	31.58	0.95	0.37

Samples	Th	Tm	U	V	W	Y	Yb	Zn	Zr
LBF	9.22	0.33	1.85	103.1	1.73	20.00	2.22	33.98	196.6
LRF	9.88	0.24	1.95	108.8	1.94	13.63	1.72	37.1	207.2
LMF	0.81	0.67	0.39	259.3	n.d	44.02	4.39	91.68	213.4
LVF	13.00	0.18	2.28	150.2	1.98	9.89	1.26	34.05	194.6
LJPF	8.31	0.37	2.11	151.7	1.40	24.56	2.35	26.07	246
LNF	9.13	0.55	1.96	97.54	1.66	36.75	3.63	39.12	259.2
LJFF	7.54	0.23	3.98	90.96	1.31	13.60	1.64	66.18	178.7

The mineralogical composition was determined on the basis of following assumptions:

The contents of illite, goethite and rutile are respectively determined from the percentage of chemical analysis of K_2O, TiO_2 and Fe_2O_3.

Kaolinite is determined by the difference between the percentage of chemical analysis of Al_2O_3 and the proportion of Al_2O_3 corresponding to illite;

In all samples, kaolinite is the main mineral with content varying from 41% to 84%. Samples contain significant amounts of quartz (14% - 27%), except Sample LVF (4.3%) in which kaolinite is great amount. Goethite is present in low concentrations (2% - 5%) for all samples except LJFF (26%). The high value for this sample is in agreement with the high amount in Fe_2O_3 obtained in the chemical analysis. All samples contain also illite (4% - 10%) associated with the rutile trace (1%).

The different colors of samples of clay materials are generally due to the iron oxidation state they contain. The red and yellow colors observed for some samples imply the presence of iron oxidized form (Fe III). The high content of goethite in the sample LJFF justifies its dark yellow color. For sample LVF, purple color (mixture of blue and red) can be due to the coexistence of ferric and ferrous iron in its composition, because the latter is responsible for the blue color. These observations correlate with the study carried out on their iron speciation by Mössbauer spectroscopy [36] . The mineral content of sample LNF is almost identical to that of most other samples. Only its appearance can justify the fact that it is not consumed internally.

3.4. Crystallinity of the Minerals

The crystallinity of the samples was evaluated using data from the X-ray diffraction, differential thermal analysis and infrared spectra. The different parameters to assess the crystallinity are shown in Table 4.

The number of layers (L) varied from 28 to 43, the crystalline index R2 (defined by Case et al. [37]) from 0.84 to 1.06 and the index of Hinkley (Ic) [38] , from 0.61 to 1.32.

These parameters calculated from the diffraction patterns indicated that the samples were moderately crystallized. Indeed, the well-crystallized kaolinites were characterized by a large number of layers ($L \approx 75$ layers), low crystallinity and Hinkley indices.

Table 3. Potential mineralogical composition of samples.

	Mineralogical composition (%)					
	Rutile	Goethite	Illite	Kaolinite	Quartz	TOTAL
LBF	1.06	2.34	9.73	66.79	17.06	96.98
LJPF	1.0	3.97	6.44	65.02	21.87	98.3
LRF	1.13	4.48	9.66	67.98	14.77	98.02
LVF	1.12	4.92	4.74	83.61	4.3	98.69
LMF	1.10	4.25	8.38	57.06	25.96	96.75
LJFF	1.0	26.02	5.28	41.86	20.67	94.83
LNF	1.03	2.23	8.11	56.77	26.71	94.85

Table 4. Crystallinity parameters.

Samples	L	R2	Ic	T_{endo} (°C)	S.R	P1	P2
LBF	39	0.92	1.32	521	2.21	-3.2×10^{-3}	0.95
LJPF	43	0.99	0.97	509	2.33	-5.7×10^{-3}	0.92
LRF	32	0.96	1.00	524	1.72	-4.5×10^{-3}	0.96
LVF	35	0.84	0.61	525	2.29	-3.4×10^{-3}	0.96
LMF	39	0.98	0.99	507	2.33	-5.1×10^{-3}	0.94
LJFF	28	0.96	1.00	515	2.31	-5.5×10^{-3}	0.73
LNF	35	1.06	1.13	517	1.97	-4.5×10^{-3}	0.94

The lower crystallinity of samples is also demonstrated by the temperature (Tendo) and the slope ratio (S.R), reflecting the asymmetry of the peak of dehydroxylation on the thermograms. The well-ordered kaolinite is characterized by a dehydroxylation temperature higher than 575°C [39] and a low slop ratio, around the unity [20].

The dehydroxylation temperature and the slope ratio determined for the samples analysed are respectively in the average 515°C and 2.15.

Stretching bands of vibration of the hydroxyl groups (domain 3500 - 3700 cm^{-1}) are also sensitive to the crystallinity of the kaolinite. When this mineral is well crystallized, the four bands are very distinct. On the other side, the two intermediate bands (3653 cm^{-1} and 3666 cm^{-1}) are summarized in a broad band [40] as it is more or less the case in the infrared spectra.

Indices P1 and P2 respectively calculated from bending modes (938 and 915 cm^{-1}) and stretching modes (3655 and 3672 cm^{-1}) of the hydroxyl groups also showed that the samples are moderately crystallized. The values are negative for P1 while those of P2 are less than one. A well crystallized kaolinite is characterized by a positive value of P1 with a P2 value closer to the unity [20].

3.5. Particle Size Distribution and Texture of the Samples

The particle size distribution curves (Figure 4) showed a magnitude in the order of meshes corresponding to 90%, 75%, 50% and 10% of the passing through particles respectively noted d_{90}, d_{75}, d_{50} and d_{10} (Table 5). The samples are characterized by relatively sharpness of grains, 90% of the passing through particles have an average diameter less than 24 microns and 50% have an average diameter less than 8 microns.

The analysis of the figure and the table shows that the samples LRF and LBF are those containing the fine particles while LJPF and LVF contain less. Generally, samples LRF and LBF are the most consumed, the sharpness of their grains could involve a high absorption capacity.

The specific surface area values obtained for the samples are generally between 35 and 43 $m^2 \cdot g^{-1}$ (Table 6) and slightly higher than the values generally observed (10 to 30 $m^2 \cdot g^{-1}$) for a kaolinite [41] [42].

These values are related in part to the presence of goethite observed by X-ray diffraction [29] [30] [42] [43]. The low value of specific surface area (26.8 $m^2 \cdot g^{-1}$) for the sample LVF is explained by its low rate of fine grains. The specific surface areas observed are large therefore these samples have a high adsorption capacity. They can be used in the medical purposes. Indeed, the clays used in protecting the intestinal mucosa or in the pharmaceutical industry must have a large specific surface and high absorption capacity [4].

3.6. Evaluation of the Edibility of the Clays

The edibility of clays on the basis of their healing properties is based among other things, on their protective action and their ability to inhibit the aggressive action of pathogens that cause dysfunctionning of the gastrointestinal mucosa.

Two modes of action govern these benefits. First, the adsorption of the "aggressors" or their toxic secretions, and also the change of the thickness and rheological properties of the mucus, reinforcing the natural defense of the gastrointestinal mucosa [8].

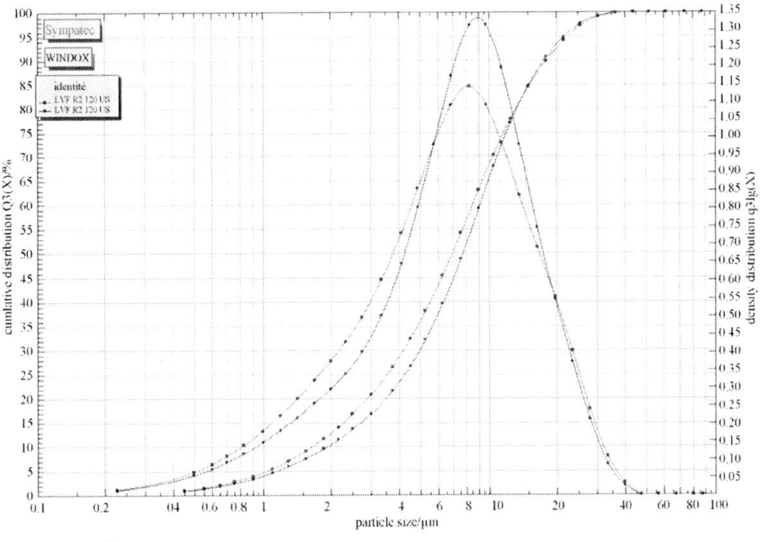

Figure 4. Granular distribution of the sample LVF.

Table 5. Value of d_{90}, d_{75}, d_{50} and d_{10}.

	Particle size (μm)				
	d_{90}	d_{75}	d_{50}	d_{25}	d_{10}
LBF	23.45	11	4.5	1.75	0.83
LJPF	14.74	8.5	5.7	3.25	1.35
LRF	13.32	7.5	4.46	2.25	1.13
LVF	17.72	11.6	7.69	4	1.94
LMF	20.85	11.6	5.12	2	0.93
LJFF	14.09	8.5	5.08	2.25	1.15
LNF	19.83	11.6	5.04	2.4	1.18

Table 6. Specific surface area (S_{BET} ($m^2 \cdot g^{-1}$)).

	LBF	LJPF	LRF	LVF	LMF	LJFF	LNF
S_{BET} ($m^2 g^{-1}$)	**35.7**	35.8	35.05	26.8	39.8	43	35.6
Fe_2O_3 (%)	2.34	3.97	4.48	4.92	4.25	26.02	2.23

The clays of the Bingerville quarry, made up of kaolinite and illite as clay minerals and other minerals including goethite, rutile and quartz, are characterized by relatively fine grain size, high specific area and thermal stability. These characteristics have already been described as suitable for healing properties therefore, their edibility [6] [7]. Indeed, all minerals constituting the clay materials are not toxic and have healing properties. Kaolinite, the main mineral is one of the most consumed clays [44] . It is used as an excipient in pharmaceutical preparations [45].

The relatively high specific surface areas (26 - 43 $m^2 \cdot g^{-1}$) should give to these clays a good adsorption capacity. They may be used in the treatment of certain poisonings by trapping toxins [46] , pesticides [47] , microorganisms [48] , gas [8] , and food allergy [49] , thus limiting their actions.

The fineness of the clay particles should give good reactivity notably strong bioadhesive properties. They can exert a stabilizing effect on the mucosal barrier, providing protection against different "aggressors" of the gastrointestinal mucosa [8].

Chemical analysis shows that samples contain numerous mineral elements including alkaline, alkaline earth metals, transition metals, lanthanides and actinides. Some of them (Cu, Al) being important for biological activity when taken up in low quantities and toxic in high dose. Others elements, such as lead, cadmium, arsenic... are toxic. However, their low level in these materials, do not induce serious dangers related to their consumption.

Radioactivity measurements performed on these clay materials [14] have shown that hazards pertaining to radionuclides are negligible.

4. CONCLUSION

Chemical analyses showed that the samples contain many elements as alkalis, heavy metal etc.; some of them are important for biological activity. Mineralogical analysis show that samples consist mainly of kaolinite, illite associated with impurities as quartz, goethite and rutile. No toxic mineral such as fibrous clay was found in the samples investigated. Samples are rich in fine-grain. The specific surface areas observed (26 - 43 $m^2 \cdot g^{-1}$) are large enough therefore these samples have a high adsorption capacity. They can be used in the treatment of certain poisonings like diarrhea. Besides the microbiological survey to be carried out, it is necessary to sift before all internal use in order to extract possible coarse particles including quartz grains.

ACKNOWLEDGEMENTS

This study was carried out in the "Laboratoire Interdisciplinaire des Environnements Continentaux" of Nancy (France). We are therefore grateful to all members of this laboratory for their contribution.

REFERENCES

1. Haussonne, J.-M., Carry, C., Bowen, P. and Barton, J. (2005) Céramiques et verres; Traité des matériaux. Presses Polytechniques et Universitaires Romandes, Lausanne (Suisse), Vol. 16.
2. Delon, J.F., Liétard, O., Cases, L.M., Richard, M., Sauret, G. and Maume, J.P. (1982) Possibilités d'emploi des Kaolins des Charentes dans le couchage des papiers et cartons. Bulletin de Mineralogie, 105, 571-581.
3. Ouachem, D. and Soltane, M. (2009) L'argile: Une alternative biologique dans l'alimentation des ruminants, Filière Ovine et Caprine, n°29-3ième trimestre.
4. Carretero, M.I. (2002) Clay Minerals and Their Beneficial Effects upon Human Health. A Review. Applied Clay Science, 21, 155-163. http://dx.doi.org/10.1016/S0169-1317(01)00085-0
5. Viseras, C., Aguzzi, C., Cerezo, P. and Lopez-Galindo, A. (2007) Uses of Clay Minerals in Semisolid Health Care and Therapeutic Product. Applied Clay Science, 36, 37-50. http://dx.doi.org/10.1016/j.clay.2006.07.006
6. Gomes, C. and Silva, J. (2001) Beach Sand and Bentonite of Porto Santo Island: Potentialities for Applications in Geomedicine. In: Gomes, C. and Silva, J., Eds., O Liberal, Camara de Lobos, Madeira, 60.
7. Gomes, C. and Silva, J. (2006) Minerals and Human Health/Os Minerais e a Saúde Humana. In: Gomes, C. and Silva, J., Eds., Litografia da Maia, Maia, 300.
8. Droy-Lefaix, M.T. and Tateo, F. (2006) Clays and Clay Minerals as Drugs, In: Bergaya, F., Theng, B.K.G., and Lagaly, G., Eds., Handbook of Clay Scienc Developments in Clay Science, 1, Elsevier, Amsterdam, 743-753
9. Williams, L.B., Holland, M., Eberl, D.D., Brunet, T. and Brunet de Courrsou, L. (2004) Killer Clays! Natural Anti- bacterial Clay Minerals. Mineralogical Society Bulletin, 139, 3-8.
10. Tong, G., Yulong, M., Peng, G. and Zirong, X. (2005) Antibacterial Effects of the Cu(II)-Exchanged Montmorillonite on Escherichia coli K88 and Salmonella choleraesuis. Veterinary Microbiology, 105, 113-122. http://dx.doi.org/10.1016/j.vetmic.2004.11.003
11. Ma'or, Z., Henis, Y., Along, Y., Orlov, E., Sorensen, K.B. and Oren, A. (2006) Antimicrobial Properties of Dead Sea Black Mineral Mud. International Journal of Dermatology, 45, 504-511. http://dx.doi.org/10.1111/j.1365-4632.2005.02621.x
12. Haydel, S.E., Remineh, C.M. and Williams, L.B. (2008) Broad-Spectrum in Vitro Antibacterial Activities of Clay Minerals against Antibiotic-Susceptible and Antibiotic-Resistant Bacteria Pathogens. Journal of Antimicrobial Chemotherapy, 61, 353-361. http://dx.doi.org/10.1093/jac/dkm468
13. Magana, S.M., Quintana, P., Aguilar, D.H., Toledo, J.A., Angeles-Chavez, C., Cortes, M.A., Leon, L., Freile-Pelegrın, Y., Lopez, T. and Torres Sanchez, R.M. (2008) Antibacterial Activity of Montmorillonites Modified with Silver. Journal of Molecular Catalysis A: Chemical, 281, 192-199. http://dx.doi.org/10.1016/j.molcata.2007.10.024
14. Coulibaly, V., Sei, J., Kouamé, N., Koua, A.A., Oyetola, S. and Brun, S. (2013) Measurement of Natural Radioactivity in the Clays Consummated in Côte d'Ivoire Using Gamma-Ray Spectrometry. Journal of Applied Sciences, 13, 140- 146.

15. Taste, J.P. (1979) Environnements sédimentaires et structuraux quaternaires du littoral du golfe de guinée (Côte d'Ivoire, Togo et Benin). Thèse Doctorat d'Etat, Université Bordeaux I, Bordeaux, 175 p.
16. Farmer, V.C. (1974) The Layer Silicates. In: Farmer, V.C., The Infrared Spectra of Minerals, Mineralogical Society, London, 331-363.
17. Russell, J.D., Farmer, V.C. and Velde, B. (1970) Replacement of OH by OD in Layer Silicates and Identification of the Vibrations of These Groups in Infrared Spectra. Mineralogical Magazine, 37, 869-879. http://dx.doi.org/10.1180/minmag.1970.037.292.01
18. Mendelovici, E., Yariv, S.H. and Villalba, R. (1979) Iron-Bearing Kaolinite in Venezuelan Laterite. I. Infrared Spectroscopy and Chemical Dissolution Evidence. Clay Minerals, 14, 323-331. http://dx.doi.org/10.1180/claymin.1979.014.4.08
19. Giese, R.F. and Datta, P. (1973) Hydroxyl Orientation in Kaolinite, Dickite and Nacrite. American Mineralogist, 58, 471-479.
20. Cases, J.-M., Liétard, O., Yvon, J. and Delon, J.-F. (1982) Etude des propriétés cristallochimiques, morphologiques, superficielles des kaolinites désordonnées. Bullettin Minéralogique, 105, 439-455.
21. Delineau, T., Allard, T., Muller, J.-P., Barres, O., Yvon, J. and Cases, J.M. (1994) FTIR Reflectance vs. EPR Studies of Structural Iron in Kaolinites. Clays and Clay Minerals, 42, 308-320. http://dx.doi.org/10.1346/CCMN.1994.0420309
22. Fan, H., Song, B. and Li, Q. (2006) Thermal Behavior of Goethite during Transformation to Hematite. Materials Chemistry and Physics, 98, 148-153. http://dx.doi. org/ 10.1016/j.matchemphys.2005.09.005
23. Singh, B. and Gilkes, R.J. (1992) Properties of Soil Kaolinite from South-Western Australia. Journal of Soil Science, 43, 654-667. http://dx.doi.org/10.1111/j.1365-2389.1992.tb00165.x
24. Melo, V.F., Singh, B., Schaefer, C.E.G.R., Novais, R.F. and Fontes, M.P.F. (2001) Chemical and Mineralogical Properties of Kaolinite-Rich Brazilian Soils. Soil Science Society of America Journal, 65, 1324-1333. http://dx.doi.org/10.2136/sssaj2001.6541324x
25. Hart, R.D., Gilkes, R.J., Siradz, S. and Singh, B. (2002) The Nature of Soil Kaolins from Indonesia and Western Australia. Clays and Clay Minerals, 50, 198-207. http://dx.doi.org/10.1346/000986002760832793
26. Toussaint, F., Fripiat, J.J. and Gastuche, M.C. (1963) Dehydroxylation of Kaolinite. I: Kinetics. The Journal of Physical Chemistry, 67, 26-30. http://dx.doi. org/10. 1021/j100795a007
27. Ortega, A., Rouquérol, F., Akhouayri, S., Laureiro, Y. and Rouquérol, J. (1993) Kinetical Study of the Thermolysis of Kaolinite between 30 °C and 1000 °C by Controlled Rate Evolved Gas Analysis. Applied Clay Science, 8, 207-214. http://dx.doi.org/10.1016/0169-1317(93)90038-3
28. Bellotto, M., Gualtieri, A., Artioli, G. and Clark, S.M. (1995) Kinetic Study of the Kaolinite-Mulliteréaction Sequence. Part I: Kaolinite Dehydroxylation. Physics and Chemistry of Minerals, 22, 207-214. http://dx.doi.org/10.1007/BF00202253
29. Jouenne, C.A. (1990) Traité de céramiques et Matériaux Minéraux. Edition Septima, Paris.
30. Brown, G. and Brindley, G.W. (1984) Crystal Structures of Clay Minerals and Their X-Ray Identification. 2nd Edition, Mineralogical Society, London, 305-360.
31. Okada, K., Otsuka, N. and Ossaka, J. (1986) Characterization of Spinel Phase Formed in the Kaolinite-Mullite Thermal Sequence. Journal of the American Ceramic Society, 69, 251-253. http://dx.doi.org/10.1111/j.1151-2916.1986.tb07353.x
32. Srikrishna, K., Thomas, G., Martinez, R., Corral, M.P., De Aza, S. and Moya, J.S. (1990) Kaolinite-Mullitereactionseries: A TEM Study. Journal of Materials Science, 25, 607-612. http://dx.doi.org/10.1007/BF00714083
33. Gualtieri, A., Belloto, M., Artioli, G. and Clark, S.M. (1995) Kinetic Study of the Kaolinite-Mullite Reaction Sequence. Part II: Mullite Formation. Physics and Chemistry of Minerals, 22, 215-222. http://dx.doi.org/10.1007/BF00202254
34. Njopwouo, D. (1984) Minéralogie et physico-chimie des argiles de Bomkoul et de Balengou (Cameroun). Utilisation dans la polymérisation du styrène et dans le

renforcement du caoutchouc naturel. Thèse de doctorat, d'État de l'Université de Yaoundé, Yaoundé.
35. Yvon, J., Baudracco, J., Cases, J.M. and Weiss, J. (1990) Eléments de minéralogie quantitative en micro-analyse des argiles. Matériaux argileux, structures, propriétés et applications. SFMC., Paris.
36. Coulibaly, V., Sei, J., Oyetola, S., Sougrati, M.T. and Jumas, J.C. (2012) Iron Speciation in the Clays Consummated in Côte d'Ivoire: A Transmission Mössbauer Spectroscopy Study. Asian Journal of Applied Sciences, 5, 460-472. http://dx.doi.org/10.3923/ajaps.2012.460.472
37. Liétard, O. (1977) Contribution à l'étude des propriétés physicochimiques cristallographiques et morphologiques des kaolins. Thèse doctorat ès-sciences, INPL, Nancy, 345 p.
38. Hinckley, D.N. (1962) Variability in "Crystallinity" Values among the Kaolin Deposits of the Coastal Plain of Georgia and South Carolina. Proceedings of the 11th National Conference on Clays and Clay Minerals, Ottawa, 13-17 August 1962, 229-235.
39. Smykatz-Kloss, W. (1974) Differential Thermal Analysis: Application and Result in Mineralogy. Springer Verlag Berlin Heidelberg, New York. http://dx.doi.org/10.1007/978-3-642-65951-5
40. Russell, J.D. and Fraser, A.R. (1994) Infrared Methods In-Clay Mineralogy: Spectroscopic and Chemical Determinative Methods. Chapman and Hall, London.
41. Guyot, J. (1969) Mesure des surfaces spécifiques des argiles par adsorption. Annales Agronomiques, 20, 333-359.
42. Arias, M., Barral, M.T. and Diaz-Fierros, F. (1995) Effect of Iron and Aluminium Oxides on the Colloidal and Surface Properties of Kaolin. Clays and Clay Minerals, 43, 406-416. http://dx.doi.org/10.1346/CCMN.1995.0430403
43. Borggaard, O.K. (1982) The Influence of Iron Oxides on the Surface Area of Soil. Journal of Soil Science, 33, 443-449.
44. Eaton, J.R. and Eaton. T.M. (1995) Bentonite: Public Ressearch Project: An Educational Compilation of Related Com- mentaries and Articles. http://eytonsearth.org/bentonite.html
45. Carretero, M.I. and Pozo, M. (2009) Clay and Non-Clayminerals in the Pharmaceutical Industry. Part I. Excipients and Medical Applications. Applied Clay Science, 46, 73-80. http://dx.doi.org/10.1016/j.clay.2009.07.017
46. Carnoy, C., Muller Alouf, H., Mullet, C., Droy-Lefaix, M.T. and Simonet, M. (2000) Oral Infection of Mice with Superantigenic Toxic Producing Yersinia pseudo tuberculosis. Effect of Diosmectite. International Journal of Medical Microbiology, 290, 477-482.
47. Meredith, T.J. and Vale, J.A. (1987) Treatment of Paraquat Poisoning in Man. Methods to Prevent Absorption. Human & Experimental Toxicology, 6, 49-57. http://dx.doi.org/10.1177/096032718700600108
48. Lipson, S.M. and Stotzky, G. (1984) Effect of Proteins on Reovirus Adsorption to Clay Minerals. Applied and Environmental Microbiology, 8, 525-530.
49. Theodorou, V., Fioramonti, J., Droy-Lefaix, M.T., Plique, O., Bueno, L. (1994) Protective Action of Diosmectite Treatment on Digestive Disturbances Induced by Intestinal Anaphylaxis in the Guinea Pig. Alimentary Pharmacology & Therapeutics, 8, 295-299. http://dx.doi.org/10.1111/j.1365-2036.1994.tb00291.x

CHAPTER 5

Zincobotryogen, $ZnFe^{3+}(SO_4)_2(OH)\cdot 7H_2O$: validation as a mineral species and new data

Zhuming Yang[1], Gerald Giester[2], Qian Mao[1], & Yuguang Ma[1], Di Zhang[1], He Li[1]

1 Institute of Geology and Geophysics, Chinese Academy of Sciences,

P.O. Box 9825, Beijing 100029, China

2 Institut für Mineralogie und Kristallographie, Universität Wien, Althanstraße 14, 1090 Wien, Austria

ABSTRACT

Zincobotryogen occurs in the oxidation zone of the Xitieshan lead-zinc deposit, Qinghai, China. The mineral is associated with jarosite, copiapite, zincocopiapite, and quartz. The mineral forms prismatic crystals, 0.05 to 2 mm in size. It is optically positive ($2V_{calc} = 54.1°$), with Z || *b* and X ∧ $c = 10°$. The elongation is negative. The refractive indices are $n_\alpha = 1.542(5)$, $n_\beta = 1.551(5)$, $n_\gamma = 1.587(5)$. The pleochroism scheme is X = colorless, Y = light yellow, Z = yellow. Microprobe analysis gave (in wt%): $SO_3 = 38.04$, $Al_2O_3 = 0.04$, $Fe_2O_3 = 18.46$, $ZnO = 13.75$, $MgO = 1.52$, $MnO = 1.23$, $H_2O = 31.06$ (by calculation), Total = 104.10. The simplified formula is $(Zn,Mg)Fe^{3+}(SO_4)_2(OH)\cdot 7H_2O$. The mineral is monoclinic, $P12_1/n1$, $a = 10.504(2)$, $b = 17.801(4)$, $c = 7.1263(14)$ Å, and $\beta = 100.08(3)°$, $V = 1311.9(5)$ Å^3, Z = 4. The strongest lines in the powder

X-ray diffraction pattern d(I)(hkl) are: 8.92 (100)(110), 6.32 (77)(−101), 5.56 (23)(021), 4.08 (22)(−221),3.21 (31)(231), 3.03 (34)(032), 2.77 (22)(042). The crystal structure was refined using 2816 unique reflections to R1(F) = 0.0355 and wR2(F^2) = 0.0651. The refined formula is $(Zn_{0.84}Mg_{0.16})Fe^{3+}(SO_4)_2(OH)\cdot 7H_2O$. The atomic arrangement is characterized by chains with composition $[Fe^{3+}(SO_4)_2(OH)(H_2O)]^{2-}$ and ~ 7 Å repeat distance running parallel to the c-axis. The chain links to a $[MO(H_2O)_5]$ octahedron (M = Zn, Mg) and an unshared H_2O molecule, and forms a larger chain building module with composition $[M^{2+}Fe^{3+}(SO_4)_2(OH)(H_2O)_6(H_2O)]$. The inter-chain module linkage involves only hydrogen bonding.

Keywords: Zincobotryogen, New mineral, Hydrated sulfate, Crystal-structure refinement, Hydrogen bonds

1. INTRODUCTION

Botryogen, $MgFe^{3+}(SO_4)_2(OH)\cdot 7H_2O$, is a hydrated sulfate of magnesium and ferric iron. Recently, Majzlan et al. (2016) investigated samples from Nuevo Cuyo, Argentina with chemical composition $MFe^{3+}_{1.010}(SO_4)_2(OH)\cdot 7H_2O$ with $M = Mg_{0.773}Zn_{0.165}Mn_{0.047}$. However, botryogen from Rammelsberg mine, Germany, had been reported to have a formula as $(Zn,Mn,Mg,Fe^{2+})Fe^{3+}(SO_4)_2(OH)\cdot 7H_2O$ (Zemann 1961). The crystal structure of this mineral was determined with all non-H atoms positions using the sample from the Rammelsberg mine by Süsse (1967, 1968) in the space group $P2_1/n$ with unit cell dimensions of a = 10.526(4), b = 17.872(7), c = 7.136(4) Å, and β = 100.13(4)°. On the basis of chemical composition the formula for this mineral was given as $(Zn_{0.47}Mn_{0.25}Mg_{0.20}Fe^{2+}_{0.08})_{\Sigma 1.00}Fe^{3+}(SO_4)_2(OH)\cdot 7H_2O$. Although zinc is the predominant element in the M site, this mineral species was labeled as botryogen, $MgFe^{3+}(SO_4)_2(OH)\cdot 7H_2O$. Furthermore, the name of botryogen-Zn is used to describe the sample from Mina Quetena, Calama, Chile (Lane et al. 2015).

Zincobotryogen,$(Zn_{0.65}Mg_{0.28}Mn_{0.11}Fe_{0.05}Na_{0.01})_{\Sigma 1.10}Fe^{3+}_{1.03}(SO_4)_{2.02}(OH)_{1.24}\cdot 6.68H_2O$, was firstly reported as a variety of botryogen from the Xitieshan lead-zinc deposit, Qinghai Province, China (Tu et al. 1964a,b). Mössbauer and Infrared (IR) spectra as well as investigations by thermogravimetry (TG) and differential thermal analysis

(DTA) for zincobotryogen were given by Yang and Fu (1988a). The crystal structure of zincobotryogen was determined, but without refinement of the (Zn, Mg) site populations and without hydrogen atoms locations (Yang and Fu 1988b). The unit cell reported was: $a = 10.52(1)$, $b = 17.85(2)$, $c = 7.133(5)$ Å, and $\beta = 100.14(6)^\circ$ with space group $P2_1/n$. However, according to the list of mineral species compiled by the Commission on New Minerals, Nomenclature and Classification, International Mineralogical Association (CNMNC, IMA), zincobotryogen was not regarded as a valid mineral species so far.

In the present study, using the original sample of zincobotryogen from the Xitieshan lead-zinc deposit, chemical analyses were performed by an electron microprobe analyzer and the crystal structure was reinvestigated. All atoms including hydrogen atoms have been located and the hydrogen bounding system is discussed. The (Zn, Mg) site population has been refined, confirming that zincobotryogen is deserved as a mineral species. The new mineral and its name have been approved by the CNMNC, IMA (IMA No. 2015–107). The mineral is named after its chemical composition and relationship to botryogen. The type specimen of zincobotryogen has been deposited in the mineralogical collection of the Museum, Institute of Geology and Geophysics, Chinese Academy of Sciences with registration number, KDX076.

2. GENERAL DESCRIPTION OF ZINCOBOTRYOGEN

2.1 Mineral occurrence

The Xietieshan lead-zinc deposit is located at the northern margin of the Qaidam Basin, Qinghai Province, China, and distributed in the Upper Ordovician Tanjianshan Group volcanic-sedimentary rocks. The Tanjianshan Group is separated by a fault from the Meso- to Neo-Proterozoic Dakendaben Group mica-quartz schists to the northeast and unconformable overlain by Devonian-Carboniferous purple conglomerates. The Tanjianshan Group is divided into three formation units from bottom upward: (1) volcanic-sedimentary rocks, (2) purple sandstone, and (3) intermediate-basic volcanic rocks. The Pb-Zn ore bodies are hosted in the marble and greenschists of the lower formation unit. The minerals in the deposit are mainly sphalerite, galena, pyrite and calcite; the minor minerals are quartz, dolomite, Parameters of the most intense powder diffraction lines are quoted bold chlorite, pyrrhotite, chalcopyrite, marcasite and arsenopyrite (Wang et al. 2008).

In the area of the Xietieshan lead-zinc deposit the climate is very arid. The average annual precipitation is below 100 mm and the evaporation capacity is usually up to 2000 mm. The oxidation zone of the lead-zinc deposit is well developed, and the thickness varies from 4 to 20 m. The oxidation zone can be divided into three vertical subzones in the profile from top downward: (1) Limonite-hematite subzone with the thickness of 2–5 cm, (2) Jarosite subzone with the thickness of 3–20 m, consisting of jarosite, quartz, gypsum, sulfur, anglesite, copiapite, zincocopiapite, sideronatrite and fibroferrite, and (3) Gypsum-sulfur subzone with the thickness of 1–4 m, consisting of gypsum, sulfur, anglesite, quartz, melanterite, RigakuFour-circle diffractometer equipped with a Saturn 724+ CCD detector, Mo tube, graphite monochrometor, φ-scans for distinct ω-angles, $\Delta\varphi = 0.5°$/frame, frame size: binned mode, 34 μm/2048 × 2048 pixels, detector-to sample distance: 45 mm. Unit-cell parameters were obtained by least-squares refinements of 2θ values roemerite and halotrichite. Zincobotryogen occurs in the Jarosite subzone, associated with jarosite, copiapite, zincocopiapite, and quartz (Tu and Li 1963; Tu et al. 1964a, b).

The spatial distribution of sulphate minerals shows a rather definite pattern in the oxidation zone of the Pb-Zn deposit: ferric sulphates are observed in the Jarosite subzone containing copiapite, sideronatrite and fibroferrite, while ferrous sulphates have their prominent development in the Gypsum-sulfur subzone containing melanterite, roemerite and halotrichite. Zn-bearing sulphate minerals, such as zincocopiapite and zincobotryogen, mainly occur in the Jarosite subzone, Mg-bearing sulphate minerals, such as pickeringite, in the Gypsum-sulfur subzone.

2.2 Physical and optical properties

The mineral forms prismatic crystals elongated in [001] from 0.5 to 2 mm in length and 0.05 to 0.2 mm in diameter, and commonly occurs in radial or globular aggregates (Fig. 1). The crystals are transparent; their colors are light to dark orange-red. The streak of zincobotryogen is light yellow, and its luster is vitreous. The density of crystal aggregates measured by using a micro-torsion balance is 2.20(1) g/cm^3 and the calculated crystal density (for Z = 4) is 2.266 g/cm^3. The mineral is soluble in hot water similar to botryogen.

Zincobotryogen is prismatic, with observed forms: {010}, $\{\bar{1}01\}$, {120}, and {110}. The a:b:c ratio calculated from the single-crystal unit cell parameters is 0.5901:1:0.4003.

Zincobotryogen is optically positive ($2V_{calc} = 54.1°$), with Z $\parallel \bar{b}$ and X ∧ $c = 10°$. The elongation is negative. The refractive indices, measured in Na-light, are $n_\alpha = 1.542(5)$, $n_\beta = 1.551(5)$, $n_\gamma = 1.587(5)$. The pleochroism scheme is: X = colorless, Y = light yellow, Z = yellow. The dispersion is strong with r > v. The compatibility factor calculated from the Gladstone-Dale rule (Mandarino 1981) is superior (0.006).

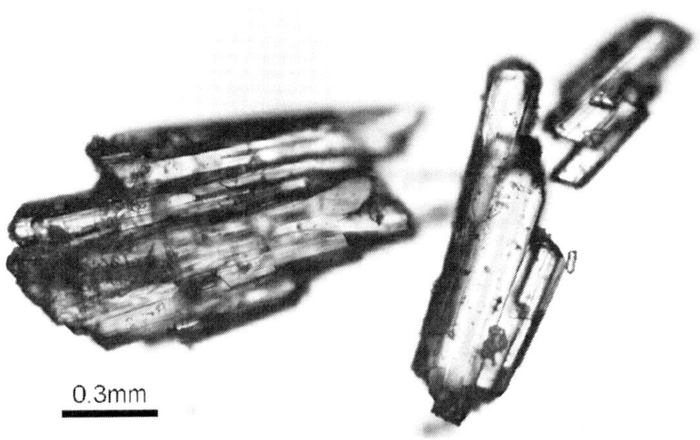

Figure. 1 Photomircograph (cross-polarized transmitted light) of several groups of zincobotryogen crystals

2.3 Chemical composition

The quantitative chemical composition of zincobotryogen was measured with a JXA-8100 electron microprobe analyzer in wavelength-dispersive spectroscopic (WDS) mode. Accelerating voltage and specimen current were kept at 15 kV and 10 nA. The beam diameter was 5 μm. The chemical composition was determined from ten electron microprobe analyses shown in Table 1. The H_2O content was calculated from charge balance and theoretical content of water molecules. The high analytical total is attributed to partial dehydration under vacuum either during carbon coating or in the microprobe chamber. This H_2O loss results in higher concentrations for the remaining constituents than are to be expected for the fully hydrated phase.

The empirical formula, based on 16 O *apfu*, is $(Zn_{0.73}Mg_{0.16}Mn_{0.08})_{\Sigma 0.97}Fe^{3+}_{0.99}(SO_4)_{2.04}(OH)_{0.82}\cdot 7H_2O$. The simplified formula is $(Zn,Mg)Fe^{3+}(SO_4)_2(OH)\cdot 7H_2O$. The ideal one is $ZnFe^{3+}(SO_4)_2(OH)\cdot 7H_2O$, with a theoretical chemical composition (in wt%): $SO_3 = 35.08$, $Fe_2O_3 = 17.49$, $ZnO = 17.83$, $H_2O = 29.60$, Total = 100.00. The chemical composition of the same sample including the concentration of H_2O and the Fe valence state was obtained by wet chemical analysis (Tu et al. 1964b) (in wt%): $SO_3 = 36.03$, $Al_2O_3 = 0.01$, $Fe_2O_3 = 18.34$, $FeO = 0.85$, $ZnO = 11.77$, $MgO = 2.50$, $MnO = 1.75$, $Na_2O = 0.05$, $H_2O = 29.35$, Total = 100.65. The empirical formula, based on 16 O *apfu*, is $(Zn_{0.65}Mg_{0.28}Mn_{0.11}Fe_{0.05}Na_{0.01})_{\Sigma 1.10}Fe^{3+}_{1.03}(SO_4)_{2.02}(OH)_{1.24}\cdot 6.68H_2O$.

Table 1 Chemical composition of zincobotryogen (10 analyses)

Constituent	Mean [wt%]	Range [wt%]	SD	EPMA calibrant
SO_3	38.0	37.24–38.46	0.34	Baryte
Al_2O_3	0.04	0–0.10	0.03	Corundum
Fe_2O_3	18.5	18.22–18.82	0.24	Hematite
ZnO	13.8	12.95–14.82	0.60	Synthetic ZnO
MgO	1.52	1.37–1.86	0.15	Diopside
MnO	1.23	1.03–1.96	0.27	Rhodonite
H_2O[a]	31.1			
Total	104.1			

[a]Calculated on the basis of $(H_2O) = 7$ *pfu* and O in (SO_4) + (OH) groups = 9 *apfu*

SD standard deviation

Our TG and DTA data indicate that zincobotryogen loses most of (H_2O) at 149 °C i.e., 28.2 wt% (Yang and Fu 1988a). The loss of (OH) is initiated at 475 °C and is complete at 578 °C associated with a weight loss of 2.0 %. The sharp absorption peak of the IR spectrum for zincobotryogen at 3550 cm^{-1} could be attributed to the OH-stretching mode, the wide peak at 3420 cm^{-1} and the sharp peak at 1635 cm^{-1} to H_2O shown in Fig. 2 (Yang and Fu 1988a).

A Mössbauer spectrum had been measured by Yang and Fu (1988a) to determine the Fe valence state, revealing that there are two quadrupole doublets.

The refined hyperfine parameters are IS (isomer shift) = 0.390 mm/s, QS (quadrupole splitting) = 1.131 mm/s for inner peaks; IS = 0.374 mm/s, QS = 1.629 mm/s for outer peaks. It indicates that iron atoms in this mineral belong to Fe^{3+}on two independent atomic sites. The ratio of areas of inner peaks to outer peaks is 36.3 : 63.7.

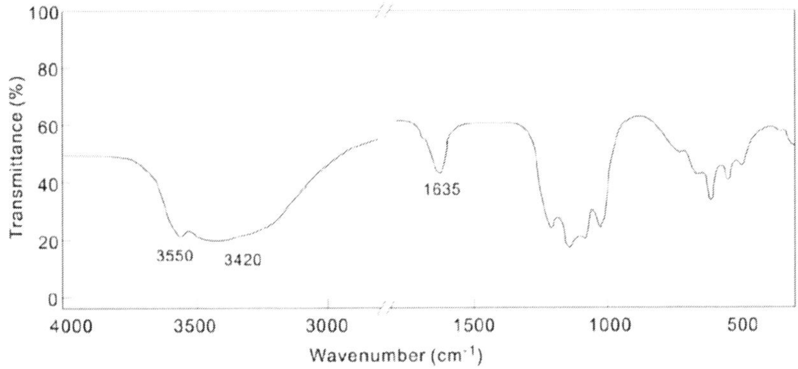

Figure. 2 Infrared-absorption spectrum of zincobotryogen

3. X-RAY CRYSTALLOGRAPHY AND CRYSTAL-STRUCTURE REFINEMENT

3.1 Generalities

The X-ray powder diffraction data on zincobotryogen were obtained using a Bruker Smart APEX instrument with a CCD detector, monochromatized Mo*K*α (0.7107 Å) radiation at 45 kV and 35 mA, and using the GADDS program (Häming 2000). Observed *d* spacings are given in Table 2. Monoclinic unit cell parameters refined from the powder data are given as: a = 10.49(1) Å, b = 17.81(1) Å, c = 7.187(9) Å, β = 100.8(2)°, V = 1318(2) Å^3. The strongest lines in the powder X-ray diffraction pattern d(I)(hkl) are: 8.92 (100)(110), 6.32 (77)(−101), 5.56 (23)(021), 4.08 (22)(−221), 3.21 (31)(231), 3.03 (34)(032), and 2.77 (22)(042).

A single anhedral crystal was examined under a polarizing microscope with no indication of twinning. Single-crystal X-ray data for zincobotryogen were collected using monochromatic Mo*K*α-radiation on a Rigaku RA-Micro7HF diffractometer with a Saturn 724+ CCD detector. A total of 1000 frames were recorded by a combination of several ω and φ rotation sets with a 0.5° scan width.

Data reduction, including intensity integration, correction for Lorentz and polarization effects, and absorption correction, was done using the software CrystalClear (Rigaku). Subsequent analysis of the intensity data by XPREP (Sheldrick 2003) indicated the centrosymmetric distribution of the normalized structure factors, allowing assignment of the unique space group $P12_1/n1$, with a = 10.504(2), b = 17.801(4), c = 7.1263(14) Å, and β = 100.08(3)°, V = 1311.9(5) Å³, Z = 4. The crystal structure, based on the model of Süsse (1967, 1968) was refined using SHELX-97 (Sheldrick 1997). The refinement procedure was conducted by full-matrix least-squares techniques on F^2.

Table 2 X-ray powder diffraction data of zincobotryogen

I	d_{meas}(Å)	d_{Calc}(Å)	h	k	l
100	8.92	8.92	1	1	0
77	6.32	6.41	−1	0	1
23	5.56	5.53	0	2	1
45	5.14	5.14	1	1	1
13	4.40	4.36	−1	3	1
22	4.08	4.08	−2	2	1
20	3.76	3.77	0	4	1
17	3.54	3.53	0	0	2
10	3.37	3.37	2	4	0
31	3.21	3.22	2	3	1
34	3.03	3.03	0	3	2
22	2.77	2.77	0	4	2
4	2.57	2.57	2	2	2
11	2.48	2.48	−4	2	1
6	2.37	2.37	−4	3	1
2	2.25	2.25	3	6	0
7	2.17	2.17	−3	5	2
12	2.06	2.06	1	8	1
1	1.97	1.97	−5	3	1
2	1.90	1.90	4	1	2
8	1.79	1.79	−1	1	4
3	1.72	1.73	−3	8	2
5	1.59	1.59	−2	5	4
2	1.52	1.52	−5	8	1
6	1.45	1.45	−7	1	2
1	1.41	1.41	3	11	1
2	1.28	1.28	−1	13	2
1	1.22	1.22	3	2	5
1	1.18	1.18	6	11	0
3	1.12	1.12	−2	10	5
1	1.04	1.04	6	10	3
1	0.95	0.95	9	3	3

Table 3 Crystal data, data collection information and refinement details for zincobotryogen

Formula	$(Zn_{0.84}Mg_{0.16})Fe(SO_4)_2(OH)\cdot 7H_2O$
Formula weight	449.73
Space group	$P12_1/n1$
a (Å)	10.504(2)
b (Å)	17.801(4)
c (Å)	7.1263(14)
β (°)	100.08(3)
V (Å^3), Z	1311.9(5), 4
μ (mm^{-1})	3.052
Crystal dimensions (mm^3)	0.14 × 0.11 × 0.10
F(000), ρ_{calc}(g·cm^{-3})	912, 2.277
λ (Mo$K\alpha$)(Å), T(K)	0.71073, 293(2)
θ range for collection	3.41 to 27.50°
Number of frames	1000
Scan time (s/°)	5
h,k,l ranges	−13 → 13, −23 → 23, −9 → 9
Total reflections measured	14234
Unique reflections	2995 [R(int) =3.88 %]
Reflections used	2816 with $I > 2\sigma(I)$
Refinement on	F^2
$R1(F)$, $wR2(F^2)$[I>2σ(I)]	3.55 %, 6.51 %
No. of refined parameters	232
GoF on F^2	1.197
$\Delta\rho$min, $\Delta\rho$max (e/Å^3)	−0.40, 0.37

As the Mn and Fe^{2+} contents are low, only the Zn:Mg ratio at the octahedral *M* site was refined, assuming full site occupancy; the respective formula is $(Zn_{0.84}Mg_{0.16})Fe^{3+}(SO_4)_2(OH)\cdot 7(H_2O)$.

The approximate positions of the H atoms of the H_2O and (OH) groups could be localized in difference-Fourier maps and were included in the final refinement, with isotropic atomic displacement parameters. Anisotropic displacement parameters were used for all other atoms. The crystal data, data-collection information and refinement details for zincobotryogen are listed in Table 3. The atomic coordinates, displacement parameters and site occupancy factors are shown in Tables 4 and 5. The relevant bond lengths and angles are shown in Table 6.

Table 4 Atomic coordinates and isotropic atomic displacement parameters (in Å²) with estimated standard deviations (e.s.d.'s) in parentheses for zincobotryogen

Atom	*x*	*y*	*z*	U_{eq}
Fe(1)	0	0	0	0.01180(13)
Fe(2)	0	0	0.5	0.01451(14)
M[a]	0.40859(4)	0.18173(2)	0.34904(5)	0.01873(13)
S(1)	0.09275(7)	0.13894(4)	0.27867(9)	0.01330(15)
S(2)	0.70889(7)	0.06827(4)	0.88554(10)	0.01473(16)
O(1)	0.00992(19)	0.10571(11)	0.1082(3)	0.0155(4)
O(2)	0.21619(19)	0.16132(12)	0.2300(3)	0.0212(5)
O(3)	0.1173(2)	0.08029(12)	0.4307(3)	0.0187(4)
O(4)	0.0253(2)	0.20244(12)	0.3439(3)	0.0244(5)
O(5)	0.7426(2)	0.11666(12)	0.7336(3)	0.0240(5)
O(6)	0.80630(19)	0.00762(11)	0.9301(3)	0.0182(4)
O_W(4)	0.8331(2)	0.06166(14)	0.4193(3)	0.0262(5)
H(4A)	0.799(4)	0.079(2)	0.499(5)	0.039
H(4B)	0.796(4)	0.077(2)	0.317(5)	0.039
O_W(5)	0.4132(2)	0.27464(13)	0.1733(4)	0.0287(5)
H(5A)	0.451(4)	0.280(2)	0.088(5)	0.043
H(5B)	0.392(4)	0.315(2)	0.210(6)	0.043
O_W(6)	0.2050(3)	0.16113(16)	0.8252(4)	0.0333(6)
H(6A)	0.259(4)	0.145(2)	0.906(6)	0.05
H(6B)	0.249(4)	0.178(2)	0.750(6)	0.05
O_W(7)	0.4046(3)	0.08851(15)	0.5275(4)	0.0356(6)
H(7A)	0.460(4)	0.070(2)	0.617(6)	0.053
H(7B)	0.339(4)	0.067(2)	0.521(7)	0.053

[a] Occupancy: *M* = 0.836(3)Zn + 0.164(3)Mg

Table 5 Anisotropic displacement parameters (in Å^2) with estimated standard deviations (e.s.d.'s) in parentheses for zincobotryogen

Atom	U_{11}	U_{22}	U_{33}	U_{23}	U_{13}	U_{12}
Fe(1)	0.0132(3)	0.0138(3)	0.0084(3)	0.0001(2)	0.0017(2)	0.0003(2)
Fe(2)	0.0168(3)	0.0173(3)	0.0095(3)	0.0010(2)	0.0025(2)	0.0000(2)
M	0.0179(2)	0.0202(2)	0.0175(2)	0.00000(15)	0.00165(15)	−0.00026(16)
S(1)	0.0149(3)	0.0147(3)	0.0102(3)	−0.0008(2)	0.0020(3)	−0.0017(3)
S(2)	0.0131(3)	0.0159(3)	0.0149(3)	0.0007(3)	0.0016(3)	0.0012(3)
O(1)	0.0164(10)	0.0161(10)	0.0131(9)	−0.0015(8)	0.0000(8)	−0.0004(8)
O(2)	0.0153(10)	0.0297(12)	0.0176(10)	0.0036(9)	0.0006(8)	−0.0090(9)
O(3)	0.0211(11)	0.0216(11)	0.0127(10)	0.0041(8)	0.0014(8)	−0.0043(9)
O(4)	0.0316(13)	0.0200(11)	0.0221(11)	−0.0054(9)	0.0058(10)	0.0062(9)
O(5)	0.0230(11)	0.0253(12)	0.0246(11)	0.0092(9)	0.0067(9)	0.0027(9)
O(6)	0.0136(10)	0.0182(10)	0.0221(11)	0.0022(8)	0.0011(8)	0.0022(8)
O(7)	0.0254(12)	0.0295(12)	0.0227(11)	−0.0069(9)	0.0071(9)	0.0050(10)
O(8)	0.0126(10)	0.0242(11)	0.0318(12)	0.0030(10)	−0.0043(9)	−0.0024(9)
O_H	0.0196(11)	0.0179(11)	0.0107(9)	−0.0001(8)	0.0032(8)	−0.0046(8)
$O_W(1)$	0.0250(12)	0.0232(12)	0.0360(13)	−0.0066(10)	0.0144(11)	−0.0016(10)
$O_W(2)$	0.0256(13)	0.0305(14)	0.0351(14)	0.0070(11)	−0.0044(11)	−0.0033(10)
$O_W(3)$	0.0228(13)	0.0237(12)	0.0289(13)	−0.0069(10)	0.0028(10)	−0.0017(9)
$O_W(4)$	0.0277(13)	0.0354(14)	0.0157(11)	0.0039(10)	0.0040(10)	0.0142(10)
$O_W(5)$	0.0356(14)	0.0226(12)	0.0317(13)	0.0054(10)	0.0161(11)	0.0068(10)
$O_W(6)$	0.0316(14)	0.0462(16)	0.0216(13)	0.0040(11)	0.0033(11)	−0.0076(12)
$O_W(7)$	0.0288(14)	0.0331(15)	0.0394(15)	0.0156(11)	−0.0089(12)	−0.0022(11)

Table 6 Relevant bond lengths (Å) and angles (°) in zincobotryogen

$Fe(1)O_6$		$Fe(2)O_6$	
Fe(1)–O_H#1	1.9648(19)	Fe(2)–O_H#1	1.9349(19)
–O_H#2	1.9648(19)	–O_H	1.9350(19)
–O(6)#3	2.013(2)	–O(3)	2.005(2)
–O(6)#4	2.013(2)	–O(3)#1	2.005(2)
–O(1)	2.0294(19)	–O_W(4)#3	2.063(2)
–O(1)#5	2.0294(19)	–O_W(4)#6	2.063(2)
Mean	2.002	Mean	2.001
MO_6			
M–O_W(2)	2.049(3)	O_W(1)-M-O(2)	84.76(10)
–O_W(1)	2.068(2)	O(2)-M-O_W(5)	90.94(10)
–O(2)	2.080(2)	O_W(2)-M-O_W(5)	86.84(11)
–O_W(5)	2.082(2)	O_W(2)-M-O_W(1)	94.22(11)
–O_W(7)	2.096(3)	O_W(1)-M-O_W(5)	87.02(10)
–O_W(3)	2.112(2)	O_W(1)-M-O_W(7)	93.25(11)
Mean	2.081	O_W(7)-M-O_W(3)	87.12(11)
$S(1)O_4$		$S(2)O_4$	
S(1)–O(4)	1.453(2)	S(2)–O(8)	1.462(2)
–O(2)	1.455(2)	–O(7)	1.471(2)
–O(1)	1.487(2)	–O(5)	1.475(2)
–O(3)	1.494(2)	–O(6)	1.483(2)
Mean	1.472	Mean	1.473
H_2O			
H(1A)-O_W(1)-H(1B)	104(4)	H(5A)-O_W(5)-H(5B)	110(4)
H(2A)-O_W(2)-H(2B)	113(4)	H(6A)-O_W(6)-H(6B)	102(4)
H(3A)-O_W(3)-H(3B)	101(4)	H(7A)-O_W(7)-H(7B)	109(4)
H(4A)-O_W(4)-H(4B)	108(4)		

Symmetry code for equivalent positions: #1 = −x, −y, −z + 1; #2 = x, y, z-1; #3 = −x + 1, −y, −z + 1; #4 = x-1, y, z-1; #5 = −x, −y, −z; #6 = x-1, y, z

3.2 Structure description

The crystal structure of zincobotryogen corresponds to the model reported by Süsse (1967, 1968), Yang and Fu (1988a) and Majzlan et al. (2016). The structure of zincobotryogen is shown in a projection along the *b*-axis in Fig. 3. The atomic arrangement is characterized by chains of corner sharing $[Fe(1)O_4(OH)_2]$ and $[Fe(2)O_2(OH)_2(H_2O)_2]$ octahedra running parallel to the *c*-axis. Both atoms of ferric iron have site symmetry 1--1--, Fe(1) is coordinated to four oxygen atoms of sulfate groups and to two of hydroxyl groups, Fe(2) to two oxygen atoms of sulfate groups, two of hydroxyl groups and two of H_2O molecules. Two independent sulfate tetrahedra, $S(1)O_4$ and $S(2)O_4$, attached via corners on alternate sides of the chain, provide further intra-chain linkages to constitute a structural building unit with composition $[Fe^{3+}(SO_4)_2(OH)(H_2O)]^{2-}$ and ~7 Å repeat distance, similar to that in sideronatrite-2*M* (Yang et al. 2015). But, the chain in zincobotryogen links to a $[MO(H_2O)_5]$ octahedron through one of its vertices and the $[S(1)O_4]$ tetrahedron with an unshared H_2O molecule, and forms a larger chain building a module with composition $[M^{2+}Fe^{3+}(SO_4)_2(OH)(H_2O)_6(H_2O)]$. The packing of the chain modules in the structure is shown in Fig. 4; inter-chain module linkage involves only hydrogen bonding.

Both $[Fe(1)O_4(OH)_2]$ and $[Fe(2)O_2(OH)_2(H_2O)_2]$ octahedra are nearly regular. The average Fe–O distances, 2.002 and 2.001 Å, are close to those found in the structure of botryogen (2.002 and 1.998 Å, Majzlan et al. 2016), in sideronatrite-*2 M* (2.006 Å, Yang et al. 2015), in the orthorhombic polytype (1.999 Å, Scordari and Ventruti 2009), in the structure of metasideronatrite (2.008 Å, Ventruti et al. 2010), in chaidamuite (2.008 and 2.011 Å, Li and Wang 1990), or in $Fe^{3+}_2Fe^{2+}(SO_4)_4{\cdot}2(H_2O)$ (1.984 Å for ferric iron, Wildner and Giester 1991). The distances of Fe(2)–O_H are shorter than those to O(1), O(6), O(3) and $O_w(4)$ as similarly observed in the sideronatrite polytypes (Yang et al. 2015). The distances of Fe(2)–O_W from iron to H_2O molecules are longer [2.063(2) Å].

The average S–O distances, 1.472 and 1.473 Å, fall in the range 1.47–1.48 Å reported for most hydrated sulfates (Palmer et al. 1972; Hawthorne et al. 2000). In the structure of botryogen, the average S–O distances are 1.471 and 1.470 Å (Majzlan et al. 2016).

The (MO_6) octahedron is nearly regular. The average M–O distance, 2.081 Å, can be compared with those found in the structure of botryogen (2.083 Å, Majzlan et al. 2016) and in chaidamuite, $ZnFe^{3+}(SO_4)_2(OH)\cdot 4H_2O$, [2.081 Å, (originally mistyped as 2.008 Å) for $Zn(1)O_6$, 2.114 Å for $Zn(2)O_6$, Li and Wang 1990].

3.3 System of hydrogen bonds

All hydrogen atoms have been approximately located in good agreement with the model proposed by Majzlan et al. (2016). Bond-valence theory allows estimating the reliability of a structure model. The bond-valence sums including contributions of hydrogen bonds, calculated from the O···O distances as suggested by Ferraris and Ivaldi (1988) (Tables 7 and 8), show satisfactory agreement with the valence-sum rule (Brown 2002). The valence sums of the bond strengths reaching each oxygen are quite satisfactory. The range of variation in S–O bond-valences is 1.421–1.588 *v.u.* (valence units), which is in accord with the bond-valence curve for S–O bond-valences (1.13–1.92 *v.u.*) given by Brown (1981) and Hawthorne et al. (2000).

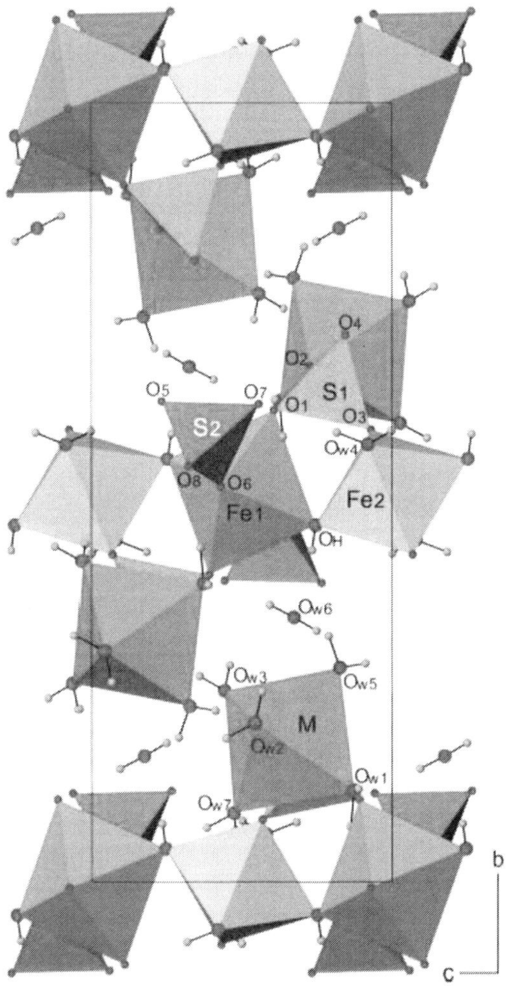

Figure. 3 The open-branched octahedral-tetrahedral chain in zincobotryogen viewed down the *a*-direction. $[Fe^{3+}(SO_4)_2(OH)(H_2O)]^{2-}$ chains with open-branched MO_6-polyhedra run parallel to the *c*-axis and are linked by hydrogen-bonds. All crystal structure drawings were done with ATOMS (Dowty 2013)

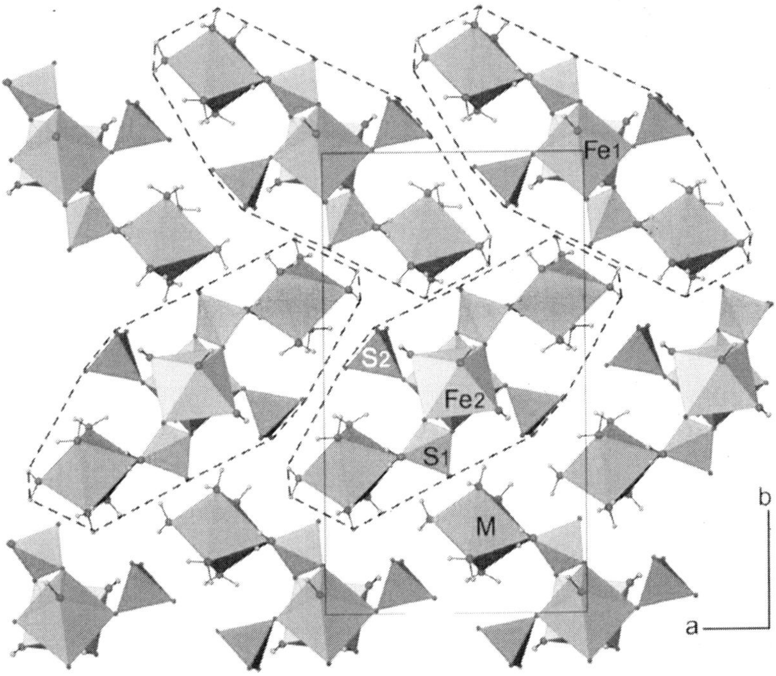

Figure. 4 Packing of chains in the zincobotryogen structure viewed down the *c*-direction. The chain modules are shown in broken lines

The O_H atom shared between two consecutive Fe atoms is part of a hydroxyl group, whereas $O_w(1)$, $O_w(2)$, $O_w(3)$, $O_w(4)$,$O_w(5)$, $O_w(6)$, and $O_w(7)$ belong to H_2O molecules. The hydrogen bonding system is illustrated in Figs. 5 and 6. $O_w(6)$ is the only oxygen atom not linked to any cations, it solely takes part in the hydrogen bonding system through four types of hydrogen bonds, $O_w(6)$-H(6A)···$O_w(1)$ and $O_w(6)$-H(6B)···$O_w(3)$, as well as $O_w(6)$···H, and $O_w(6)$···H(2A). $O_w(4)$ is the only not cross-linked oxygen atom between chain modules, and belongs to the hydrogen bonding system of the intra-chain module.

Adjacent chain modules in the structure are cross-linked through eight types of hydrogen bonds of five oxygen atoms of H_2O molecules, $O_w(1)$-H(1A)···O(7), $O_w(2)$-H(2A)···$O_w(6)$, $O_w(2)$-H(2B)···O(5), $O_w(3)$-H(3A)···O(4), $O_w(3)$-H(3B)···O(7), $O_w(5)$-H(5A)···O(4), $O_w(5)$-H(5B)···O(5), and $O_w(7)$-H(7A)···O(8) as shown in Figs. 5 and 6. The total strength of these eight types of hydrogen bonds from a chain building module with composition $[M^{2+}Fe^{3+}(SO_4)_2(OH)(H_2O)_6(H_2O)]$ to the adjacent ones is 1.508 *v.u.* (Table 8), which may represent the effective charge for zincobotryogen (Hawthorne and Schindler 2008). The ideal effective charge for this structural formula charge is 0 (the formal charge of the structural unit) −8 × 0.20 (the charge transferred by hydrogen bonding, assuming a hydrogen bond-valence of 0.20 *v.u.*) =1.60⁻(Hawthorne and Schindler 2008), which is approximately in agreement with the calculated value (1.508 *v.u.*).

Table 7 Hydrogen-bond geometry (Å, °) for zincobotryogen

O–H···O	*d*(O–H)	*d*(H···O)	O–H···O	*d*(O···O)
O_H–H···O_W(6)	0.77(3)	2.074	174.09	2.841
O_W(1)–H(1A)···O(7)	0.81(3)	2.011	173.17	2.822
O_W(1)–H(1B)···O(8)#1	0.83(3)	1.905	176.33	2.733
O_W(2)–H(2A)···O_W(6)#2	0.82(3)	2.06	170.18	2.868
O_W(2)–H(2B)···O(5)	0.83(3)	1.949	163.91	2.753
O_W(3)–H(3A)···O(4)#3	0.84(3)	1.906	173.55	2.739
O_W(3)–H(3B)···O(7)#4	0.77(3)	2.128	160.69	2.862
O_W(4)–H(4A)···O(5)	0.79(3)	1.978	168.92	2.761
O_W(4)–H(4B)···O(7)	0.81(3)	2.027	178.13	2.84
O_W(5)–H(5A)···O(4)#5	0.79(3)	2.05	172.62	2.834
O_W(5)–H(5B)···O(5)#6	0.80(3)	2.023	145.65	2.722
O_W(6)–H(6A)···O_W(1)#7	0.79(3)	2.446	114.7	2.867
O_W(6)–H(6B)···O_W(3)	0.82(3)	2.144	167.26	2.954
O_W(7)–H(7A)···O(8)	0.85(3)	1.887	177.62	2.738
O_W(7)–H(7B)···O(3)	0.78(3)	2.316	142.93	2.978

Symmetry code for equivalent positions: #1 = −x + 1, −y, −z + 1;#2 = x + 1/2, −y + 1/2, z-1/2; #3 = x + 1/2, −y + 1/2, z + 1/2;#4 = x-1/2, −y + 1/2, z + 1/2;#5 = x + 1/2, −y + 1/2, z-1/2; #6 = x-1/2, −y + 1/2, z-1/2; #7 = x, y, z + 1

3.4 Relationships to other minerals, and concluding remarks

Zincobotryogen belongs to the botryogen group (Strunz and Nickel: 07.DC.25; Dana: 31.09.06), which consists of two isotypic members: zincobotryogen is the Zn-end member, while botryogen, $MgFe^{3+}(SO_4)_2(OH)\cdot 7(H_2O)$, is the Mg-end member.

Table 8 Bond valences for zincobotryogen

	O(1)	O(2)	O(3)	O(4)	O(5)	O(6)	O(7)	O(8)	O_H	Ow(1)	Ow(2)	Ow(3)	Ow(4)	Ow(5)	Ow(6)	Ow(7)	Σ
Fe(1)	$0.482^{\times2\rightarrow}$					$0.503^{\times2\rightarrow}$			$0.573^{\times2\rightarrow}$								3.12
Fe(2)			$0.514^{\times2\rightarrow}$						$0.621^{\times2\rightarrow}$				$0.440^{\times2\rightarrow}$				3.15
M		0.360								0.372	0.391	0.330		0.358		0.345	2.16
S(1)	1.448	1.579	1.421	1.588													6.04
S(2)					1.464	1.496	1.512	1.549									6.02
H									0.830						0.170		1.00
H(1A)							0.176			0.824							1.00
H(1B)								0.211		0.789							1.00
H(2A)											0.838				0.162		1.00
H(2B)					0.202						0.798						1.00
H(3A)				0.208								0.792					1.00
H(3B)							0.163					0.837					1.00
H(4A)					0.199								0.801				1.00
H(4B)							0.170						0.830				1.00
H(5A)				0.172										0.828			1.00
H(5B)					0.216									0.784			1.00
H(6A)										0.162					0.838		1.00
H(6B)												0.140			0.860		1.00
H(7A)								0.209								0.791	1.00
H(7B)			0.135													0.865	1.00
	1.93	1.94	2.07	1.97	2.08	2.00	2.02	1.97	2.02	2.15	2.03	2.10	2.07	1.97	2.03	2.00	

The bond valences were calculated from the curves of Brese and O'Keeffe (1991); hydrogen bond strengths were calculated from the curves of Ferraris and Ivaldi (1988)

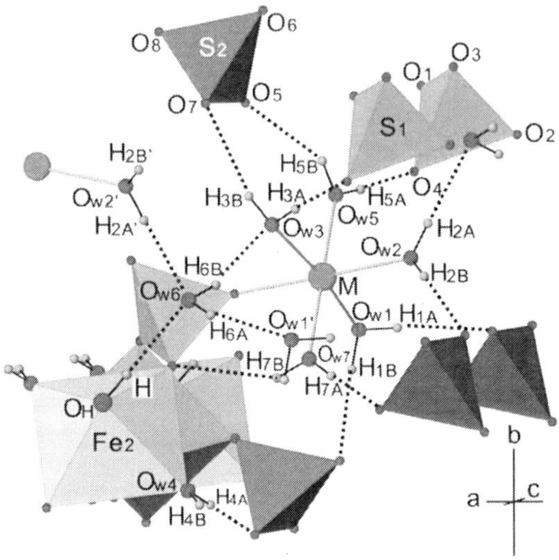

Figure. 5 The hydrogen-bond system in zincobotryogen seen along the *c*-direction

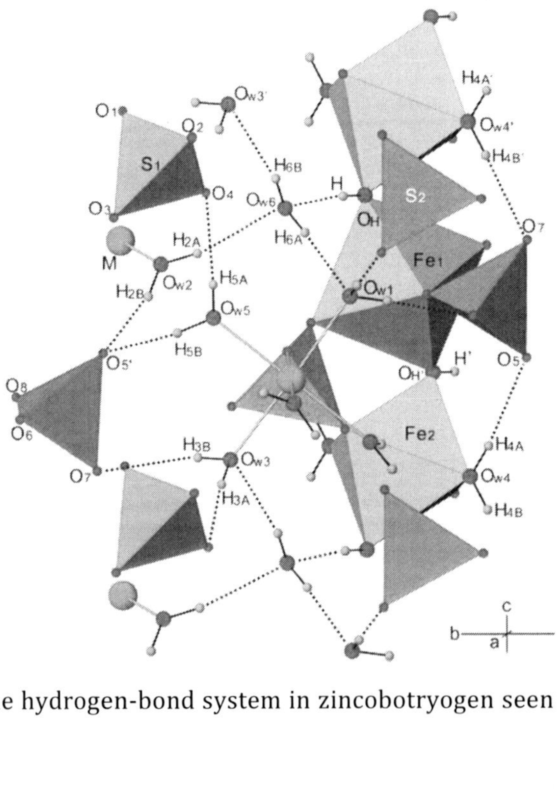

Figure. 6 The hydrogen-bond system in zincobotryogen seen along the *a*-direction

In the structure of zincobotryogen, zinc, magnesium, manganese, and ferrous iron were assigned to the *M* site, ferric iron fully occupies the Fe sites. The calculated site scattering value (26.62 *epfu*) is in good agreement with the refined one (27.12 *epfu*), confirming the reliability of the model. "Botryogen" from the Rammelsberg mine, Germany, should be renamed as zincobotryogen, since zinc is the predominant element in the M site (Zemann 1961; Süsse 1967, 1968). The samples from Al-caparrosa locality, Antofagasta Province, Chile, and from Nuevo Cuyo, Argentina are botryogen, as magnesium is the predominant element in the *M* site (Frost et al. 2011; Majzlan et al. 2016).

ACKNOWLEDGMENTS

Open access funding provided by University of Vienna. The authors acknowledge the help of Hao Xiang and Liang Tongling with the single-crystal structure measurement. The project was supported by the National Natural Science Foundation of China (No. 41272063).

REFERENCES

1. Brese NE, O'Keeffe M (1991) Bond-valence parameters for solids. Acta Cryst B47:192–197
2. Brown ID (1981) The bond-valence method: an empirical approach to chemical structure and bonding. In: O'Keeffe M, Navrotsky A (eds) Structure and bonding in crystals. Academic, New York, pp 1–30
3. Brown ID (2002) The chemical bond in inorganic chemistry: the bond valence model. Oxford Scientific Publication, UK
4. Dowty E (2013) Atoms V6.4.1 for atomic structure display. Shape Software, Kingsport
5. Ferraris G, Ivaldi G (1988) Bond valence vs. bond length in O···O hydrogenbonds. Acta Cryst B44:341–344
6. Frost RL, Palmer SJ, Cejka J, Sejkora J, Plasil J, Jebava I, Keeffe E (2011) A Raman spectroscopic study of $M^{2+}M^{3+}$sulfate minerals, römerite $Fe^{2+}Fe_2{}^{3+}(SO_4)_4 \cdot 14H_2O$ and botryogen $Mg^{2+}Fe^{3+}(SO_4)_2(OH) \cdot 7H_2O$. J Raman Spectrosc 42:825–830
7. Häming L (2000) GADDS like experiments using SMART CCD detectors, quick guide, application scientist, single crystal diffraction. Bruker AXS GmbH, Karlsruhe
8. Hawthorne FC, Schindler M (2008) Understanding the weakly bonded constituents in oxysalt minerals. Z Kristallogr 223:41–68
9. Hawthorne FC, Krivovichev SV, Burns PC (2000) The crystal chemistry of sulfate minerals. In: Alpers CN, Jambor JL, Nordstrom BK (eds) Sulfate minerals: crystallography, geochemistry, and environmental significance. Rev Mineral Geochem l40:1–112
10. Lane MD, Bishop JL, Dyar MD, Hiroi TH, Mertzman SA, Bish DL, King PL, Rogers DR (2015) Mid-infrared emission spectroscopy and visible/near-infrared reflectance spectroscopy of Fe-sulfate minerals. Am Mineral 100:66–82
11. Li WM, Wang QG (1990) Determination and refinement of the crystal structure of chaidamuite. Sci China Chem 33:623–630

12. Majzlan J, Plasil J, Dachs E, Benisek A, Koch CB (2016) Crystal chemistry, Mössbauer spectroscopy, and thermodynamic properties of botryogen. J Mineral Geochem 193:147–159
13. Mandarino JA (1981) The Gladstone-Dale relationship: part IV the compatibility concept and its application. Can Mineral 19:441–450
14. Palmer KJ, Wong RY, Lee KS (1972) The crystal structure of ferric ammonium sulfate trihydrate, $FeNH_4(SO_4)_2{\cdot}3H_2O$. Acta Cryst B28:236–241
15. Scordari F, Ventruti G (2009) Sideronatrite, $Na_2Fe(SO_4)_2(OH){\cdot}3H_2O$: crystal structure of the orthorhombic polytype and OD character analysis. Am Mineral 94:1679–1686
16. Sheldrick GM (1997) SHELXL-97, Program for the refinement of crystal structures. University of Göttingen, Germany
17. Sheldrick GM (2003) XPREP. Bruker-Nonius AXS, Madison
18. Süsse P (1967) Die Kristallstruktur des Botryogens. Naturwissenschaften 54:139
19. Süsse P (1968) Die Kristallstruktur des Botryogens. Acta Cryst B24:760–767
20. Tu KC, Li HL (1963) Studies on the characteristic features of the oxidation zone of the sulphide deposits in arid to extremely arid regions (with special reference to observations obtained from five sulphide deposits in the Northwestern China). Acta Geol Sin 43(4):361–377 (in Chinese with English abstract)
21. Tu KC, Li HL, Hsieh HD, Yin SS (1964a) Zincobotryogen and zincocopiapite–two new varieties of sulphate minerals. Acta Geol Sin 44:99–101 (in Chinese with English abstract)
22. Tu KC, Li HL, Hsieh HD, Yin SS (1964b) Zincobotryogen and zincocopiapite–two new varieties of sulphate minerals. Sci Geol Sin 5:313–330 (in Chinese)
23. Ventruti G, Stasi F, Scordari F (2010) Metasideronatrite: crystal structure and its relation with sideronatrite. Am Mineral 95:329–334
24. Wang LJ, Zhu XY, Wang JB, Deng JN (2008) A preliminary study on fluid inclusions and mineralization of Xitieshan sedimentary-exhalative (SEDEX) lead-zinc deposit. Acta Geol Sin-Engl 82:838–844
25. Wildner M, Giester G (1991) $Fe(II)Fe(III)_2(SO_4)_4{\cdot}2H_2O$: a new Fe(II)-Fe(III) compound. Synthesis and crystal structure. Z Krist 196:269–277
26. Yang HX, Fu PQ (1988a) A further study of zincobotryogen. Acta Mineral Sin 8:119–224 (in Chinese with English abstract)
27. Yang HX, Fu PQ (1988b) Crystal structure of zincobotryogen. Acta Mineral Sin 8:1–12 (in Chinese with English abstract)
28. Yang ZM, Giester G, Ding KS, Li H (2015) Crystal structure of sideronatrite-2*M*, $Na_2Fe(SO_4)_2(OH)(H_2O)_3$, a new polytype from Xitieshan lead-zinc deposit, Qinghai Province, China. Eur J Mineral 27:427–432
29. Zemann J (1961) Über den Botryogen vom Rammelsberg. Fortschr Mineral 39:84

CHAPTER 6

Mineralogy and the Texture of the Basalt in Hail Region, Saudi Arabia

Sana'a Odat

Department of Earth and Environmental Sciences, Faculty of Science, Yarmouk University, Irbid, Jordan

ABSTRACT

The study is carried out to list mineralogy and texture of the basalt in Hail region. The basalt flows belong to the Arabian Harrat volcanism. Cenozoic volcanism has produced 13, vast, basaltic fields in western Saudi Arabia, covering a total area of about 100,000 sq. km. The volcanism starts in the late Cretaceous time and continues periodically into the Holocene. The basaltic field presently investigated is known as Harrat Al-Hamad (or Harrat As-Shamah) which lies in northwestern Saudi Arabia, and continues northwards into Jordan. Mineralogical analysis of 20 rock samples indicates that the rocks are products of continental intra-plate magmatism; mostly as coarse-grained silica—under saturated olivine-rich alkali basalt. Mineralogical study shows that rocks are olivine, plagioclase, pyroxene, Nepheline and Opaque minerals. They correspond to alkali olivine basalts. Moreover, the main well-defined texture is intergranular, vesicular and amygdaloidal texture.

Keywords: Mineralogy, Texture; Hail, Basalt, Alkali Olivine, Saudi Arabia

1. INTRODUCTION

Basalts are the most abundant extrusive igneous rocks, which have erupted on earth throughout its history. Their most extrusive forms are lava flows and pillows in which they constitute the bulk of the major lava plateau on the continents, ocean floor and significant parts of volcanic in rift valleys. Basalt as a general term is a finegrained, mafic volcanic rock containing (44 - 54)% SiO_2. It consists essentially of calcic plagioclase and abundant mafic minerals mainly Ca-rich clinopyroxene, but may contain quartz, orthopyroxene, olivine, feldspathoid, small proportions of alkali feldspar (<10% total feldspar), Fe-Ti oxides and apatite. Basalt rocks may give many types according to their mineralogical compositions such as tholeiite contains quartz, hypersthene basalt contains hypersthene, olivine tholeiite contains hypersthene and olivine, olivine basalt; contains olivine, and alkali basalt which contains nepheline [1].

Basalts are often associated with lava flows and narrow dykes and sills. Most of basalts occur as lava flows either in volcanoes or as extensive sheets building up a lava plateau, which may cover hundreds of thousands of square kilometers, and may be fed by numerous fissures. The surface forms of lavas are of two principal types; smooth or ropy (the surface looks like a rope) which is known by the Hawaiian term of pahoehoe, and scoriaceous which is rough and clinker and has the Hawaiian name A'A. Another common form is pillow lava which consists of pillows or balloon-like masses of basalt—usually with a very fine-grained or glassy outer layer. They are the major constituent of the upper layer of the ocean floors (usually as pillow lava), and hot spot volcanoes such as the Hawaiian Islands. Basalt commonly forms on the continents too, usually the result of hot spot activity. Here it may also exist as intrusive dikes and sills, or extrusive cinder volcanoes [2].

Harrat Al-Shaam is a large intra-continental basalt terrain, which covers an area of about 50,000 km^2, about 25% of the Arabian Harrat. It stretches over 700 km in a NW-SE direction and extends over the southern rim of the Damascus Basin in Syria. It is marked by the folded mountain of Antilebanon, southwards through NorthJordan up to the northern rim of Al-Azraq depression, and then to the "Al-Nofuth" desert in the northwestern part of Saudi Arabia [3].

The basalt originated during several phases of eruptions, which were closely associated in space and time with: 1) the Cainozoic evolution of the Red Sea through two-stages of spreading; the first was before 30 - 15 Ma and the second was initiated over the past 5 Ma-recent; 2) collision of the Arabian and Eurasian plates and 3) the uplift of the Afro-Arabian dome [4].

Previous studies on the North Arabian volcanic province suggest that the derivation of this basalt from deep mantles source material and that it is also characterized by a low degree of partial melting of upper mantle peridotites with the minor secondary differentiation [5] [6] . The depth of the basalt source material is 37 - 60 km [7].

2. MATERIALS AND METHODS

2.1. The Study Area

Volcanic activity that accompanied the opening of the Red Sea from the Miocene (25 Ma) to the present, resulted in the formation of vast fields of subaerial basaltic flows in the western part of Saudi Arabia, referred to by the Arabic term "Harrat". These harrats cover an area over about 90,000 km^2, and extending over parts of the Proterozoic Arabian shield and adjacent Phanerozoic rocks of the Arabian Platform and Red Sea basin [4] [8] . The principal harrats are Harrat Rahat, Harrat Uwayrid, Harrat al Hutaymah, Harrat Kishb, Harrat Khaybar, and Harrat al Birk. The lava flows are commonly composed of picritic to ankaramitic basalt and may contain peridotite nodules (Figure 1) [9].

The study area is located in the southern part of Hail region. It is located between 27°13'N & 27°31'N and 42°09'E & 42°35'E. The principal features of the region are the two great mountain ranges of Aja (granites) and Salma (basalts) and the immense rolling sand dunes of An-Nafūd [10].

In order to determine the mineralogical and texture characteristics of basaltic rocks of Hail region, twenty representative samples (as fresh as possible) were collected and studied under the polarizing microscope.

2.2. Methodology and Techniques

Twenty rock samples have been collected from the study area. Thin sections were studied under polarizing microscope.

2.2.1. Thin Section Preparation

Thin sections were prepared as follows: A slab of basalt with 30 mm × 20 mm × 10 mm thick was cut, polished using (200, 400, 800, 1000) silicon carbide powder, in order to get a smooth surface easy to stack with glass slide of (26 × 42 mm), using Canada palsam adherent. An Automated thin section machine, called Multiplate Grinder housed was used for making the slides thinner in thickness of approximately 45 microns. Then, Heavy duty polishing table and hand polishing using (1000) silicon carbide, grinded the slide down to 30 microns [11].

2.2.2. Polarizing Optical Microscope

Nikon optical microscope, are used particularly to identify mineral constituents of the studied basaltic rocks, and to determine the mineralogical properties and textures.

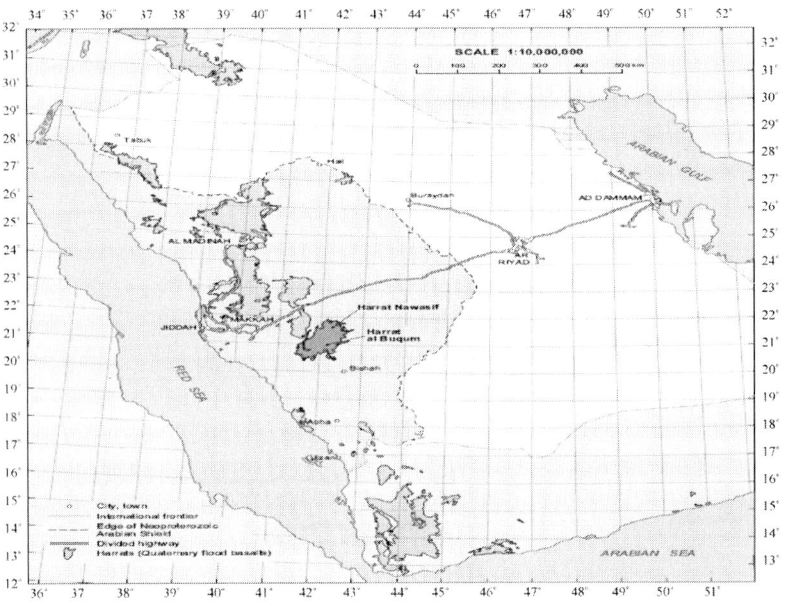

Figure 1. Map showing the major Cenozoic lava fields of Saudi Arabia.

3. RESULT

3.1. Minerals Description

3.1.1. Primary Minerals

1) Olivine

Olivine is the major mineral phase in the studied basalts, rates between 8 vol% and 11 vol%. It usually present as subhedral to anhedral, light grey to colorless crystals. Olivine crystals are characterized by high relief, parallel extinction crystals that exhibit two generations of size, which rather variable reflecting a seriate texture, as phenocrysts and groundmass . They range in size between 0.01 mm and 2 mm in all samples (Figure 2).

Olivine minerals of different flows don't exhibit twining. However, all basalt flows show corroded olivine crystals, due to resorbtion, the shaping of the resorbtion of olivine can be noted through the bands of iddingsite that forms a long their embayment rims (Figure 3).

Olivine minerals had been subjected to alteration processes, such as hydration and oxidation, the processes represented by a slight chlorite and much iddingsite that occur at the rims, along the fractures of the crystals core and rarely completely at the whole olivine crystal. Iddingsite appears as a dark red, slightly pleochroic brown color.

2) Plagioclase

Plagioclase is the most abundant mineral phase in the studied basalts, which ranges from 20 vol% to 44 vol%, this variation is due to the cooling rate, as well as the thickness of the flow where the lowest flow is the thickest. Plagioclase phenocrysts occur as lath-like Clusters of phenocrysts up to 7 crystals forming a golmeroporpheritic texture. The second flow is crystallized by one generation as phenocrysts (Figure 4), they are commonly lath-like shape.

Figure 2. Microphotograph showing of olivine crystals under PPL (×10).

Plagioclases phenocrysts are subhedral, but some are euhedral and tabular in shape and colorless crystals exhibit two sets of cleavage intersects at 90°. Groundmass plagioclase occurs as skeletal like microlites. The crystals are slightly fractured. Polysenthetic or multiple twining is common, simple twining are also noted, the multiple twining lamellae are wide, using the MICHEAL-LEVY method [12] . Plagioclase laths were found to have extinction angle between 38° to 42°, indicates a labradorite (An 59-An68) and anorthite (An 100-An 80) range of composition. Sericitisation of plagioclase is common, that appear as pale yellow color band at the outer most rims of the crystals, some of them affect the inner most parts of the crystals (Figure 5).

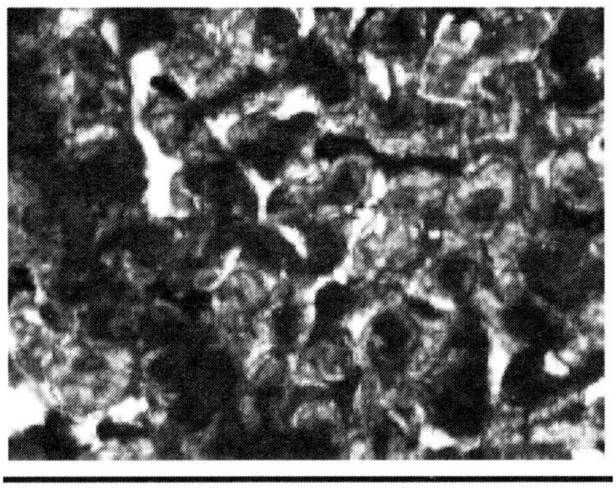

(a)

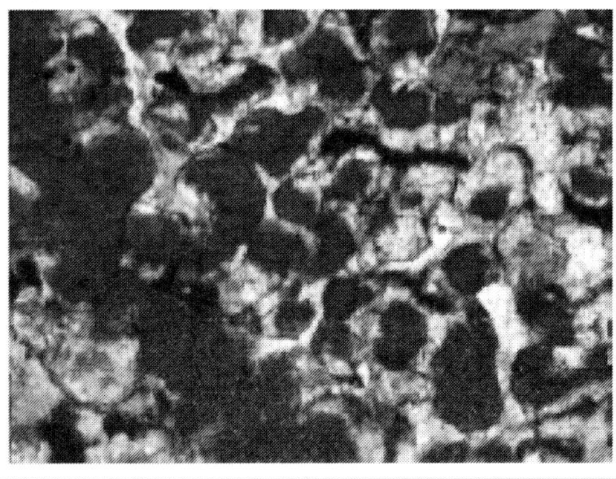

(b)

Figure 3. Microphotograph showing resorbtion of olivine crystals: (a) Under PPL; (b) CN (×10).

3) Pyroxene

Pyroxene are grayish brown in color with anhedral crystals and can be distinguished by two sets of cleavage intersects in 90°. The abundance of pyroxene is ranging between 12 vol% to 20 vol%, as one generation (phenocrysts) in all samples. Pyroxene crystals have an inclined extinction which represents clinopyroxene minerals, this inclination is measured and found to be between (36° - 42°), that indicates diopside mineral of composition. These crystals have subhedral to anhedral oxides inclusions of magnetite type. Pyroxene crystals also had been affected by the alteration processes which appear as a green chlorite on the fractures, crystal rims and on the cleavage sets (Figure 6).

(a)

(b)

Figure 4. Microphotograph of plagioclase crystals (a) Under PPL; (b) CN (×10).

4) Nepheline

Nepheline crystals are colorless under plane polarized light and forms about 2 vol% of the rock. The crystals are euhedral to subhedral in shape and exhibit two sets of cleavage that are intersected at 90°. They have parallel extinction and biaxial interference figure. It is noticed that they are slightly fractured and slightly altered along rims and cleavage planes, (Figure 7).

5) Opaque minerals

Opaque minerals is abundant—as inclusions within olivine and pyroxene crystals. Generally, they are black, and show homogeneous optical properties. Most of them are anhedral to subhedral (rod-like), rarely magnetite shows four sided octagonal shape (Figure 8).

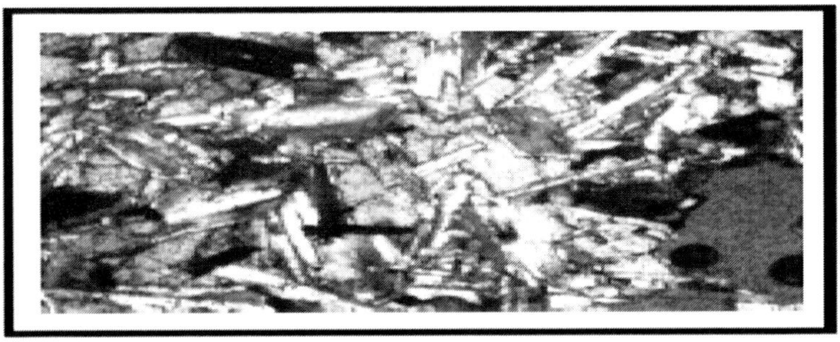

Figure 5. Microphotograph showing the serisitsation of plagioclase crystals under PPL (×10).

Figure 6. Microphotograph showing the pyroxene crystals under PPL (×10).

Figure 7. Microphotograph showing the Nepheline crystals under PPL (×10).

3.2. Secondary Minerals

Several secondary (authogenic) minerals are formed due to the post eruption processes such as hydration and oxidation of the original pyrogenic minerals. These processes include oxidation of olivine, albitization and seritisation of plagioclases and chloritisation of pyroxene. The microscopic investigations show that the second flow was suffered from alteration.

Sericite: pale yellow to turbid color, along the rims and twining planes of the crystals, formed about 5% of the highly weathered, and mainly distributed the original characteristics of the plagioclase, this disturbing affecting the extinction of the crystals and the original color of the mineral.

Iddingsite: the studied thin sections suffered from highly to slight iddingtisation, however particularly this psedomorphism of olvine to iddengsite is noted mostly as a dark brown to reddish brown to yellowish orange and mostly recorded along the rims and fractures of crystals and occasionally forming about (5 - 15 vol%) of crystals.

Chlorite: Chloritization is seen by the occurrences of chlorite along the edges of clino pyroxene (augite) and occasionally along cleavage sets. They are green. Some pyroxenes are completely chloritized.

3.3. Textures Description

The texture for all basalt flows is found to be intergranular texture (Figure 9), vesicular texture (Figure 10) and amygdaloidal textures (Figure 11).

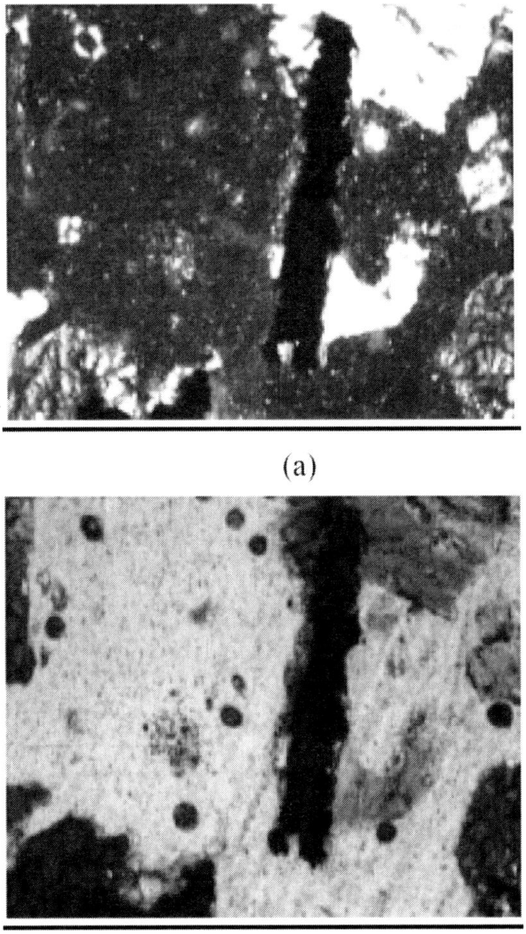

(a)

(b)

Figure 8. Microphotograph showing rod like oxides shape, (a) Under PPL; (b) CN (×10).

4. DISCUSSION

4.1. Mineralogy

The basalt samples in hand specimen are melanocratic, holocrystalline, medium-grained, and porphyritic. Vesicles have an elongated and oval shape. Moreover, the main mineral constituents are olivine, plagioclase, pyroxene, and opaque minerals (mainly magnetite). The secondary minerals include iddingsite, sericite, and chlorite. Common textures occurring are intergranular, vesicular and amygdaloidal textures.

4.1.1. Olivine

Olivine occurs as phenocrysts and in groundmass. The phenocrysts are euhedral to subhedral. They are colourless to palę greys. Olivine in the groundmass has subhedral to rounded shape. The larger crystals are fractured slightly—to-moderately. Iddingtization is common, particularly along fractures and along edges of the crystals. Some crystals are partially to completely pseudomorphed to dull brown iddingsite. Groundmass olivine is also iddingsitized.

4.1.2. Plagioclase

Plagioclase occurs in two generations as larger phenocrysts and as small tabular to elongated microlites in the groundmass. The phenocrysts are subhedral laths. The extinction angles on several plagioclase phenocrysts range from 28° to 32°, indicating a labradorite composition (An50-An70) by using the method described by Michel-Levy's [12].

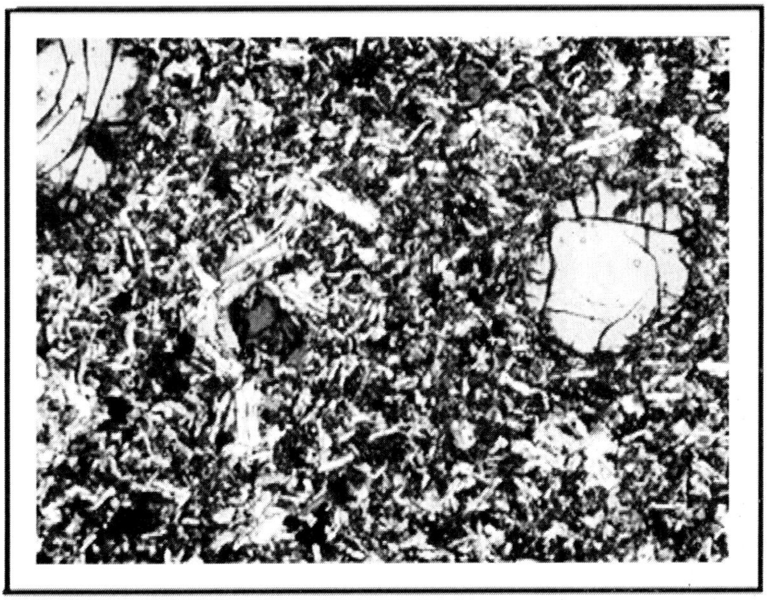

Figure 9. Microphotograph showing be intergranular texture.

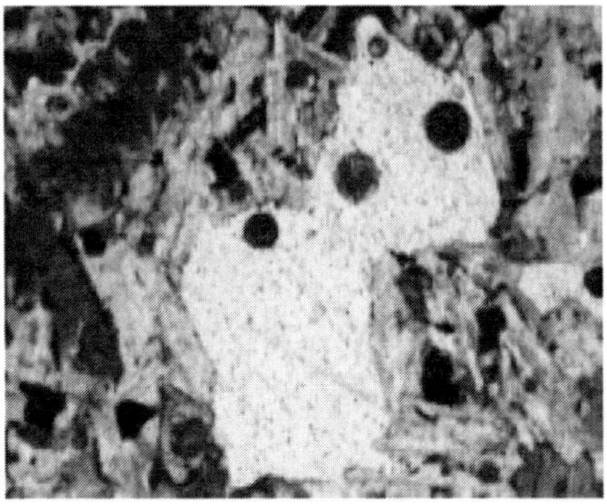

(b)

(a)

Figure10. Microphotograph showing vesicle net shape, (a) Under PPL; (b) CN (×10).

4.1.3. Pyroxene

Pyroxene occurs as brownish anhedral crystals. The pyroxene crystals have an inclined extinction of between (36° - 42°), that indicates diopside mineral of composition. Pyroxene crystals are affected by chloritization. Green chlorite is present along fractures and along crystal rims.

4.1.4. Opaque Minerals

Opaque minerals are mostly magnetite phenocrysts scattered throughout the rock and throughout inclusions within olivine and pyroxene crystals. Generally, magnetite is black; and shows homogeneous optical properties.

Most of it is anhedral to subhedral, but crystals with square outline are also presented.

Figure 11. Microphotograph showing be amygdaloidal texture.

4.1.5. Groundmass

The groundmass consists mainly of plagioclase (labradorite), olivine, pyroxene (diopside), and opaque minerals (mainly magnetite). Sericite, iddingsite and chlorite are secondary minerals.

5. CONCLUSION

Minerological studies of selected basalt samples from Hail region show that the basalts are generally composed of primary minerals such as olivine,plagioclase and pyroxene. Olivine is considered the most common essential mineral phase, followed by plagioclase pyroxene, Nepheline and opaque minerals. Secondary minerals are sericite, iddingsite and chlorite. Quartz is totally absent. Presence of olivine as phenocrysts as well as in the groundmass (more iron rich, similar to one described by [13] -[15] is a common feature of alkali olivine basalts.

ACKNOWLEDGEMENTS

The author would like to express her deep thanks to Dr. Ahmad Alshammari from Hail University for his help in providing the geological information and samples from Hail region in Saudi Arabia.

REFERENCES

1. Shalley, D. (1993) Igneous and Metamorphic Rocks under the Microscope. Springer, London, 234 p.
2. Solyom, Z., Andreasson, P. and Johansson, I. (1985) Petrochemistry of Late Proerozoic rift Volcanism in Scandinavia, Mafic Dike Swarms in Constructive and Abortive Arms. Abstract Book, 164-171.
3. Al-Malabeh, A., El-Hasan, T., Lataifeh, M. and O'Shea, M. (2002) Geochemicaland Mineralogical-Related Magnetic Characteristics of the Tertiary-Quaternary (Umm A-Qutein) Basaltic Flows from the Basaltic Field of Harra El-Jabban, Northeast Jordan. Physica B: Condensed Matter, 321, 396-403. http://dx.doi.org/10.1016/S0921-4526(02)01083-9
4. Camp, V. and Roobol, M., (1989) The Arabian Continental Alkali Province: Part I Evolution of Harrat Rahat, Kingdom of Saudi Arabia. Geological Society of. America Bulletin, 101, 71-95. http://dx.doi.org/10.1130/0016-7606(1989)101<0071:TACABP>2.3.CO;2
5. Barberi, F., Capaldi, G., Gasperini, P., Marinelli, G., Santacroce, R., Scandone, R., Treuil, M. and Varet, J., (1979) Recent Basaltic Volcanism of Jordan and Its Implications on the Geodynamic Evolution of the Afro-Arabian Rift System. Rome, 667-683.
6. Saffarini, G, Nassir, S. and Abed, A. (1985) A Contribution to the Petrology and Geochemistry of the QuaternaryNeogene Basalts of Central Jordan. Dirasat, 12, 133-144.
7. Ibrahim, K. (1996) The Regional Geology of Al-Azraq Area. NRA. Geol. Dir. Map Div. Bull, 36, 67.
8. Coleman, R. and McGuire, A. (1983) Magma Systems Related to the Red Sea Opening. Tectonophysics, 150, 77-100. http://dx.doi.org/10.1016/0040-1951(88)90296-X
9. Thornber, C.R. (1990) Geologic Map of Harrat Hutaymah, with Petrologic Classification and Distribution of Ultramafic Inclusions. Saudi Arabia: US Geological Survey Map MF-2129.
10. Al-Turki, T.A. and Al-Qlayan, H.A. (2003) Contribution to the Flora of Saudi Arabia: Hail Region. Saudi Journal of Biological Sciences, 10, 190-222.
11. Hughes, C. (1982) Igneous Petrology. Development in Petrology. Elsevier, New York.
12. Kerr, P. (1977) Optical Mineralogy. John Wiley and Sons, New York.
13. Brown, G.M. (1968) Mineralogy of Basaltic Rocks. In: Hess, H.H. and Poldervaart, A., Eds., Basa1ts, Vol., I., Interscience, John Willey and Sons, New York, 103, 160-162.
14. Nagao T. and Sakaguchi, Y. (1990) Ultramafic Xenoliths in the Cenozoic Kawashimo Alkali Basalt from Masuda, Shimane Prefecture SW Japan. Journal of Mineralogy, Petrology and Economic, 17, 21-27. http://dx.doi.org/10.2465/ganko.85.469
15. Shoji, A. and Natsue, A. (1995) Reaction of Orthopyroxene in Peridotite Xenoliths with Alkali Basalt Melt and Its Implication of Genesis of Alpine Type Chromitite. American Mineralogist, 80, 1041-1047.

CHAPTER 7

An Alternative Source for Ceramics and Glass Raw Materials: Augen-Gneiss

Murat Olgaç Kangal, Gülay Bulut, Zeynep Yeşilyurt, Onur Güven and Fırat Burat *

Mineral Processing Engineering Department, Mining Faculty, Maslak, Istanbul Technical University, 34469 Istanbul, Turkey

AUTHOR CONTRIBUTIONS

M.O.K., G.B. and F.B. conceived and designed the experiments; all researchers performed the experiments; O.G. and Z.Y. analyzed the data, M.O.K. and F.B. wrote the paper.

ABSTRACT

Due to the depletion of well-known feldspar reserves in the forms of granite, trachyte and pegmatite, many investigations are focused on new and economic alternatives to these resources. Of these, gneiss is a metamorphic rock type, which is characterized by dark and light colored minerals including feldspar and quartz. In this study, Augen-Gneiss rock was subjected to flotation tests to produce alkali (Na + Ca) feldspar and quartz concentrates by adapting different flotation conditions with/without hydrofluoric acid (HF). During these tests, the effect of particle size and different reagent combinations were investigated. The results of these tests showed that, in the presence of HF, a saleable feldspar concentrate with

0.04% Fe_2O_3 content and a saleable quartz concentrate with 98.45% SiO_2 content were obtained. Products with similar qualities to a feldspar concentrate with 0.13% Fe_2O_3 and a quartz concentrate with 92.53% SiO_2 content were produced without HF. The results of these tests clearly showed that, compared to other types of origin, saleable quality feldspar and quartz could well be enriched with/without **Keywords**: Augen-Gneiss; feldspar; quartz; flotation

1. INTRODUCTION

Gneiss is a widely distributed type of metamorphic rock originating by either igneous or sedimentary formations. Layered forms as planar structures in gneiss rock are generally formed by alternating dark and light colored silicates. The most common minerals of gneiss rock are quartz, potassium and sodium feldspar, as well as mica and other dark colored minerals such as tourmaline and hornblende. Thus, the majority of the aforementioned components are classified as gneiss rocks. Of these, "Augen-Gneiss" is a coarse-grained gneiss rock, which contains feldspar and microcline within the layering of the quartz and biotite. Many investigations have been devoted to the geology of gneiss rock, however, the enrichment characteristics of its content as feldspar has been given little attention [1,2]. Therefore, the processing of feldspar of gneiss rock origin will provide an alternative and economically beneficial new source of feldspar production.

Apart from its origin, feldspar minerals are classified based on their Na, K or Ca ratios as albite ($NaAlSi_3O_8$) known as Na-feldspar, orthoclase/microcline ($KAlSi_3O_8$), or anorthite ($CaAl_2Si_2O_8$). Albite mineral is generally used in glass production, while microcline and orthoclase are used in ceramic and porcelain production [3]. Thus, the alkaline and alumina contents of these minerals are very important for usage in these industries. Therefore, in order to ensure these qualifications, the removal of iron and titanium-bearing minerals (mica, iron oxides, titanium oxides, tourmaline) and also the removal of quartz from feldspar is required. In the literature, numerous papers have been devoted to investigating the effect of different parameters such as methods, reagent type, pH adjusters, different flowsheets with/without the usage of hydrofluoric acid (HF) and other flotation-related parameters, in order to obtain higher quality feldspar and quartz products [4,5,6].

In previous studies, the removal of iron and titanium-bearing minerals (tourmaline, hornblende, ilmenite or mica, etc.) was reported with either magnetic separation or flotation depending on the characteristics of the mixture [4]. These colored impurities are found as iron-bearing silicate minerals, which can be removed by sulfonate type collectors at acidic pH values (3–4). However, based on their similar crystalline structures and physicochemical properties, flotation becomes the only option for separating, in particular, quartz and feldspar [4,6,7,8,9,10,11]. Apart from other flotation-related parameters such as reagent type or electrolyte concentration, material-based characteristics also play an important role in selecting the pH range, collector and frother dosage, among other parameters. When considered, due to the close values of the iso-electric point for both quartz and feldspar mineral within the pH range of 1.5–2, the difference in charge between these materials can well explain the electrostatic interactions during the adsorption of different reagents such as amine and sulfonate-based collectors on feldspar and quartz. Thus, at that pH range, when using an anionic collector, feldspar is not floatable in the absence of activating cations [11,12].

Therefore, in most industrial plants, the separation of feldspar from quartz is obtained by using a long-chain primary cationic amine collector (12, 14 or 16 carbons) in acidic pH conditions. In general, the pH adjustment is carried out with sulfuric acid (H_2SO_4), while the usage of HF is due to its important role as a fluoride source in the separation of feldspar and quartz. This is because the effect of HF on pH values is negligible compared to other acids, according to the acid ionization constant in dilute solutions. In other words, the lower interaction of fluorine ions with hydrogen is due to its covalent structure. In contrast to its low solubility, the usage of HF causes environmental and health problems. For that reason, research has leaned toward new schemes without the usage of HF for feldspar/quartz separation [10,13,14,15]. Besides other types of collectors such as anionic, cationic or non-ionic collectors, the efficiency of mixed systems such as tallow propylene di-amine-dioleate (Duomeen T series), polyethylene glycol hexadecyl ether, and sodium dodecyl sulfate (Brij 58 + SDS) was found to be higher compared to single systems [13,14,16].

Considering these findings in the literature, this study aimed to investigate the applicability of different flowsheets including different flotation conditions with/without the usage of HF in order to obtain a feldspar concentrate with higher alkali and a quartz concentrate with high silica content from Augen-Gneiss rock.

2. EXPERIMENTAL

2.1. Material

The representative gneiss sample used in the experimental studies was obtained from Muğla-Milas region, Turkey. According to the chemical analysis results with the XRF (X-ray Fluorescence) method, the sample contained 79.48% SiO_2, 13.52% Al_2O_3, 0.21% Fe_2O_3, 0.18% TiO_2, 2.70% CaO, 4.54% Na_2O, and 0.47% K_2O.

The microscopic and XRD analyses employing a Rigaku X-ray diffractometer (model 2028, Cu KR, Tokyo, Japan) (Figure 1) revealed that the ore mainly contained feldspar and quartz with minor amounts of biotite, muscovite, hornblende, tourmaline, apatite, zircon and mica minerals. Feldspar minerals were albite (Na-feldspar) and anorthite (Ca-feldspar), and the product quality was investigated in terms of alkali content (Na_2O + CaO). In addition, because of decomposition, schorlite type minerals in tourmalines were found to color the quartz minerals. The microscopic images of the Augen-Gneiss sample are illustrated in Figure 2 and Figure 3.

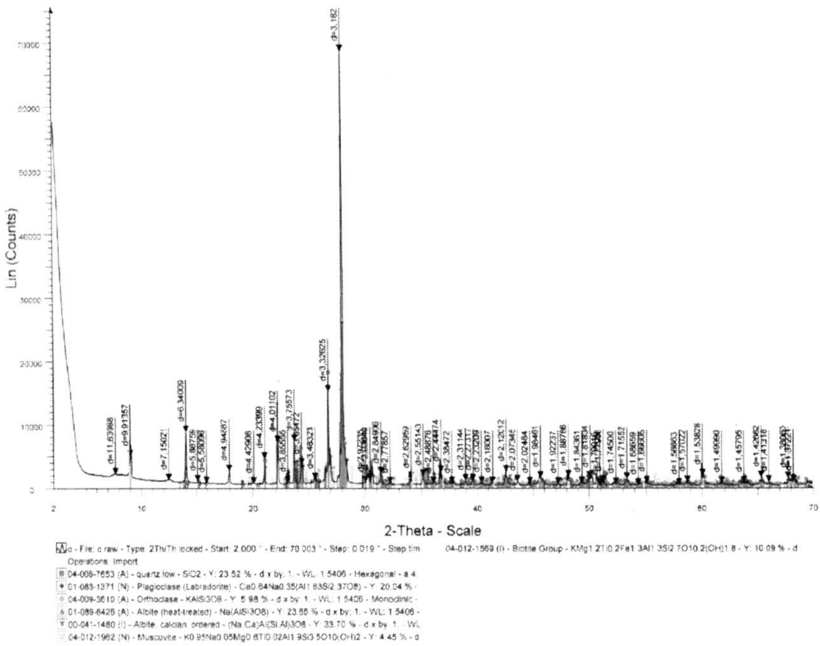

Figure 1. XRD analysis of the representative Augen-Gneiss sample.

2.2. Reagents

Dodecylamine hydrochloride (DAHC) from Acros Organics (Geel, Belgium), as well as derivatives of petroleum sulfonates, namely R801-R825 from Cytec (Woodland Park, NJ, USA), Flotigam 4343, a derivative of alkyl diamine from Clariant (Muttenz, Switzerland), sodium dodecyl sulfate (SDS), tallow propylene di-amine-dioleate namely Duomeen TDO (Akzo Nobel), were used as flotation collectors. MIBC, namely, methyl isobutyl carbinol (DOW Chemical Company, Midland, MI, USA) was used as a frother. To depress the quartz particles, sodium silicate ($Na_2SiO_3 \cdot 9H_2O$) was added to the pulp at the conditioning stage. Hydrofluoric acid (HF) was used as a depressant regulator to prevent quartz flotation. Sulfuric acid (H_2SO_4) and sodium hydroxide (NaOH) were used as pH adjusters.

(a)

(b)

Figure 2. Microscopic images of (a) feldspar (fe), quartz (qu) and (b) quartz (qu), biotite (bi) in the Augen-Gneiss sample.

(a)

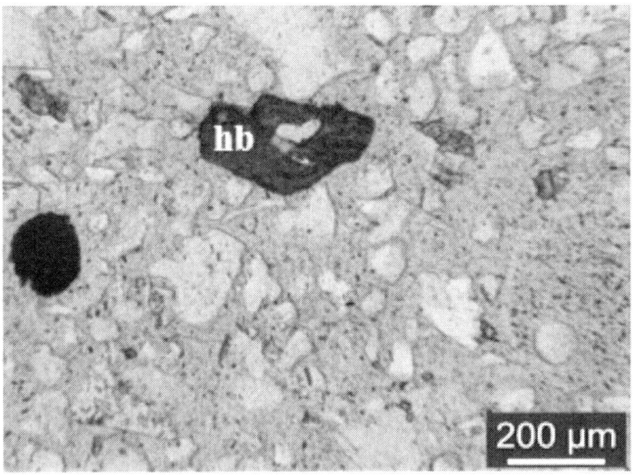

(b)

Figure 3. Microscopic images of (a) feldspar (fe) and schorlite (sc) and (b) hornblende (hb) in the Augen-Gneiss sample.

2.3. Flotation Experiments

A mixture of ceramic balls with diameters of 35, 25, 20 and 18 mm, which weighed about 2600 g, was adapted for the grinding process in a ceramic mill under wet grinding conditions. To prevent the negative effect of slime materials [6,9,17], material under 0.063 mm was removed and the feed size was selected as 0.063 < r < 0.300 mm. The results of the sieve analysis showed that d_{80} size of the ground material was 0.250 mm. In addition, to investigate the effect of particle size, series of tests were adapted with the material under −0.212 mm where the d_{80} size was found to be 0.150 mm.

Flotation experiments were conducted in a self-aerated Denver flotation machine, equipped with a 1.5-L cell. A sample weighing about 300 g was mixed with 300 mL tap water at a constant impeller speed of 1250 rpm. The percentage of the solids at the conditioning stage was selected as 50 wt % in order to provide better interaction of reagents with material, while it decreased to 20 wt % by the addition of water during flotation. As it is well known that in feldspar flotation the gangue minerals associated with feldspar such as mica and other iron-bearing minerals were initially floated by the addition of a suitable amine at an acidic pH range of 2.5–3.5. After the flotation of mica, other iron-bearing oxide minerals were floated using sulfonate type collectors at an acidic pH range of 3–4. Following these stages, the flotation of feldspar was performed in an acidic circuit with/without the addition of HF.

In the experimental studies, the initial pH of the mixture was around 8.0, while it was decreased to 2.5 for mica flotation, 3.5 for oxide flotation and 2.4 for feldspar flotation by the addition of H_2SO_4 to the medium. Thus, based on the results of previous studies on the flotation of similar materials, DAHC for mica flotation, R801 and R825 for oxide flotation and Flotigam 4343 and SDS were used as collectors for colored impurities in feldspar flotation. Additionally, series of tests were also performed without HF, using Duomeen TDO as a collector and MIBC as a frother. Following the aforementioned stages and conditions for each type of mineral, the effects of different parameters on the flotation process such as particle size and reagent type on the separation of quartz and feldspar from gneiss origin ore were investigated. In addition, the dosage of sodium silicate on the separation of feldspar and quartz was examined. The chemical analysis of raw material and flotation products were performed by X-ray florescence (XRF-1700, Kyoto, Japan).

3. RESULTS AND DISCUSSION

3.1. Determination of the Optimum Particle Size

As is well known, particle liberation plays an important role in flotation processes, as demonstrated by many flotation studies of various types of ores [4,5,6,7,8,9]. This issue in turn is of significant importance for determining the grinding cost for an industrial-scale process. On the other hand, the removal of fine material, so-called slime, is another important process due to its adverse effect of coating the surfaces by electrostatic interaction and hence reducing the flotation recovery [18,19]. In addition, very fine particles will both increase the collector consumption and HF amount. Thus, prior to flotation tests, material under 63 μm was removed in order to enhance the flotation recovery by preventing the negative effects of very fine particles. The effect of particle size was evaluated by considering two sizes of fraction ranges; $0.063 < r < 0.300$ mm and $0.063 < r < 0.212$ mm, for determining the optimum liberation size and thus acquiring the maximum yields during flotation tests. The flowsheet for these series of experiments is shown in Figure 4. The conditions of the experiment are given in Table 1 as Test No. 1, and the results are shown in Table 2.

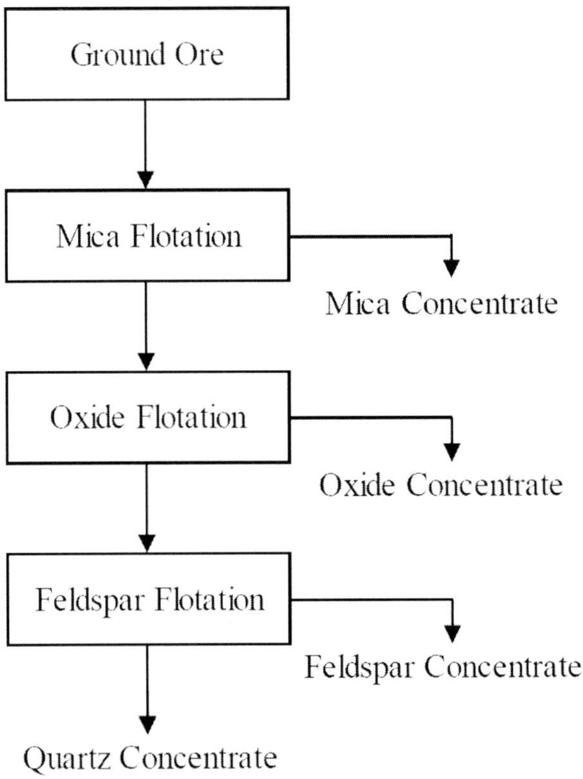

Figure 4. The bench flotation flowsheet for the determination of the optimum particle size.

Table 1. Conditions for different flotation tests.

Test No.	1	2	3	4	5
		Mica Flotation			
Collector	Amine (DAHC)	Amine (DAHC)	Amine (DAHC)	Amine (DAHC)	Amine (DAHC)
Concentration, g/t	100 × 3	100 × 3	100 × 3	100 × 3	100 × 3
Na_2SiO_3 concentration g/t	0	1000	2000	1000	1000
Conditioning Time, min	5 × 3	5 × 3	5 × 3	5 × 3	5 × 3
Flotation Time, min	3 × 3	3 × 3	3 × 3	3 × 3	3 × 3
		Oxide Flotation			
Collector	Sulfonates (R801 + R825)	Sulfonates (R801 + R825)	Sulfonates (R801 + R825)	Sulfonates + SDS	Sulfonates (R801 + R825)
Concentration, g/t	1600 × 3	1600 × 3	1600 × 3	800 × 3 + 100 × 3	1600 × 3
Conditioning Time, min	5 × 3	5 × 3	5 × 3	5 × 3	5 × 3
Flotation Time, min	3 × 3	3 × 3	3 × 3	3 × 3	3 × 3
		Feldspar Flotation			
Collector	Amine (Flotigam 4343)	Amine (Flotigam 4343)	Amine (Flotigam 4343)	Amine (Flotigam 4343)	Duomeen TDO
HF Concentration, g/t	1000	1000	1000	0	0
Collector Concentration, g/t	200 × 5	200 × 5	200 × 5	200 × 5	500 × 5
Conditioning Time, min	5 × 5	5 × 5	5 × 5	5 × 5	4 × 5
Flotation Time, min	3 × 5	3 × 5	3 × 5	3 × 5	4 × 5

("×" indicates the number of stagewise additions).

Table 2. The effect of particle size on flotation (Ct: concentrate).

Size Range, mm	Products	Weight, %	Content		
			Fe_2O_3, %	SiO_2, %	Total Alkali, %
$0.063 < r < 0.300$	Feldspar Ct.	45.2	0.04	80.01	7.21
	Quartz Ct.	22.1	0.01	98.45	0.16
$0.063 < r < 0.212$	Feldspar Ct.	29.2	0.04	79.45	7.52
	Quartz Ct.	24.4	0.01	97.25	0.11

The results and observations during these series of tests clearly indicated that, due to the long grinding times needed to obtain a size range below 0.212 mm, the over-grinding of feldspar (moreso than quartz) leads to a significant amount of feldspar loss in the oxide flotation stage. A comparison of the flotation results with the two size fractions showed that the $0.063 < r < 0.300$ mm size range showed better performance than the $0.063 < r < 0.212$ mm range, especially for feldspar yield, as 45.2% of the total feed was concentrated.

3.2. Determination of the Optimum Depressant Concentration

Additional series of tests were adapted with different amounts of Na_2SiO_3 addition in order to show the influence of quartz depression on the yield of oxide products with the conditions shown in Table 1 (Test No. 2 and No. 3), while the results of these tests are shown in Table 3. As seen in Table 3, the addition of sodium silicate did not make any significant change to the quartz content. With the addition of Na_2SiO_3, the total alkali content of the feldspar concentrate was increased from 7.21% to 10.8%, having similar yields, which was favorable in terms of sale conditions. Doubling the depressant concentration to 2000 g/t did not show any improvements, therefore, 1000 g/t Na_2SiO_3 addition was noted as the optimal concentration.

3.3. Effect of Different Collectors on the Efficiency of Flotation

The effect of different reagents and reagent combinations were tested to obtain products with lower impurities for feldspar and quartz concentrates. In the oxide flotation stage, the efficiency of SDS over sulfonates in the last step of feldspar flotation, and Duomeen TDO efficiency over HF and amine combinations were investigated.

Table 3. The results of flotation tests for evaluating the influence of sodium silicate addition (Ct: concentrate).

Na_2SiO_3, g/t	Products	Weight,%	Content		
			Fe_2O_3, %	SiO_2, %	Total Alkali, %
0	Feldspar Ct.	45.2	0.04	80.01	7.21
	Quartz Ct.	22.1	0.01	98.45	0.16
1000	Feldspar Ct.	29.9	0.07	68.75	10.8
	Quartz Ct.	42.7	0.02	96.36	0.33
2000	Feldspar Ct.	30.7	0.07	68.76	10.9
	Quartz Ct.	43.9	0.02	96.09	0.37

To this aim, following the mica flotation with DAHC in acidic pH conditions, 250 g/t sodium dodecyl sulfate (SDS) instead of R-825 was adapted along with R-801 for the flotation of oxide minerals under the same conditions as our previous series of tests. It is worth to note that conditions such as the dosage of HF and Flotigam 4343 for feldspar flotation were kept the same as the previous tests. The conditions and results of these tests are shown in Table 1 (Test No. 4) and Figure 5, Figure 6, Figure 7 and Figure 8, respectively. According to the test results for separating oxide minerals using SDS, a feldspar concentrate with 0.11% Fe_2O_3 content and a quartz concentrate with 96.43% SiO_2 content were obtained, as seen in Figure 6 and Figure 7. Very similar values were obtained for oxide concentrates in terms of both the contents and yields of different components by the substitution of R-825 with SDS, however, the Fe_2O_3 content of feldspar was found to be almost double compared to previous series of tests performed with the combination of sulfonates (R801 + R825). Thus, the results and our observations during the experiments also proved that most of the iron-bearing colored minerals remained in the feldspar product, due to the ineffectiveness of SDS usage in the oxide stages. Considering the overall results of these series of tests, besides the efficiency of the separation stages for mica and oxide products, DAHC was found to give better results in mica flotation and the usage of petroleum sulfonates, so-called "R-801 + R-825" was found to be effective for oxide flotation.

The effect of different reagents during feldspar-quartz separation were also investigated. For decades, many investigations have been adapted to find an alternative reagent or reagent combinations in order to separate quartz and feldspar without HF, due to the hazardous environmental and health problems caused by HF over long periods of time. To this aim, mixtures of cationic and anionic reagents were used. Katayanagi et al. reported that the mixture of cationic alkyltrimethylene diamine and anionic petroleum sulphonate can be used for this separation [20]. Likewise, Vidyadhar et al. also suggested a similar mixture composed of N-Tallow-1.3 propylene diamine/dioleate for feldspar ore [16]. Thus, in our experimental studies, the effect of the same reagent, namely "Duomeen TDO" N-tallow 1,3-propane diamine dioleate was used in order to float feldspar with gneiss origin instead of relying on activation by HF. The experimental conditions and the results are shown in Table 1 (Test No. 5) and Figure 5, Figure 6, Figure 7 and Figure 8, respectively.

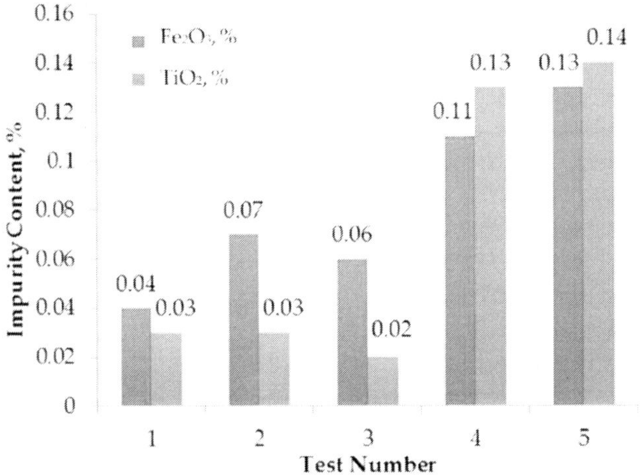

Figure 5. Impurity content of feldspar concentrates for each test.

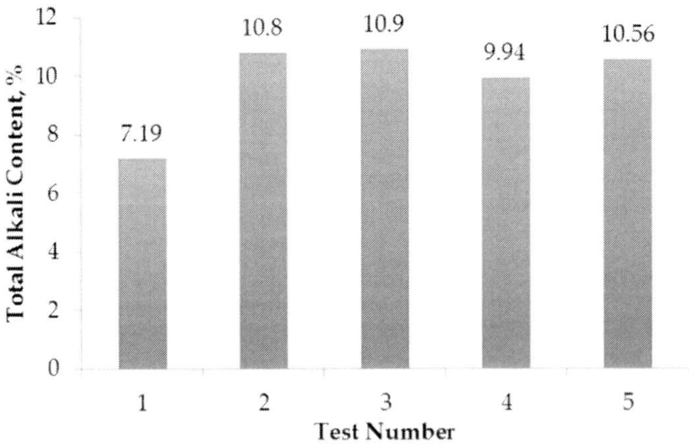

Figure 6. Total alkali (Na_2O + CaO) content of feldspar concentrates for each test.

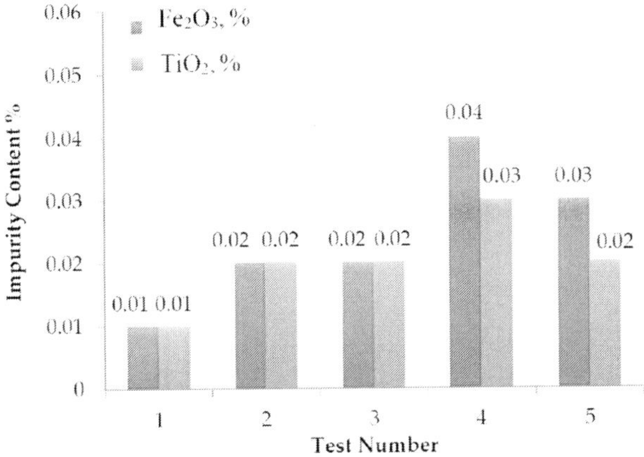

Figure 7. Impurity content of quartz concentrates for each test.

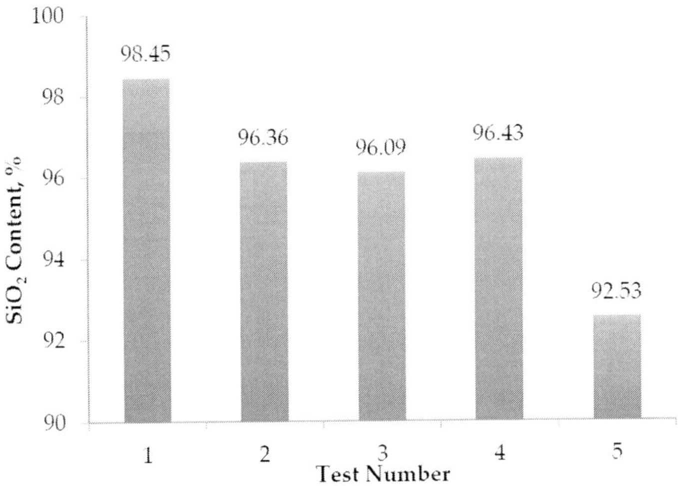

Figure 8. SiO_2 content of quartz concentrates for each test.

According to the results of these tests, a feldspar concentrate with 0.13% Fe_2O_3 content was obtained with 38.8 wt %, while a quartz concentrate with 92.53% SiO_2 content was obtained with 51.1 wt %. If a comparison is made between the optimum results with HF, it is clear that the usage of HF is advantageous for the Fe_2O_3 and total alkali content of feldspar as well as for the quality of the quartz concentrate. The results are shown in Table 4 and Table 5.

Table 4. The results of the optimized flotation test using hydrofluoric acid + sulfonates + 1000 g/t Na_2SiO_3.

Products	Weight, %	Fe_2O_3, %		TiO_2, %		SiO_2, %		Al_2O_3, %		Na_2O, %		CaO, %	
		C	R	C	R	C	R	C	R	C	R	C	R
Feldspar Ct. 1	29.86	0.07	7.2	0.03	5.5	66.75	25.3	20.33	47.5	7.31	59.4	3.46	44.3
Feldspar Ct. 2	3.01	0.08	0.9	0.05	0.9	76.60	2.9	13.89	3.3	5.67	4.6	2.45	3.2
Quartz Ct.	42.69	0.02	2.3	0.02	5.6	96.36	51.0	4.33	14.4	0.01	0.1	0.32	5.9
Oxide Ct.	21.84	0.65	52.6	0.55	72.1	68.38	18.9	17.64	30.1	5.83	34.6	4.36	40.9
Mica Ct.	2.61	3.85	37.0	1.02	15.9	56.56	1.9	22.90	4.7	1.78	1.3	5.13	5.7
Total	100.0	0.27	100.0	0.17	100.0	78.92	100.0	12.79	100.0	3.68	100.0	2.33	100.0

(C: content, R: recovery, Ct: concentrate).

Table 5. The results of the flotation test using Duomeen TDO (without using hydrofluoric acid).

Products	Weight, %	Fe_2O_3, %		TiO_2, %		SiO_2, %		Al_2O_3, %		Na_2O, %		CaO, %	
		C	R	C	R	C	R	C	R	C	R	C	R
Feldspar Ct.	38.8	0.13	20.4	0.14	33.1	69.37	33.7	17.96	58.0	7.09	70.3	3.47	58.4
Quartz Ct.	51.1	0.03	6.5	0.02	6.9	92.53	57.9	6.33	26.9	1.30	17.0	0.75	16.6
Oxide Ct.	5.7	0.74	17.4	0.44	15.2	70.95	5.1	15.87	7.6	5.96	8.7	4.33	10.7
Mica Ct.	4.4	3.03	55.7	1.65	44.8	58.85	3.3	20.20	7.5	3.52	4.0	7.39	14.3
Total	100.0	0.24	100.0	0.16	100.0	79.80	100.0	12.00	100.0	3.91	100.0	2.30	100.0

(C: content, R: recovery, Ct: concentrate).

Therefore, based on the yield of products, the results of the tests with the addition of HF were found to be more effective than the tests without HF for acquiring feldspar and quartz from gneiss origin. Thus, in future studies, the effect of different reagent combinations and different collectors should be evaluated without the usage of HF in order to obtain feldspar and quartz products with a saleable degree.

4. CONCLUSIONS

Based on the mineralogical investigations on selected Augen-Gneiss samples, primary minerals consisted of quartz and feldspar, while secondary minerals were identified as biotite, muscovite, hornblende, tourmaline (schorlite type), apatite, and zircon, which indicates that Augen-Gneiss rock can be an alternative source for glass and ceramic raw materials. Flotation tests were conducted in order to obtain high quality quartz and feldspar concentrates from the samples.

In the series of tests to identify the effect of particle size on flotation, it was found that the usage of either $0.063 < r < 0.300$ mm or $0.063 < r < 0.212$ mm size ranges did not produce significant differences in terms of metal content or yield. Therefore, all the tests were adapted with the coarse size range, taking into consideration the economic concerns of industrial-scale plants.

The results of tests to identify the effect of reagent type on flotation suggested that the optimum values were obtained for all products in terms of content and yield in the case of usage of DAHC for mica flotation, R-801 and R-825 for oxide flotation and Flotigam 4343 for feldspar flotation at a $0.063 < r < 0.300$ mm size range.

Sodium silicate addition did not make any significant difference on the SiO_2 content of the quartz concentrate, but it increased the total alkali content of the feldspar concentrate up to 11%.

The usage of a cationic fatty amine Duomeen TDO (Tallow Di-Oleat) was investigated in order to eliminate HF usage, but it is clear that the usage of HF is more advantageous for the Fe_2O_3 and total alkali content of the feldspar as well as for the quality of the quartz concentrate obtained from Augen-Gneiss rocks.

ACKNOWLEDGMENTS

This work was funded by Straton Mining Company of Turkey.

REFERENCES

1. Yoshihide, O. On the formation of augen structure. Lithos 1969, 2, 109–132.
2. Burianek, D.; Hrdlickova, K.; Hanzl, P. Geological position and origin of augen gneisses from the Policka Unit, Eastern Bohemian Massif. J. Geosci. 2009, 54, 201–219.
3. Potters, M. Feldspars. In Industrial Minerals and Rocks-Commodities, Markets, and Uses, 7th ed.; Kogel, J.E., Trivedi, N.C., Barker, J.M., Krukowski, S.T., Eds.; SME Press: Englewood, CO, USA, 2006; pp. 451–460.
4. Celik, M.S.; Can, I.; Eren, R.H. Removal of titanium impurities from feldspar ores by new flotation collectors. Min. Eng. 1998, 11, 1201–1208.
5. Karaguzel, C.; Çobanoğlu, G. Stagewise flotation for the removal of colored minerals from feldspathic slimes using laboratory scale Jameson cell. Sep. Purif. Technol. 2010, 74, 100–107.
6. Burat, F.; Kokkilic, O.; Kangal, O.; Gurkan, V.; Celik, M.S. Quartz-feldspar separation for the glass and ceramic industries. Min. Metall. Process. 2007, 24, 75–80.
7. Kangal, O.; Guney, A. Beneficiation of low-grade feldspars using free jet flotation. Min. Process. Extr. Metall. Rev. 2002, 23, 129–140.
8. Sekulic, Z.; Canic, N.; Bartulovic, Z.; Dakovic, A. Application of different collectors in the flotation Addition of feldspar, mica and quartz sand. Min. Eng. 2004, 17, 77–80.
9. Gülsoy, Ö.Y.; Can, N.M.; Bayraktar, I. Production of potassium feldspar concentrate from a low-grade pegmatitic ore in Turkey. Min. Process. Extr. Metall. 2005, 114, 80–86.
10. Vidyadhar, A.; Hanumanta, R. Adsorption mechanism of mixed cationic/anionic collectors in feldspar-quartz flotation system. J. Colloid Interface Sci. 2007, 306, 195–204.
11. Demir, C.; Bentli, I.; Gulgonul, I.; Çelik, M.S. Differential separation of albite from microcline by monovalent salts in HF medium. Min. Metall. Process. 2003, 20, 120–124.
12. Bayat, O.; Arslan, V.; Cebeci, Y. Combined application of different collectors in the flotation Addition of Turkish feldspars. Min. Eng. 2006, 19, 98–101.
13. Klyachin, V.V. The selective flotation of quartz-feldspar sands by a new method. Glass Ceram. 1969, 26, 174–176.
14. Kalyon, D.; Gulsoy, Ö. The comparison of non-hydrofluoric acid flotation methods in feldspar quartz separation. Earth Sci. 2004, 26, 49–59.
15. Ibrahim, I.; Hussin, H.; Azizli, K.; Alimon, M. A study on the interaction of feldspar and quartz with mixed anionic/cationic collector. J. Fundam. Sci. 2011, 7, 101–107.
16. Vidyadhar, A.; Rao, H.; Forssberg, K.S.E. Separation of feldspar from quartz: Mechanism of mixed cationic/anionic collector adsorption on minerals and flotation selectivity. Min. Metall. Process. 2002, 19, 128–136.
17. Gaied, M.E.; Gallala, W. Beneficiation of feldspar ore for application in the ceramic industry: Influence of composition on the physical characteristics. Arab. J. Chem. 2015, 8, 186–190.
18. Karaguzel, C. Selective separation of fine albite from feldspathic slime containing colored minerals (Fe-Min) by batch scale dissolved air flotation (DAF). Min. Eng. 2010, 23, 17–24.
19. Walker, G.F. Interactions of N-alkylammonium ions with mica-type layer lattices. Clay Miner. 1967, 7, 129.
20. Katayanagi, K. A new feldspar flotation process without hydrofluoric acid. Ceram. Jpn. 1973, 8, 58–61.

CHAPTER 8

Chemical and Mineralogy Characteristics of Dust Collected Near the Phosphate Mining Basin of Gafsa (South-Western of Tunisia)

Mohamed Raja*, Taieb Dalila and Ben Brahim Ammar

Research Unit: Applied Thermodynamics (UR11 ES 80), National Engineering School of Gabes, Gabès University, Tunisia

Mohamed Raja, Research Unit: Applied Thermodynamics (UR11 ES 80), National Engineering School of Gabes, Omar Ibn El Khattab Street 6029 Gabès, Gabès University, Tunisia

ABSTRACT

The study aimed at chemical and mineralogical characterization of whole particulate matter (PM) in the vicinity of a mining phosphate basin at urban area in Gafsa. Heavy metals concentrations in PM samples (Cd, Fe, Cu, Mn, Cr, Zn, Ni, Al, Pb), MgO and SiO_2 were determined by atomic absorption spectrometry. Calcium and phosphorus oxides (CaO, P_2O_5) were analyzed by Technicon Auto-analyzer. Potassium and sodium oxides (K_2O, Na_2O) were analyzed by flame photometer. Ionic species (SO_4^{2-}, Cl^-, NO_3^-) were quantified by ion chromatography while organic carbon (OC) content was determined using carbon sulfur Analyzer. The mineral phase identified by X-Ray powder diffraction technique and SEM-EDX provides information about morphology and chemical composition of PM. As regards the chemical characterization it was found that samples were enriched predominantly in SiO_2, CaO and P_2O_5 which were detected only at mining area (S2 and S3) Weak concentrations of K_2O, Na_2O and MgO were also found. In addition the minerals phases identified in the samples were carbonate fluorapatite, Calcite, Heulandite, Gypsum and Dolomite. The values of ionic composition were in the flowing order $SO_4{}^{2-}$>Cl->$NO_3{}^{-}$. Results also showed that heavy metals pollution increased by the proximity to the mining area. SEM-EDX revealed that particles are almost spherical

Keywords: Dust; Mining pollution; Metals; Mineralogy; Mining basin; Gafsa.

1. INTRODUCTION

Air pollution is a major problem nowadays. Exposure to ambient air pollution has been associated to several health outcomes, starting from modest transient changes in the respiratory tract and impaired pulmonary function, continuing to restricted activity, emergency room visits and hospital admissions and to mortality [1].

According the World Health Organisation (WHO), 4-8% of deaths occurring annually in the world are related to air pollution [2].

In recent decades, there has been a growing concern for the potential contribution of dust as one of major sources of atmospheric pollution. The interest in atmospheric particulate matter (PM) is mainly due to its effect on health and its role in climate change [3]. Dust can also affect biogeochemical cycling and air temperature because of the absorption and scattering of solar radiation and can influence sulphur dioxide levels in the atmosphere [4]. Fine particles in the air are most efficient in scattering light because they are similar in size to wavelength of visible light and have the largest surface area, so reducing visibility [5]. In daily life, dust presence in atmosphere dramatically reduces air quality leading to consequences such as respiratory and allergy diseases and aggravation in people suffering from heart diseases. Moreover, metals present in dust can accumulate in human body via directly inhalation, ingestion and dermal contact resulting in serious health problems [6]. In addition, the particulate material or its soluble components may be transported to organs some distance from the lungs and have a detrimental effect on these organs [7]. A large variety of dust is found in the atmosphere which originates from a variety of sources. Dust can have natural or anthropogenic origins. Maximal concentrations of PM10 are found in mining area and the concentrations are gradually diminished with increase in distance due to transportation, deposition and dispersion of particles [8].

Currently, the knowledge of the chemical composition of particulate matter (PM) has gathered increasing importance in the scientific community as the necessity to differentiate PM components and their influence to both health effects and role in the climate change [9]. Also, research on the physic-chemistry properties of particulate matter is intense and toxicological studies attempt to identify which adverse biological responses (e.g. particle number, size, surface, chemical composition), and suggest that the chemical composition of PM (which reflects differences in source contributions) plays an important role in these responses [9]. In this framework, this study presents a comparative analysis of particulate matter at three sites located in Gafsa and identified minerals and trace metals, ionic components, and carbonaceous material.

2. THE STUDY AREA

2.1 Geology

Gafsa is the capital of Gafsa Governorate of Tunisia. It lies 369 Km by road southwest of Tunis. The geographical coordinates of the city are 34°25'N8°47'E. It is located in one perforated in the middle of an alignment mountainous, called "mounts of Gafsa " between DjbelBou- Ramli and DjbelOrbata which culminates to 1165 meters.

The governorate of Gafsa is composed of different towns such as: El Guettar, El Kaser, Gafsa, Mdhila, Métlaoui, Moularès, Rdyef and Sned. The city has 90.000 inhabitants (2005 estimate), the governorate has 340.000 inhabitants (20005 estimate) and an area of 8990 Km^2.

Mining basin of Gafsa is a part of the south-western dry area of Tunisia. It covers 325 Km^2, representing 42% of the entire governorate of Gafsa area and 38% of its total population [6].

Measurements were conducted at urban area of south Gafsa (S1) and (S2, S3) were chosen at Metlaoui as shown in (Figure 1) because Metlaoui is considered as a predominantly mining zone. Phosphate is the main wealth of the region. Metlaoui is located in the southwestern part of Tunisia (34°18'/34 °19'N; 8°22'/8°25'E), at 250 m above sea level. There are less than 40.000 inhabitants in the province of Metlaoui. The regional landscapes are mountains, basins and plains. It is a transition zone between steppes and desert [10].

2.2 Climate

Gafsa has a dry continental climate. It is characterized by important annual variation in temperature. Winters are cold and dry; with a mean annual rainfall of 137 mm and monthly mean temperatures ranging from 40°C in July to 7°C in January. Summers are very hot and dry because of dry sirocco blowing from the Sahara desert. Wind speed varies from 7 to 10 knots throughout the year and winds are predominately from the west or northwest [11].

2.3 Soil type

Soil in Gafsa is characterized by high salinity. The soil is stony and contains gypsums crusts and limestone. Mobility of sand is the limiting factor for the development of soil and rain-fed agriculture in general [12].

2.4 Vegetation

Vegetation includes steppe of this climate; Thymeleahirusta (menten), Artemisia eampestris (tgouft), Artemisia bleached on grass Alba (chir), Diplotaxix (harra) and Peganumharmala (harmel). Oasis of south-western Gafsa, El-Kasba, El ksar, Lala and Guettar cover about 3200 hectares. The oasis counts 32500 palm trees belonging to seven different varieties; "Degletnour", "Allique ", "Kenta","BesserHlou"," Quabrichou" and "Hammouri" [12].

2.5 Chemical analysis

Malvern Mastersizer 2000 was used for determining particle size distribution. Dust is introduced into a tank with circulation containing a dispersing liquid. The suspension is then transported towards a laser cell measurement. Themastersizer has a fully optimized optical design which allows characterizing particle in the size range 0.02-2000 micron. The results are given in function of volume of the size of particles

Frequency=50/60 Hz

Power =335 W

Capacity =3.2 m^2/min

These samples were collected during July 2011

Details of sampling are given in Table 1.

Metals concentration (Cd, Fe, Cu, Mn, C r, Zn, Ni, Al, Pb) and magnesium oxide MgO, SiO_2, were determined using the Atomic spectrometer spectra AA 220. CaO and P_2O_5 were analyzed by with Technicon Autoanlyzer. K_2O, Na_2O were characterized by using flame photometer 410 model

Mineral phases identified by X-RAY powder diffraction technique. In this technique peaks occur when the path of the diffracted X-rays is equal to an integer multiple of the path difference expressed by Bragg's equation which is given by: $N\lambda = 2d \sin\theta$ Ion chromatography (Shimadzu Degasser SCL.10ASP) was used to analyze the three anions (SO_4^{2-}, Cl^-, NO_3^-).

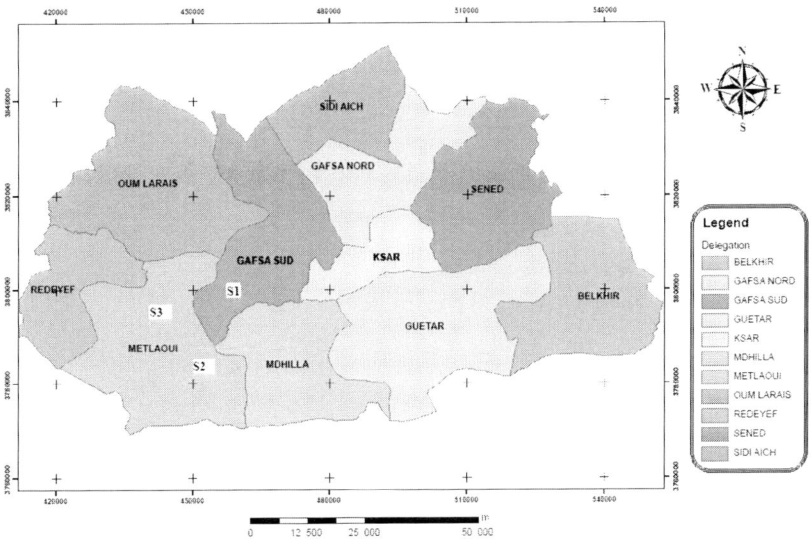

Figure 1: Study area and sampling sites.

Scanning Electron Microscopy (SEM) coupled with energy dispersive X-ray spectrometry (EDX) (FEI Quanta 200) provides information on morphology and chemical composition of individual particles down to sub-micrometer and nano-meter size. Organic carbon (OC) content was determined using Horiba EMIA-220V series carbon sulfur analyzer.

3. RESULTS AND DISCUSSION

3.1 Particle size distribution

The particle size of dust is a very important parameter. Particle size is often described by the diameter of the particle. It has already been proved that the degree of harmful effect of inhaled dust is closely related with dust particle size. The results of particle size distributions are summarized in Tables 2-4 and their particle size distribution curves (S1, S2, S3) are shown respectively in Figure 2. The particle size distribution curves of samples show more or less similar distribution patterns marked by overlapping of the curves.

Table 1: Details of sampling collection.

Station code	Location	Period sampling
S1	Urban area	17/7/2011 - 22/7/2011
S2	Near Landry of phosphate	22/7/2011 - 26/7/2011
S3	Near Breakers of phosphate	26/7/2011 - 30/7/2011

Table 2: Particle size distribution of S.

Particle size (µm)	0.010	1	6	10	20	32	40	50	71	90	100	125	160	200	315
% volume	0.21	1.38	1.03	1.43	5.41	6.62	10.12	21.67	15.95	6.64	12.1	9.36	4.87	3.18	0.02

Table 3: Particle size distribution of S2.

Particle size (µm)	0.010	1	6	10	20	32	40	50	71	90	100	125
% volume	1.21	3.9	3.06	16.67	23.1	12.55	12.04	15.48	7	2.12	2.71	0.16

Table 4: Particle size distribution of S3.

Particle size (µm)	0.010	1	6	10	20	32	40	50	71	90	100	125	160	200	315	340
% volume	0.89	5.28	3.87	10.75	15.6	10.57	11.85	18.14	9.71	3.34	5.03	2.86	1.04	0.6	0.07	0.26

Where n is an integer, λ is the wavelength, d is the inter-atomic spacing and θ is the diffraction angle

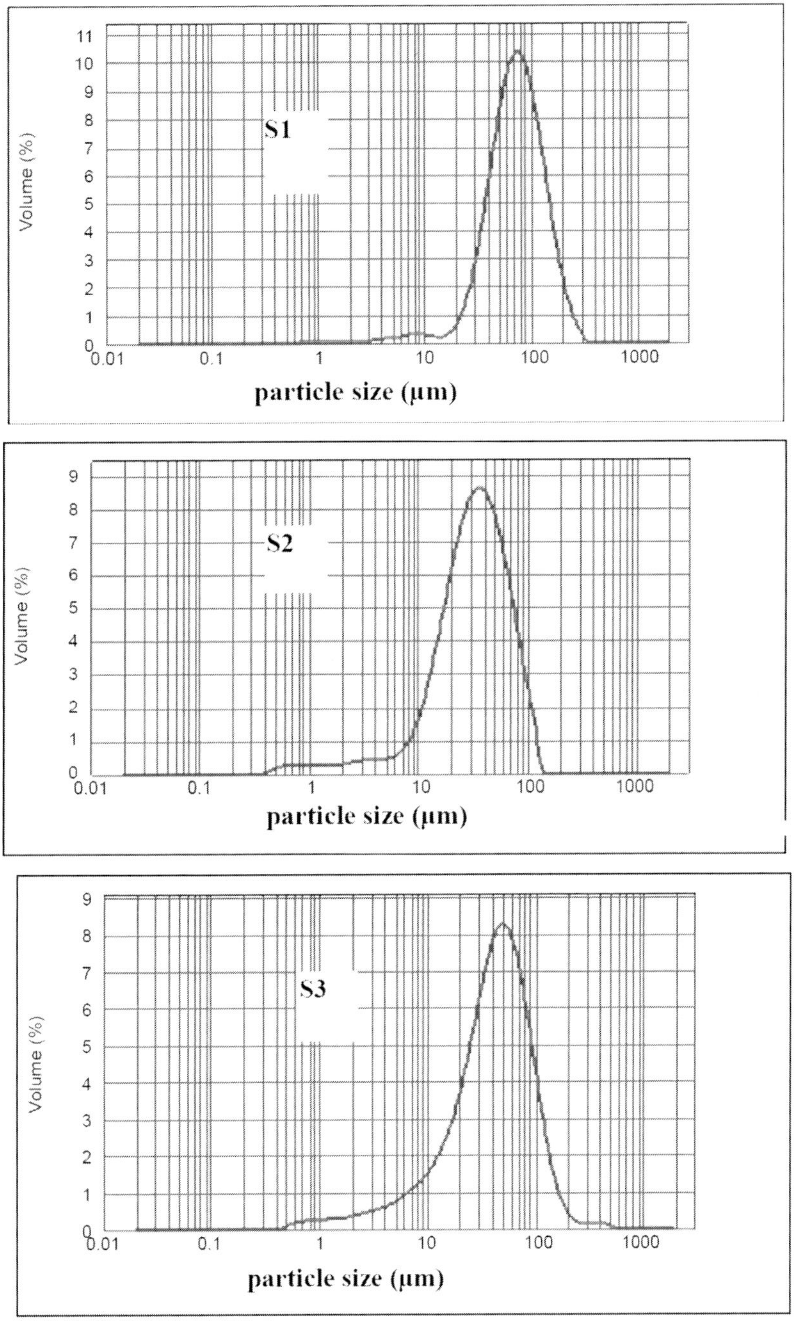

Figure 2: Particle size distribution.

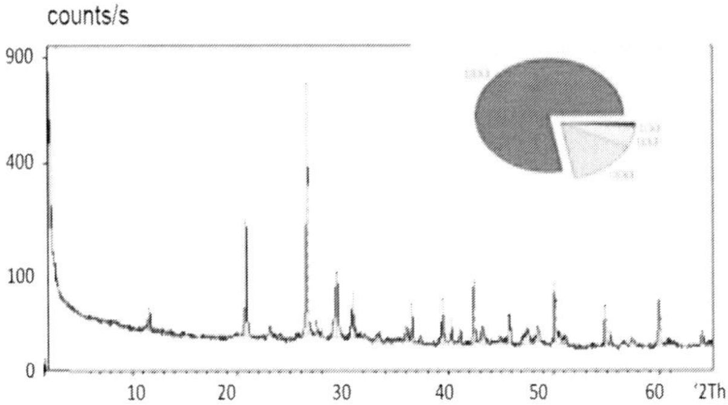

(a)

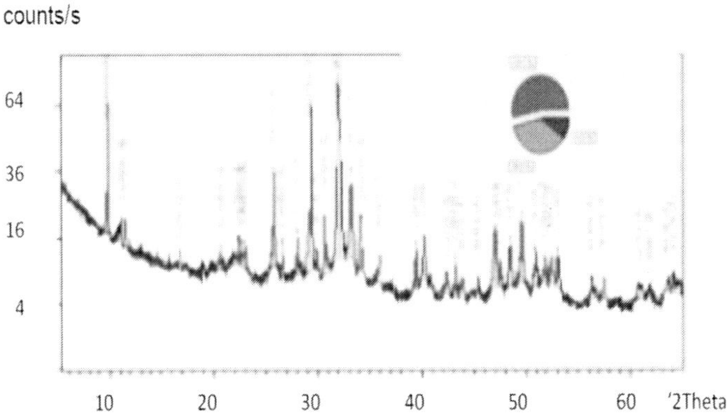

(b)

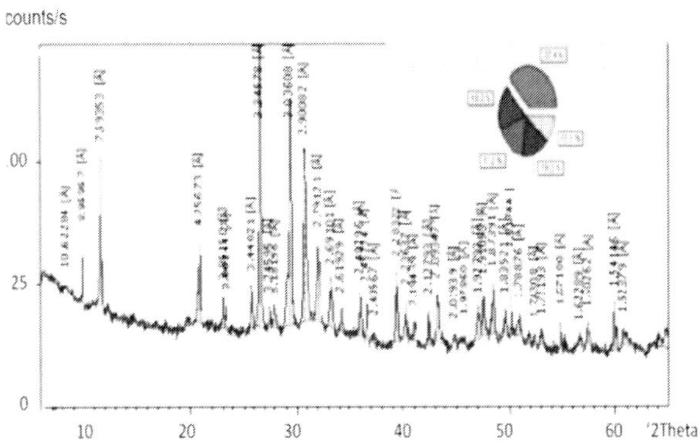

(c)

Figure 3: Representative x-ray diffraction patterns of the 3 samples; (a) urban area; (b) near Landry of phosphate; (c) near Breakers of phosphate.

Percentage of particles between 0.01 and 10 µm: The mesh that contains very fine particles between 0.01 and 10 microns is represented on Figure 3. The samples obtained (S1, S2, S3) contain (2.62 %, 8.17%, 10.04 %) of total sample volume respectively. Fraction of coarse mining dust superior 10 µm around mining area is considered to be nonbreathable, that is why is considered with less attention than breathable dust. Breathable fractions of airborne ambient particulate matter (PM10) specifically those fractions that are less than 2. 5 µm (PM2.5) in size. Fine particles have been identified as potential risk of general public. It has been one the largest occupational problem. They are small enough to penetrate into lungs, where they may exacerbate conditions such as bronchitis and asthma and have also been associated with visibility degradation and climate change. It causes lung damage such as pneumoconiosis and in particular silicosis, asbestosis, damage to the noise; throat and eyes and damage to the skin; it may cause various types of dermatitis, which are a widespread and often serious problem or even skin cancer [14].

3.2 Mineralogical and Chemical Composition

Mineralogical composition: Based on XRD analysis (Figure 3), samples of dusts collected at mentioned sites (S1, S2, S3) are composed of minerals listed in Table 5 with their formula

The characterization in XRD, S_1 shows that this sample mainly consists of quartz (78%), this sample is very rich in breathable silica crystalline in the form of quartz.

For S2 and S3 silica is not detected by diffraction on the X-ray but this phase appears later on compared to the results obtained by MEBEDX.

The other phases identified by this sample are calcite (15%), gypsum (1%) and dolomite (6%).

The S3 sample is characterized by a major phase of heulandites 65% and a minor gypsum phase 4%. The S2 sample is less rich in heulandites (37%) and presents calcite 10%.

As it is shown in Table 6, Carbonatefluorapatite is not uniform in different sites. Very high percentage of Carbonate fluorapatite at S2 and S3 and do not have in the Gafsa urban area could be explained from the distance of mining activities (landry and Breakers of phosphate) because Carbonate fluorapatite (francolite) is the principal mineral of phosphorites, the sedimentary rock phosphates which is an combination of phosphate radical PO_4 with water, calcium and the trace element fluoride [15].

Inhalation of fluoride containing phosphate dust can damage the tooth forming cells, leading to a defect in the enamel known as dental fluorosis.

Fluorosis is a serious health problem .White and yellow glistening patches on the teeth are seen which may turn brown. The brown streaks may turn black and affect the whole tooth and may get pitted perforated and chipped. Dental fluorosis not only poses cosmetic problems but has also social problems [16].

Table 5: Formula of investigated minerals

Name	Formula
Carbonate fluorapatite	$Ca9.55 (PO_4)4.96F1.96 (CO_3)1.283$
Calcite	$CaCO_3$
Heulandite	$((C_2H_5) NH_3)7.85 ((Al 8.7 Si 27.3) O72)(H_2O)6.92$
Quartz	SiO_2
Gypsum	$Ca(SO4) (H_2O)2$
Dolomite	$Ca Mg (CO_3)2$

Table 6: Major minerals determined by XRF (%).

	Carbonate fluorapatite	Calcite	Heulandite	Quartz	Gypsum	Dolomite
S1	-	15	-	78	1	6
S2	53	10	37	-	-	-
S3	31	-	65	-	4	-

3.3 Contents of heavy metals in dusts

Particulate samples collected from 3 sites located in Metlaoui around mining phosphate area and urban area Gafsa were analyzed for (Cd, Fe, Cu, Mn, Cr, Zn, Ni, Al, Pb).

Table 7 shows the results of the heavy metals. Graphs obtained using concentration of the heavy metals in 3 samples is given in Figures 4-12.

The Ni concentrations measured at 3 sample points were in the range 57-224 ppm .The highest concentration of nickel was found at site S1 (224 ppm) flowed by S2 (60 ppm) ,but the lowest value was found in S3.The data are represented in Figure 4.

Ni concentrations measured in the present study were higher than concentration at the SelebiPhikwe nickel-copper plan area in Botswana (41 ppm) [17].

The high concentrations of Ni were attributed to the pollution originating from traffic in Work place and anthropogenic activities (landry and Breakers of phosphate).

This element is mainly associated with emissions from stationary and industrial sources; it can also be emitted from vehicle exhaust as it is used as an additive in fuels. Health hazards associated with the exposure to Ni the occupational environment have resulted primarily from inhalation.

About 10% of women and 2% of men in the population are highly sensitive to Ni. A portion of these sensitive people can develop skin rash called nickel dermatitis if they are exposed to Ni through direct contact [18].

Cadmium concentration measured in all samples, presented the highest value at S2 site (67 ppm), but the lowest value was found at S1site (4 ppm).Concentration values of cadmium in those three sites were significantly higher than this measured at SelebiPhikwe nickel-copper plan area in Botswana (0,03 ppm) [17] (Figure 5). A high concentration of Cd in S1and S2 may be due to phosphate activities because Cd can be found in phosphate rocks as can be seen in Table 8 [19].

People living near mine sites, in our study especially site S2 and S3 try to keep their windows and doors closed, because after inhalation,the absorption of Cd compounds may vary greatly depending upon the particles sizes and their solubility. Cadmium is a metal, which can cause severe toxicity in humans. Prolonged exposure to Cd can affect a variety of organs with the kidney being the principal target [20]. Also it induces chlorosis and necrosis in plants and exhibits mutagenic and carcinogenic effects in animals 33%-72% of the local Cd is supplied by air and airborne cadmium is transferred predominately by large scale atmospheric transport [21].

Concerning Cu, the high concentrations of copper found at S3 may be associated with electrical and chemical working; also Cu may be derived from the mechanical abrasion of vehicles and may be associated with diesel engine and wearing of break vehicles [22].

Table 7: Metals concentrations in dust in (ppm).

Sites	Cd	Fe	Ni	Cu	Zn	Mn	Cr	Pb	Al
S1	4	1500	224	-	80	177	71	15.8	1345.33
S2	67	2400	60	47	367	30	183	6.22	9953.23
S3	49	2940	57	170	300	35	277	13.7	3848.72

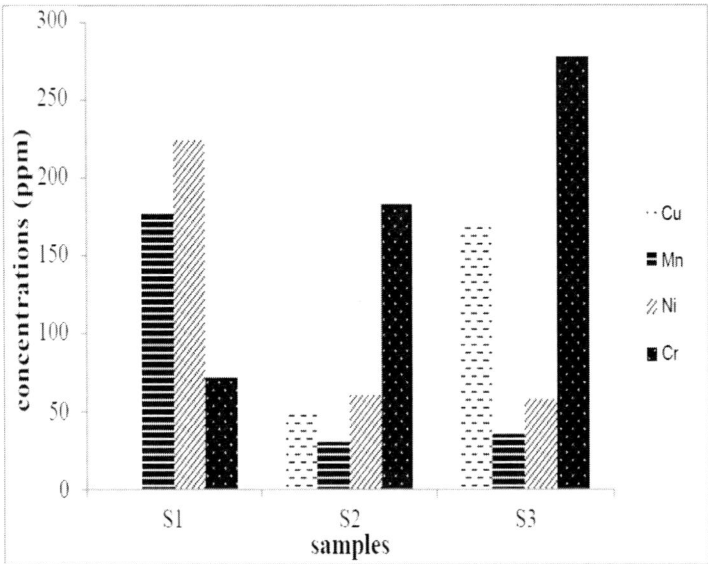

Figure 4: Representative x-ray diffraction patterns of the 3 samples; (a) urban area; (b) near Landry of phosphate; (c) near Breakers of phosphate.

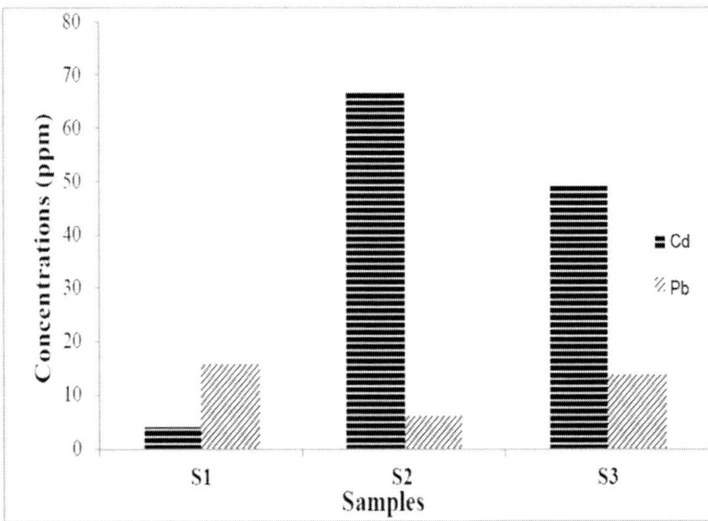

Figure 5: Cd, Pb concentrations in the dust sample.

Table 8: Global pattern of cadmium in contents in phosphate rocks.

Phosphate rock origin	Cd concentration (ppm)
USA	3-186
Morocco	3-165
Peru	2-186
Russia	0.1-<13
North Africa	60
South Africa	2-<13
Brazil	4
Jordan Aman	6<13
Togo	44-179
Tunisia ,Gafsa	38-173
Algeria	22.5-62,6
Syria	6.1-52
China(Yunam)	4
Mexico	8
Egypt (Quseir)	Aug-74

Lead is the most significant toxin of the heavy metals and affects are of toxicological and neurotoxic in nature, which include irreversible brain damage. From Table 7 and Figure 8 it can be seen that the Pb concentration in the range of 6 to 78 ppm. The highest lead concentration has been found in sample collected in urban environment S1, and the lowest at S2 sample. These may be due to the heavy traffic density that used lead such as deterioration of lead paint, home age, smoking and cooking activities.

It is noticed that important concentration of lead was found near Work place S3 can be associated to cars services site. The contents of Zinc in the dust samples are in the range of 50-367 ppm .the highest Zn concentrations were respectively in S2 and S1 as given in (Table 7 andFigure 9). This could be attributed to the use of zinc in accumulatorsof motor vehicles or in carburetors. Moreover, zinc may come from lubricating oils and tires of motor vehicles.

To better understand the pollution level of elements associated on dust in 3 sites, the concentrations of (Ni, Cd, Cu, Pb, Zn) measured at various places around the world are shown in Table 9 [20]. With regard to Fe, when comparing Fe concentrations with those measured at SelebiPhikwe nickel-copper plan area in Botswana (91 ppm) [17]

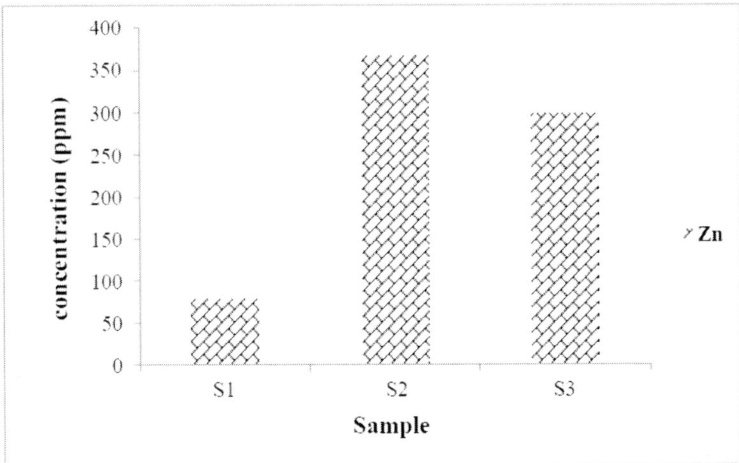

Figure 6: Zn concentrations in the dust sample.

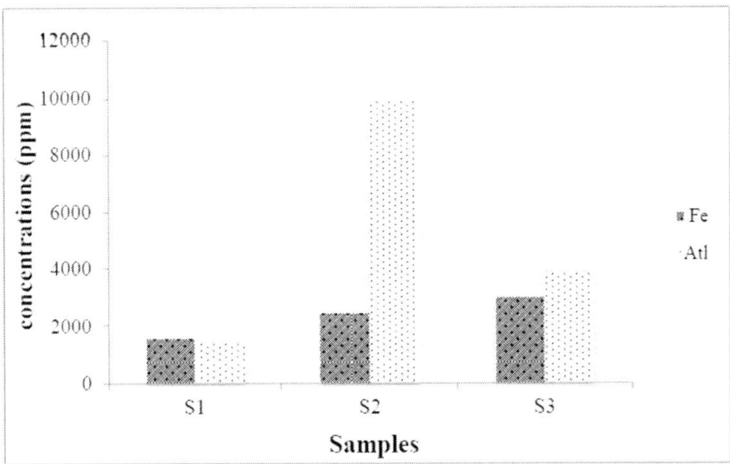

Figure 7: Fe, Al concentrations in the dust sample.

3.4 Organic Carbone content

Measurement results of OC content are given in Table 13. The % of OC ranged between 2.26% and 4.69%.

1/13/2012 10:59:01 AM
Mag 300x
HV 20.0 kV
Spot 6.6
Sig BSE
WD 10.3 mm
Det SSD
Pressure 1.00 mbar
100.0µm
INRAP

1/13/2012 10:43:19 AM
Mag 150x
HV 20.0 kV
Spot 5.4
Sig ---
WD 9.1 mm
Det Mix
Pressure 1.00 mbar
200.0µm
INRAP

1/13/2012 10:25:56 AM
Mag 1200x
HV 20.0 kV
Spot 5.4
Sig BSE
WD 10.5 mm
Det SSD
Pressure 1.00 mbar
20.0µm
INRAP

1/13/2012 10:25:56 AM
Mag 1200x
HV 20.0 kV
Spot 5.4
Sig BSE
WD 10.5 mm
Det SSD
Pressure 1.00 mbar
20.0µm
INRAP

1/13/2012 10:20:42 AM
Mag 1200x
HV 20.0 kV
Spot 5.1
Sig BSE
WD 10.5 mm
Det SSD
Pressure 1.00 mbar
20.0µm
INRAP

Figure 8: SEM images of dusts

Table 9: Global comparison of heavy metals with present study data (ppm).

City /country		Elements				
		Cd	Cu	Ni	Pb	Zn
Africa	Land /Angola	11.5	41.78	10	351.3	316.6
Asia	Amman/Jordan	3.1-11.2	66.5-350	43-88	210-1131	116-410
	Bahrain	72	-	126	697.2	151.8
Europe	Madrid/spain	-	188	44	1927	476
	Palermo/Italy	1.1	98	14	544	207
	Paris/France	1.7	1075	25	1450	840
North America	Hermosillo/ Mexico	4.25	26.34	4.7	36.15	387.98
	Ottawa/ Canada	0.37	65.84	15.2	39.05	12.5

Table 10: Comparison of the heavy metal (Cr, Mn) concentration (ppm) of the three sample points and other selected cites

	S1	S2	S3	Baoji	Xi'an	Guangzhou	Hong Kong	Ottawa	Avilés	Calcutta	Madrid
Cr	71	183	277	126.7	167.3	78.8	124	43.3	41.6	54	61
Mn	177	30	35	804	687	481	594	432	1661	619	362

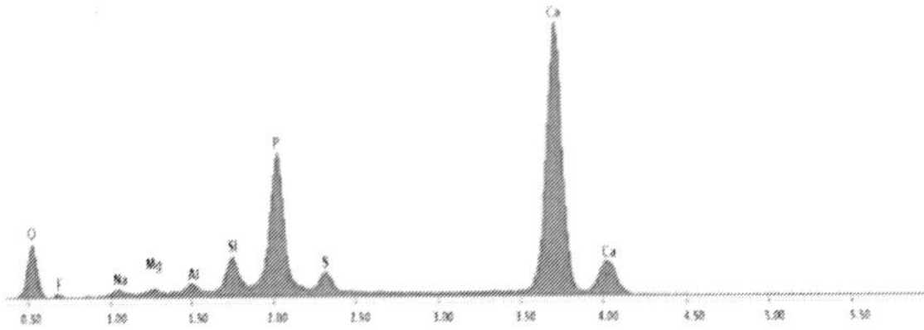

P1

Elements	atomic %
C	9.52
O	34.27
NA	0.26
Mg	0.83
Al	0.90
Si	47.87
S	0.35
Ca	6.00

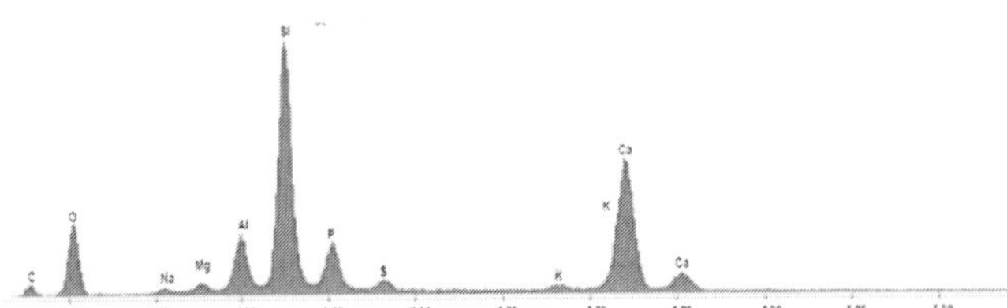

P2

Elements	atomic %
C	12.72
O	34.19
Mg	1.81
Al	0.92
Si	24.93
S	0.59
K	0.69
Ca	23.00
Fe	1.16

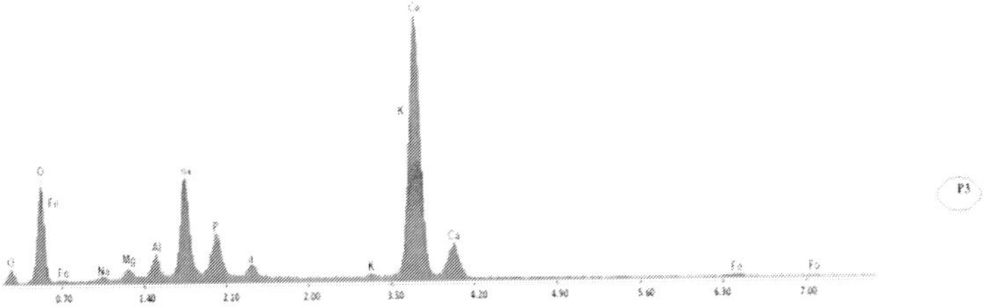

Elements	atomic %
C	10.44
O	26.34
Mg	1.42
Al	1.30
Si	26.22
S	0.30
K	0.44
Ca	14.18
Fe	19.64

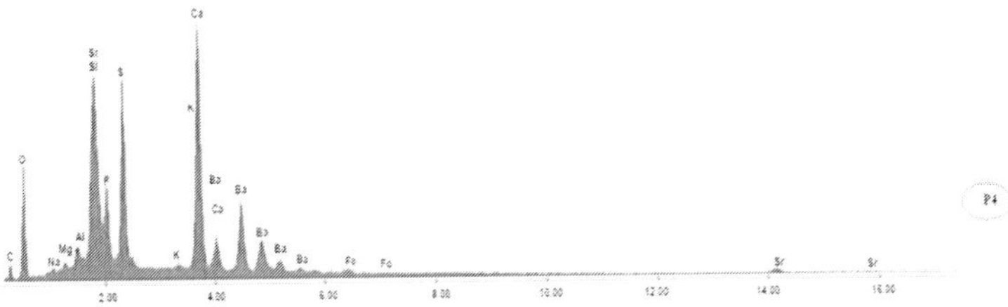

Elements	atomic %
C	6.40
O	22.55
Mg	1.38
Al	1.14
Si	14.26
S	0.33
K	0.28
Ca	5.88
Ti	3.85
Fe	43.96

Figure 9A: SEM micrographs of representative S1 sample showing the morphologies of the main constituents and their corresponding EDS spectrum of element.

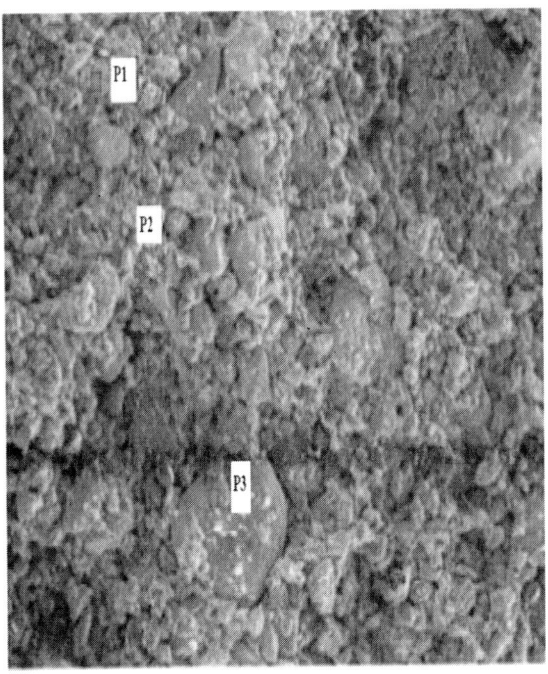

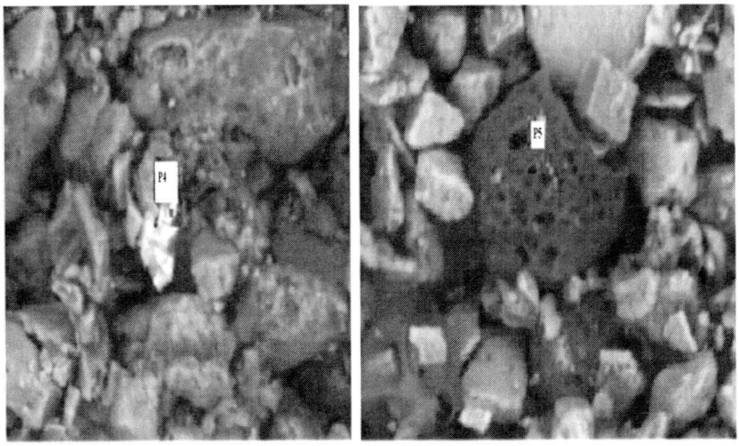

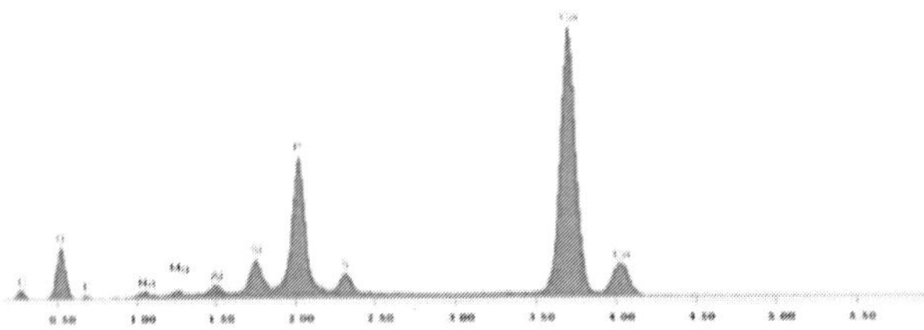

P1

Elements	atomic %
C	11.38
O	35.29
Fe	1.45
Na	1.06
Mg	0.77
Al	1.01
Si	2.94
P	12.12
S	1.88
Ca	32.10

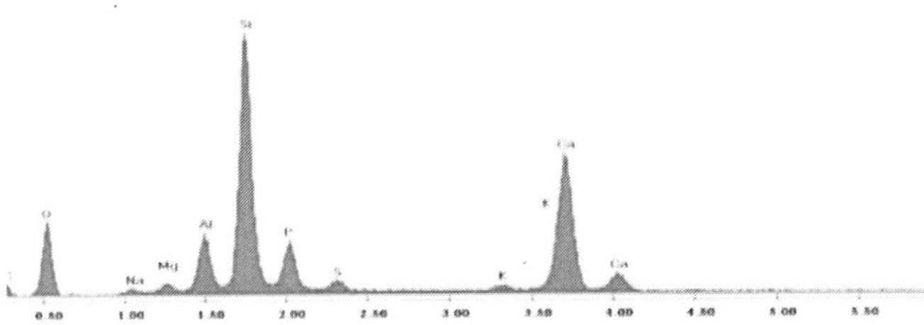

P2

Elements	atomic%
C	15.13
O	36.08
Na	0.72
Mg	0.96
Al	4.53
Si	19.97
P	4.80
S	1.08
K	0.66
Ca	16.07

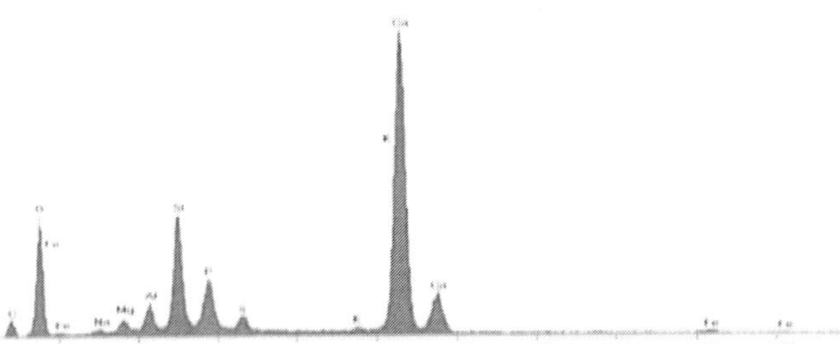

P3

Elements	atomic %
C	10.31
O	48.97
Na	0.15
Mg	0.64
Al	1.34
Si	6.61
P	3.22
S	0.89
K	0.31
Ca	26.80
Fe	0.78

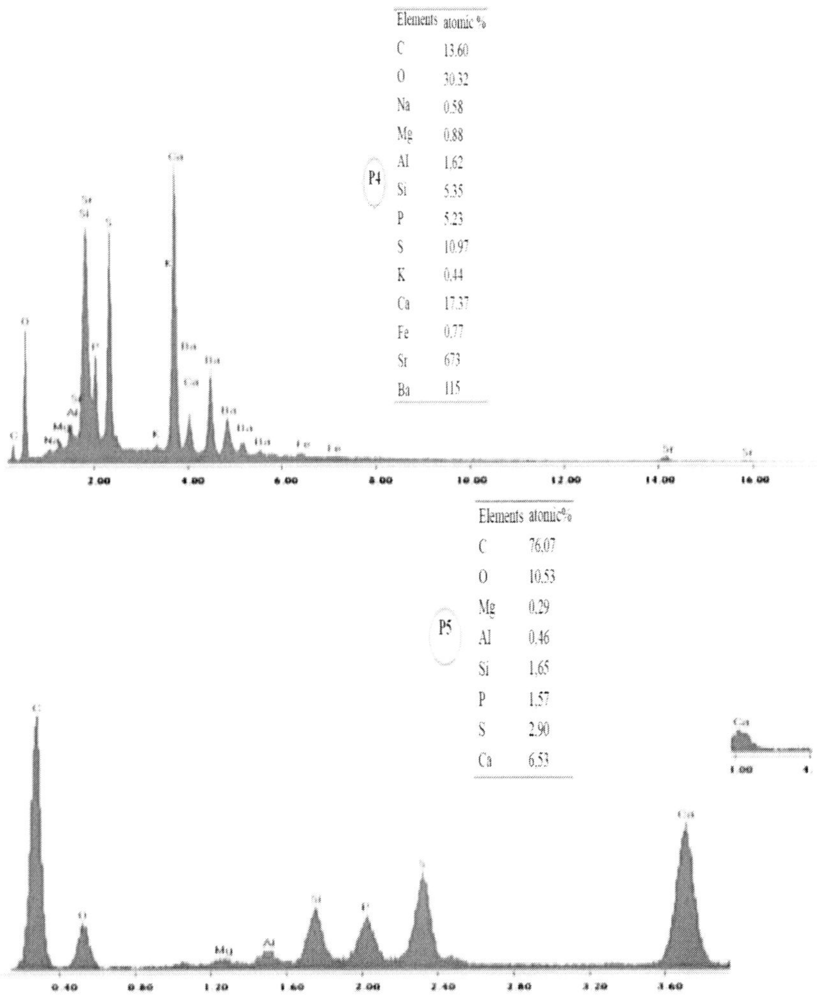

Figure 9B: SEM micrographs of representative S2 sample showing the morphologies of the main constituents and their corresponding EDS spectrum of element.

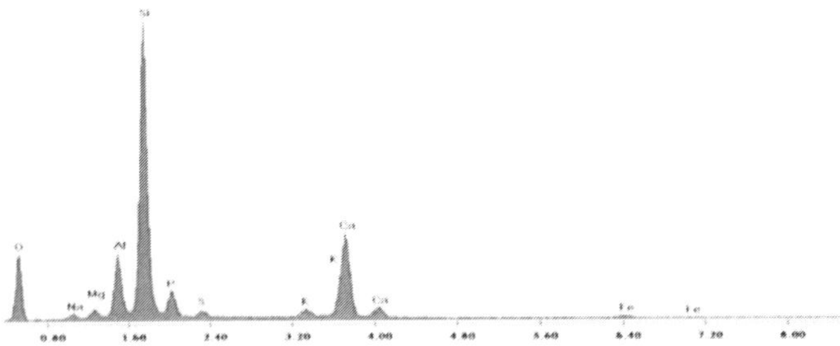

P2

Elements	atomic %
C	8.37
O	36.63
Na	0.99
Mg	1.18
Al	6
Si	28.02
P	3.44
S	0.84
K	1.15
Ca	12.29
Fe	1.10

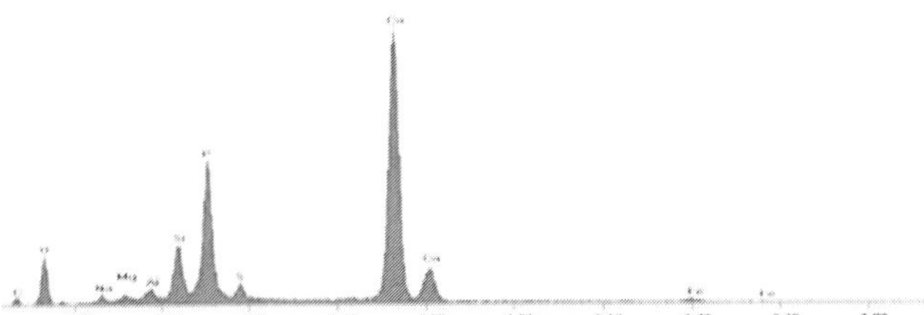

P1

Elements	atomic %
C	9.73
O	33.55
Na	1.19
Mg	0.87
Al	1.23
Si	4.66
P	12.69
S	1.45
Ca	33.42

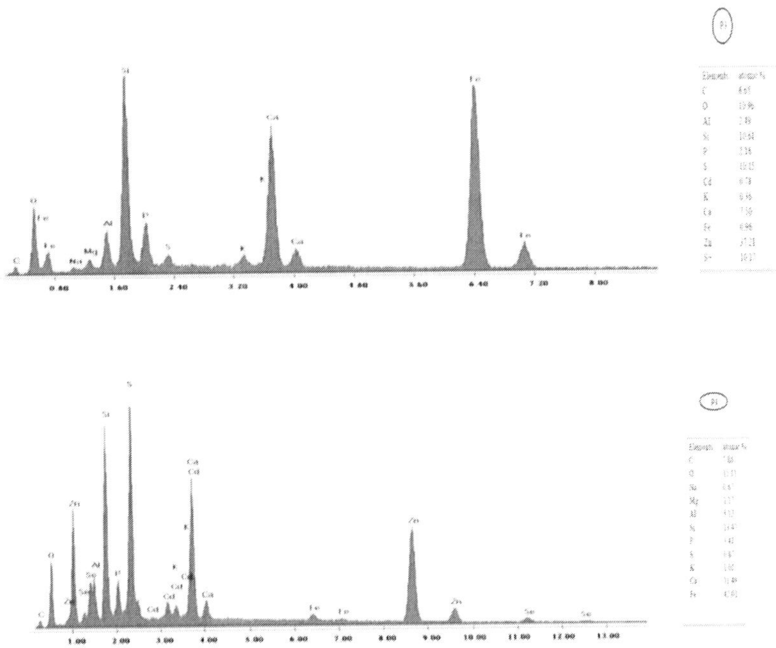

Figure 9C: SEM micrographs of representative S3 sample showing the morphologies of the main constituents and their corresponding EDS spectrum of elements

Table 11: Major oxides (%).

Oxides Sites	P_2O_5	MgO	CaO	K_2O	Na_2O	SiO_2
S1	0	8.49	15.48	4.9	5.52	65.61
S2	20.5	4.04	38.51	1.16	2.36	31.98
S3	17.48	6.4	32.06	4.48	3.86	35.72

Table 12: Ions concentration in dust (g/Kg).

Ions sites	Cl^-	NO_3^-	SO_4^{2-}
S1	2.633	1.23	5.351
S2	1.578	1.11	17.73
S3	1.43	-	10.14

Table 13: OC in sample points.

Sites	S1	S2	S3
OC (%)	2.26	4.69	2.84

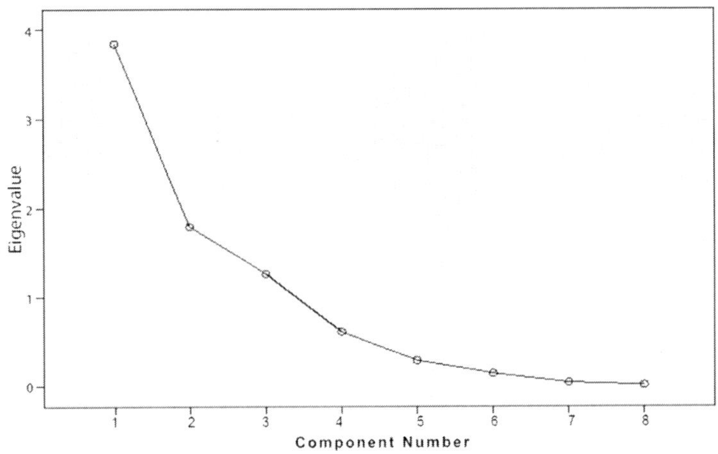

Figure 10: Scree plot.

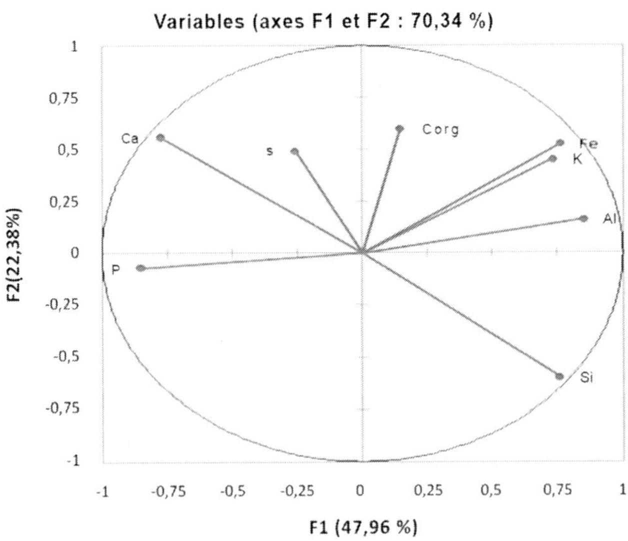

Figure 11: The results of factor analysis, showing the plots of factor loadings (scores) variables on a correlation circle.

35 Scanning electron Microscopy (SEM) and Electron dispersive x-ray (EDX) data

Morphology of dusts: SEM can provide size and morphology information of particles on the submicron scale. The size and morphological characteristics of dust particles observed by ESM are illustrated in Figure. The SEM studies indicate that samples are characterized by multi-modal particle-size distributions; spherical, irregular, long and prismatic, crystalline are the most common shapes of dusts in the samples (Figure 13).

3.6 EDX analysis

S1 sample: The three spectra obtained are shown as (P1), (P2) and (P3) in Figure 9. It must be noted that chemical composition of the particle is mainly Si and O with minor Mg, Al, S and K as expected the atomic percentages of these elements determined by analyzing the three EDX spectra. Figure shows that spherical grains (P4) have very high iron (Fe) content with small amounts of S, K and Al.

S2 sample: EDX spectrum P1 (Figure 9) indicated that the sample are predominantly composed of Ca (32.101%), O (35.29%) and P (12.12%).

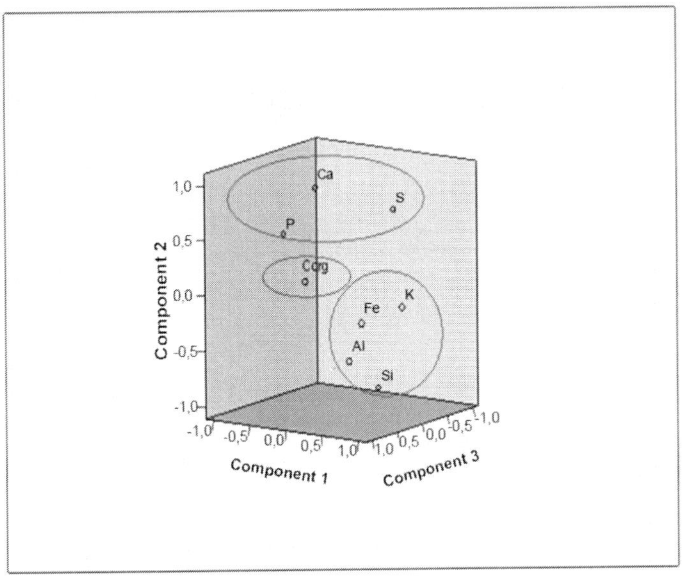

Figure 12: PCA results in three dimensional space plot of loadings of the three principal components.

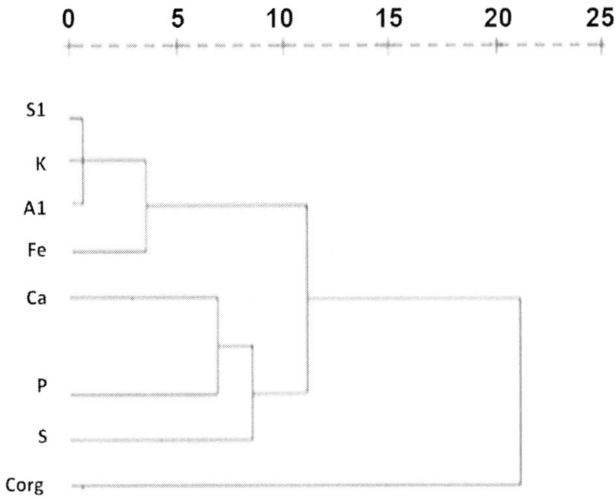

Figure 13: Dendogram resulting from the word's method of hierarchical cluster analysis.

The two spectra obtained are shown as (P2) and (P3) in Figure 9. It must be noted that both spectra exhibit characteristic peaks at the same element (C, O, Ca) although there are difference in the intensity.

The elements detected at spectrum P4 (Figure 9) are (C, O ,Na , Mg ,Al ,Si ,P ,S, K ,Ca ,Fe ,Sr ,Ba), where O, C, Ca have the largest contribution as is presented in the atomic percentages of these elements

Spectrum P5 Figure 9 clearly shows that this sample have very high carbon content (76.07%) with small amounts of (O, Mg, Al, Si, P, S, Ca).

The high fraction of C was possibly due to the mixture carbonaceous aerosols during the dust event. Figure 9 SEM micrographs of representative S2 sample showing the morphologies of the main constituents and their corresponding EDS spectrum of element.

S3 sample: Measurement in P1 (Figure 9) indicated that the dominant elements were O (33.55%), Ca (33.42%) and P (12.69%).

As shown in Figure 9 the particle P2 is rich in O (36.83%) and Si (28.02%), other elements are present with a smaller percentage. From Figure 9 we can see that the main elements of P3 are Fe (42.01%), O (15.11%), Si (13.47%) and Ca (11.49%). As it can be seen in P4 (Figure 9), sample shows the highest content in Zn (37. 21%), other element (C, O, Al, Si, P, S, Cd, K, Ca, Fe, Zn, Se) are detected with small amounts.

4. CONCLUSIONS

This paper provides a detailed update of the chemical and mineralogical composition of PM at different site types in Gafsa. There is a clear evidence of metallic trace elements contamination in the city; this contamination is much higher around mines areas. It is highly recommended to keep a healthy distance between the mining and urban areas to reduce the exposure of population to contaminants. More guidelines and standards need to be produced to specify the maximum allowable level of every mineral substance without endangering the health of the population, to put these standards in place, the cooperation of Hygienists with people from the mining sector are paramount. The hygienist can provide people with more accurate information on exposure potential, dust sources, existent prevention methodology and others.

ACKNOWLEDGMENT

The authors would like to thank the administration of biotech pole Sidi Thabet (Tunis). Thanks are also due to Mr Habib Hamid director of research center, Gafsa phosphate company (CPG) for his help during the sampling campaigns. We are particularly indebted to Professor Marie Le coq at laboratory of process engineering, environment, food (GEPEA), Nantes, France for her continuous and precious collaboration. Finally, we wish express our gratitude to Thomas Bergantz for his technical support.

REFERENCE

1. Nilamadhab S, Singhdeo AK (2008) Monitoring of sulphur dioxide, nitrogen oxides, PM10 and TSP present in the ambient air of Nit Rourkela.
2. Zheng N, Liu J, Wang Q, Liang Z (2010) Health risk assessment of heavy metal exposure to street dust in the zinc smelting district, Northeast of China. See comment in PubMed Commons below Sci Total Environ 408: 726-733.
3. Pederzoli A, Mircea M, Finardi S, Sarra A, Zanini G (2010) Quantification of Saharan dust contribution to PM10 concentrations over Italy during 2003-2005. Atmospheric Environment 44: 4181-4190.
4. Feng J (2006) characteristics and source apportionment of organic matter in PM2,5 from cities in different climatic zones of China, Hong Kong.
5. Goudie AS (2009) Dust storms: recent developments. See comment in PubMed Commons below J Environ Manage 90: 89-94.
6. Karaouli F, Touzi S, Tarhouni J, Bousselmi L (2009) Improvement potential of the integrated water resources management in the mining basin of Gafsa. Desalination 246: 478-484.
7. Kawah YE (1993) Comprehensive characterization of organic Matters in Aerosol in Hong Kong. Applied chemistry. Hong Kong Baptist college, Hong Kong.

8. Mlayah A, da Silva EF, Rocha F, Hamza CB, Charef A, et al. (2009) The Oued Mellègue: Mining activity, stream sediments and dispersion of base metals in natural environments, North-western Tunisia. Journal of Geochemical Exploration 102: 27-36.
9. Querol X, Alastuey A, Lopez-Solera A, Planaa F, Mesasb A, et al. (1999) Physico-chemical characterisation of atmospheric aerosols in a rural area affected by the aznalcollar toxic spill, south-west Spain during the soil reclamation activities. The Science of the Total Environment 242: 89-104.
10. Jaouadi K, Depaquit J, Haouasa N, Chaara D, Gorcii M, et al. (2012) Twentyfour new human cases of cutaneous leishmaniasis due to Leishmania killicki in Metlaoui, southwestern Tunisia. Probable role of Phlebotomus sergenti in the transmission. Acta Tropica 122: 276-283.
11. Medical Force protection-Additional information: Tunisia. Threat Assessment Department Navy Environmental & Preventive Medicine Unit 2 1887 Powhatan Street Norfolk.
12. Job JO (1992) les sols sales de l'oasis D'el Guetar (sud tunisien).
13. Phosphate
14. Petavratzi E, Kingman S, Lowndes I (2005) Particulates from mining operations. A review Of sources, effects and regulations. Minerals Engineering 18: 1183- 1199.
15. Mathews A, Nathan Y (1977) The decarbonation of carbonate-fluorapatite (francolite). American Mineralogist 62: 565-573.
16. Bellomo S, Aiuppa A, Alessandro WD, Parello F (2007) Environmental impact of magmatic fluorine emission in the Mt. Etna area. Journal of Volcanology and Geothermal Research 165: 87-101.
17. Ekosse G, van den Heever DJ, de Jager L, Totolo O (2004) Environmental chemistry and mineralogy of particulate air matter around Selebi Phikwe nickel–copper plant, Botswana. Minerals Engineering 17: 349-353.
18. Zarasvandi A, Carranza EJM, Moore F, Rastmanesh F (2011) Spatio-temporal occurrences and mineralogical–geochemical characteristics of airborne dusts in Khuzestan Province (southwestern Iran). Journal of Geochemical Exploration 111: 138-151.
19. Kyle JH, Breuer PL, Bunney KG, Pleysier R, May PM (2011) Review of trace toxic elements (Pb, Cd, Hg, As, Sb, Bi, Se, Te) and their deportment in gold processing. Part 1, Mineralogy, aqueous chemistry and toxicity. Hydrometallurgy 107: 91-100.
20. Faiz Y, Tufail M, Javed MT, Chaudhry MM, Siddique N (2009) Road dust pollution of Cd, Cu, Ni, Pb and Zn along Islamabad Expressway, Pakistan. Microchemical Journal 92: 186-192.
21. Komarnicki GJ (2005) Lead and cadmium in indoor air and the urban environment. See comment in PubMed Commons below Environ Pollut 136: 47-61.
22. López JM, Callén MS, Murillo R, García T, Navarro MV, et al. (2005) Levels of selected metals in ambient air PM10 in an urban site of Zaragoza (Spain). See comment in PubMed Commons below Environ Res 99: 58-67.
23. Gunawardana C, Goonetilleke A, Egodawatta P, Dawes L, Kokot S (2012) Source characterisation of road dust based on chemical and mineralogical Composition. Chemosphere 87: 163-170.
24. Lu X, Wang L, Li LY, Lei K, Huang L, et al. (2010) Multivariate statistical analysis of heavy metals in street dust of Baoji, NW China. See comment in PubMed Commons below J Hazard Mater 173: 744-749.
25. Kumar A, Mishra MK, Divkar JK, Rout S, Narayanan U, et al. (2010) Impact of particle size on distribution of major ions in acid- and water-soluble components of PM10 atmospheric aerosols in the coastal region of Mumbai. Atmospheric Research 98: 406-415.

CHAPTER 9

Clay Minerals from the Perspective of Oil and Gas Exploration

Shu Jiang

[1]*Energy & Geoscience Institute, University of Utah, Salt Lake City, USA*

1. INTRODUCTION

The clay minerals e.g. kaolinite, smectite, illite, chlorite, etc. are ubiquitous in the targeting rocks of oil and gas exploration. During the early age (1940s) of worldwide oil exploration, clay minerals were studied to predict the quality of organic rick source rock and generation mechanism when scientists tried to investigate the origin of oil and gas (Grim, 1947, Brooks, 1952). Then the clay minerals analysis was used as a tool in terms of environmental determination, stratigraphic correlation and hydrocarbon generation zone identification to find exploration target interval, which was preliminarily and generally summarized by Weaver in 1960. By the 1970s, the clay minerals began to be widely studied for diagenesis and reservoir quality prediction due to the application of petrological analysis and quantitative mineralogical analysis by X-ray diffraction (Griffin, 1971; Pettijohn, 1975; Heald and Larese, 1974; Bloch et al., 2002). Since 1980s, the clay minerals analysis has been used to determine the hydrocarbon emplacement time and petroleum system analysis (Lee et al., 1985). These intermittent clay minerals research progresses are the result of exploration demands of conventional reservoirs (sandstone and carbonate rocks) at different times.

Despite their increasing importance in fundamental geological research and oil industry, clay minerals prove difficult to study in the past. Their sheet structure results in features that can only be resolvable at the sub-micron scale. They are also subtly variable in chemical composition (Fe, Mg, K, Al, etc) and can be confused with each other and other silicates. the recent innovative analytical tools and modern analysis techniques, e.g., QEMSCAN (Automated Mineralogy and Petrography), FIB/SEM (Focused Ion Beam/Scanning Electron Microscope), EDS (Energy-dispersive X-ray spectroscopy), etc., have the capability of quantitative and qualitative characterizing nano-pore and mineralogy of fine grained shale rocks (Lemmens et al., 2011), which creates new era of studying clay minerals for facilitating unconventional (shale) reservoir exploration.

Even though there were the numerous sporadic reports about the application of clay minerals in the oil and gas exploration. So far, relatively little work has been documented on the detailed summary of clay minerals from the perspective of oil and gas exploration. This paper is to systematically summarize the important role of clay minerals in oil and gas exploration from many points of view: basin tectonic evolution, depositional environment, thermal history and maturation history of organic matter in the source rock, hydrocarbon generation, migration and accumulation process, diagenetic history and reservoir quality prediction. The traditional and cutting-edge analytical tools and techniques are also be introduced to identify and characterize the clay mineralogy, rock fabrics property and micro- to nano-scale pores both conventional and unconventional oil and gas exploration.

2. THE USES OF CLAY MINERALS IN OIL AND GAS EXPLORATION

2.1. Indication of tectonics and sedimentation

During the evolution of petroliferous sedimentary basin, the clay minerals contained in the rocks undergo a series of changes in composition and crystal structure in response to tectonics and sedimentation. The amount and type of clay minerals are a function of the provenance of clastic minerals and of diagenetic reactions at shallow and greater depth in different tectonic and sedimentary settings. Clay minerals can be used to infer tectonic/structural regime, basin evolution history and the timing of various geologic events. This may even provide useful tool in helping to unravel the histories in tectonically complex area, e.g., Schoonmaker et al. (1986) found that the depth distribution of illite/smectite (I/S) compositions showed an irregular, zig-zag trend with depth. This trend is probably the result of multi-stage reverse faultings resulted from the compressional tectonic movement. I/S data were also used to infer several kilometers of uplift and subsequent erosion of the section. The depositional facies appears to be an important factor controlling the abundance of clays in the sediments. Fluvial facies generally possesses higher clay mineral abundance. Well-sorted clean aeolian sands typically have a low clay abundance (<15%).

2.2. Indicator of hydrocarbon generation and expulsion

For oil and gas exploration, we need at least to confirm the exploration area has potential source that generates the oil and gas. This drives geologists to study the potential source rocks (usually organic rich shales) to understand if the organic matter in the source rock can generate hydrocarbons at a given depth in a specific geologic time and when the generated hydrocarbons reach the expulsion peak. Organic geochemistry is the main discipline for studying oil and gas generation and expulsion. However, clay mineralogy is also important for evaluation of these parameters since clay minerals and organic matters usually coexist in the sedimentary rocks and the ultrafine clay minerals are sensitive to the changes in the rocks accompanying the hydrocarbon generation and expulsion processes. Association of clay minerals and organic matter in shales is a significant factor in petroleum genesis. Grim (1947) emphasized the likelihood that the clay minerals in shales concentrated organic constituents by adsorption to form abundant source material, and subsequently acted as catalysts in petroleum generation (Brooks,1952).

Many authors report the transformation of clay minerals during diagenesis is from montmorillonite to mixed-layer montmorillonite/illite to illite (Hower et al, 1976) and changes in the ordering of Illite/smectite (I/S) are particularly useful in studying the hydrocarbon generation because of the common coincidence between the temperatures for the conversion from random to ordered I/S and those for the onset of peak oil generation. Percentage of expandable layers in Illite/smectite decreases sharply where the Tmax (from Rock-Eval pyrolysis) of S2 hydrocarbon production peak and indicator of thermal maturation-production index (PI) [PI=S1/(S1+S2)] indicate it is oil generation zone (Burtner and Warner,1986). The use of mixed-layer illite/smectite (I/S) as a geothermometer and indicator of thermal maturity is based on the concepts of shale diagenesis that were first described in detailed studies of Gulf Coast (Powers, 1957; Hower et al., 1976; Hoffman and Hower, 1979). The good agreement between changes in ordering of I/S and calculated maximum burial temperatures or hydrocarbon maturity suggests that I/S is a reliable semi-quantitative geothermometer and an excellent measures of thermal maturity (Waples,

1980; Bruce, 1984; Pollastro, 1993). The clay mineral association even can be used to evaluate the hydrocarbon generation degree, e.g., the presence of illite-smectite-tobelite demonstrates that oil generation has taken place and absence of tobelite layers shows that the rock has not been heated sufficiently to generate large amounts of oil (Drits et al., 2002).

The significant changes of clay minerals during burial and their relations with diagenetic stages, temperature, organic matter maturity, hydrocarbon generation and expulsion can be summarized in Figure 1. During early diagenesis, the maturity of source rock indicated by vitrinite reflectance (Ro) is low and low percentate of illitic beds in illite-smectite mixed-layer clay minerals, e.g. Ro= 0.5% approximately corresponds to around 25% illite presence. The Clay minerals mainly experience loss of pore water and little oil is generated during this period. 25 to 50% illitic beds in illite-smectite mixed-layer clay minerals correspond to major oil generating zone (R_O= 0.5 to 1.0%). When more than 75% illitic layers are present in illite-smectite mixed-layer clay minerals, cracking of hydrocarbons form dry gas (Ro> 1.5%). This general trend can be used to predict if the source rock is able to generate hydrocarbon in an area. For example, the smectite alters to illite at temperature of 80 to 120°C, which corresponds to the oil generation peak at the same temperature range. Figure 2 presents data from Liaodong Bay area in Bohai Bay Basin in Northeast China to this aspect showing the change in maturity of organic matter and reaction progress in the smectite to illite transformation, which indicates that the rapid increase in illite and decrease in smectite (montmorillonite) in I/S correspond to rapid oil generation.

The reaction of smectite to illite in these clays also can indicate the producing high pore-fluid pressures (Powers, 1967) and expulse hydrocarbons from the shales (Burst,1959; Bruce, 1984). This can be demonstrated in Figure 2 that the overpressure development interval corresponds to the transformation of smectite to illite and hydrocarbon generation zone (Figure 2C, D, E).

2.3. Indices for hydrocarbon migration and accumulation

It is critical to establish that hydrocarbon formation and migration occurred after the formation of the trap (anticline, etc.) that is to hold the oil. There is still very little known about the manner in which hydrocarbons formed in argillaceous source rocks migrate and accumulate in porous reservoirs. Some evidence exists, however, that the clay mineral-kerogen complex plays a role in modifying hydrocarbon compositions during migration. Some time ago, Legate & Johns (1964) used gas chromatography to measure the affinity of montmorillonite clays for hydrocarbons of differing polarity, and suggested that during the migration of petroleum chromatographic effects might modify their composition. Young & Melver (1977) developed the chromatographic technique further and convincingly showed that they could in numerous instances predict oil compositions following migration, where the clay-kerogen complex was the chromatographic agent. "Organo-pores" proposed by Yariv (1976) could be migration paths for hydrocarbons in argillaceous source rocks.

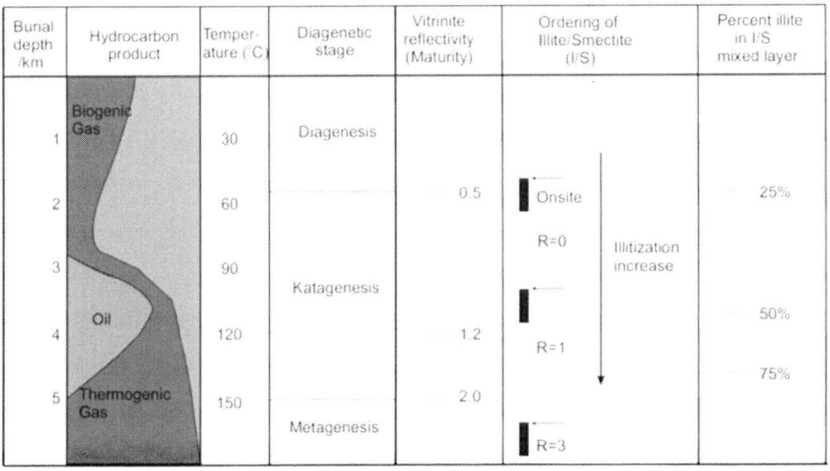

Figure 1. Generalized relationship between temperature, hydrocarbon generation, diagenesis, source rock maturity (vitrinite reflectance), changes in mixed-layer illite/smectite. Figure and data summarized from Foscolos et al (1976), Hoffman and Hower (1979), Waples (1980), Tissot and Welte (1984).

A number of investigators (Powers, 1967; Burst, 1969) have focused attention on the late-stage dehydration which accompanies smectite to illite transformation during burial diagenesis. This firstly suggests that the replacement of kaolinite by illite or direct precipitation of illite indicates fluid flow where the chemical potential of the fluids is in disequilibrium within the reservoir sandstone. The existence of secondary illite does indicate aqueous fluid flow and thus can be used as indices of fluid movement and hence signal the possible hydrocarbon migration. Secondly, it indicates that the water release could create a flushing action responsible for the migration of petroleum hydrocarbons from the source rock through the migration paths to nearby reservoirs. Also, the water liberation can build up abnormal pressures in less permeable sediments, which can provide migration dynamic for hydrocarbons (Figure 2 C, E).

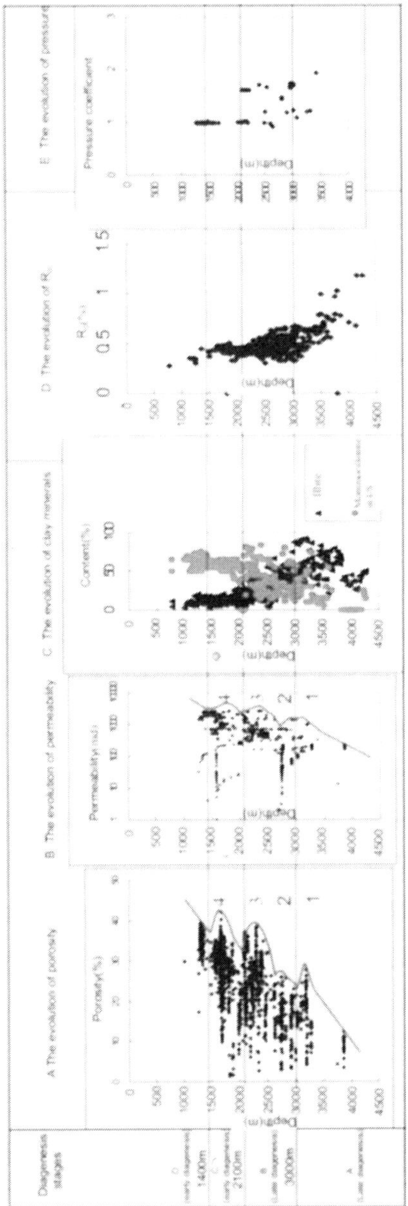

Figure 2. Five plots showing the relationships between diagenetic stages, porosity (A), permeability (B), clay minerals evolution (C), vitrinite reflectance (Ro) (D), and pressure (E) in Liaozhong depression, Liaodong bay sub-basin, Bohai Bay Basin, Northeast China. The secondary porosity zones are numbered upward from 1 to 4.

Abnormal Illite distribution has been used as an index to determine if certain rocks/strata/areas are a hydrocarbon migration pathway and its conducting capability (Zeng and Yu, 2006, Jiang et al., 2011). If there shows abnormal illite distribution, it indicates the hydrocarbon migration happened. The illite abnormal distribution of three wells from three different structure zones in Liaodong Bay Sub-basin of Bohai Bai Basin in Northeast China in Figure 3 suggests hydrocarbon migration happened in these three areas represented by three wells, but the conduiting capabilities are different in the three areas based on different abnormal magnitude of illite content. At the same depth of these three wells, illite content of well JZ25-1s-1 is the highest and the illite content of well JZ21-1-1 is the lowest, which indicates the hydrocarbon migration in the JZ25-1s-1 well area is the most active and the JZ21-1-1 area is the relatively least active area regarding to hydrocarbon migration (Figure 3). This result is consistent with current oil discoveries: The Liaoxi uplift (represented by well JZ25-1s-1) to the west of Liaodong Bay contributes to the most reserves in the Liaodong Bay sub-basin. Tan-Lu strike-slip area (represented by well JZ23-1-1) is emerging as the second largest hydrocarbon migration and accumulation area (Jiang et al., 2010, 2011). Almost no oil and gas discoveries in the rest area of Liaozhong depression (represented by well JZ21-1-1) away from strike-slip zone and Liaoxi uplift so far due to poor hydrocarbon migration pathway and poor conduiting capability.

The smearing of clay minerals can also prohibit the hydrocarbons' further migration and facilitate the hydrocarbon accumulation. When the soft clays are smeared into the fault plane during movement and they will provide an effective seal. In many cases, the presence of clay types and their proportions can even indicate if there is oil and gas accumulation.

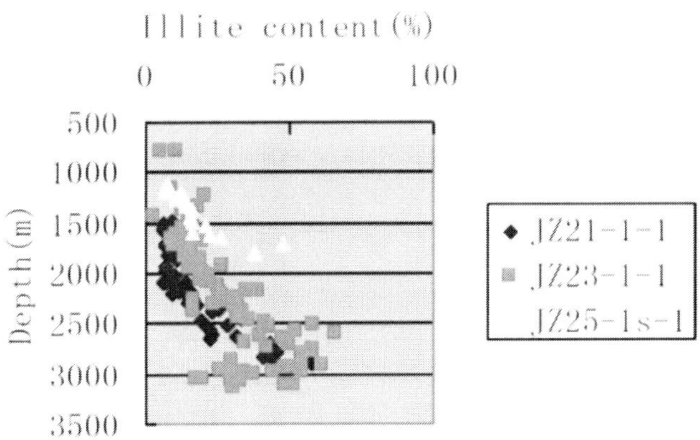

Figure 3. The illite content distribution versus depth from three wells in Liaodong Bay area, Bohai Bay Basin.

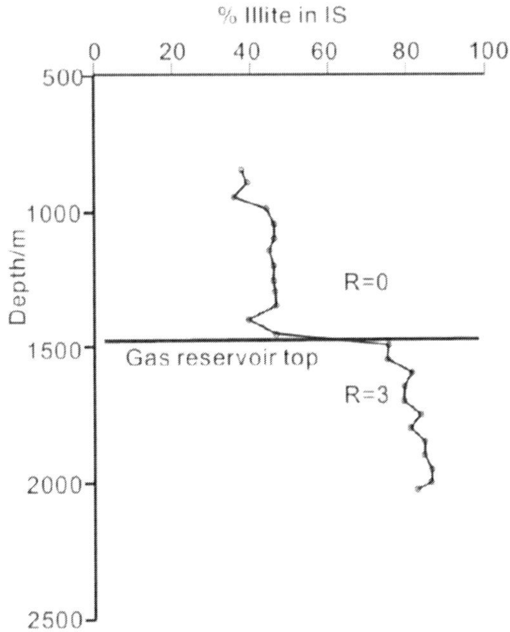

Figure 4. The abrupt changes in the percent illite in I/S and the ordering of I/S (R) in well Dongfang 6 in Dongfang Gas Field, Northern South China Sea.

Webb (1974) recorded that the Cretaceous sandstones of Wyoming generally contain abundant authigenic kaolinite where water saturated, but little if any authigenic clay is found where the sandstones are hydrocarbon saturated. Investigation on clay mineral and its relationship with gas reservoirs show that the clay mineral percentage and ordering of I/S can indicate the hydrocarbon reservoir, e.g., the high content of illite in I/S and higher ordering of I/S change indicate the gas reservoir interval in Dongfang 6 well from Dongfang gas field in Northern South China Sea (Figure 4).

2.4. Significance of clay mineralogy for reservoir quality prediction

Porosity and permeability are the most important attributes of reservoir quality. They determine the amount of oil and gas a rock can contain and the rate at which that oil and gas can be produced. Most sandstones and carbonates contain appreciable fine-grained clay material including kaolinite, chlorite, smectite, mixed layer illite-smectite and illite. These clay minerals commonly occur as both detrital matrix and authigenic cement in reservoir sandstones. The reservoirs initially have intergranular pores that are main space for oil and gas accumulation. When the reservoirs are deposited, their primary porosity is frequently destroyed or substantially reduced during burial compaction. The clay minerals are usually assumed to be detrimental to sandstone reservoir quality because they can plug pore throats as they locate on grain surface in the form of films, plates and bridge and some clay minerals promote chemical compaction. Not only in sandstone reservoir, the clay content also greatly accelerated the rate of porosity loss in limestone reservoir (Brown, 1997). Generally, the porosity loss is mainly caused by the diagenetic process including mechanical compaction, quartz and K-feldspar overgrowths, carbonate cementing and clay mineralization. Especially, the diagenetic clay minerals play a very important role in determining the reservoir quality.

Authigenic clays from diagenesis in the sandstones studied occur as illite, illite-smectite and kaolinite. They form cements around the detrital minerals. During the period of intermediate to deep burial diagenesis, Ilite and illite-smectite clays are the first cements. These early-formed clay films play an important role in reducing reservoir porosity and permeability during burial diagenesis. For example, pore-filling illite formed mainly at the expense of kaolinite.The illitic clays usually occur as pore-bridging clays to reduce the pore space and block the fluid movement by reducing permeability. For clay minerals that replaced rigid feldspar minerals are easily compacted and can be squeezed into pore throats between grains. This will also greatly influence the decrease of reservoir quality.

For oil and gas exploration, we expect the occurrences of high-quality reservoirs. Even though the porosity and permeability of reservoir generally decrease with the increase of burial depth due the diagenetic processes as state above, other diagenetic processes may enhance porosity through the forming of secondary porosity including fractures, removal of cements or leaching of framework grains, preexisting cements and clay minerals, limited compaction and/or limited cementation. The dissolution of authigenic minerals that previously replaced sedimentary constituents or authigenic cements may be responsible for a significant percentage of secondary porosity. Some micropores are found in various clays regardless of whether the clay is authigenic or detrital in origin. Also, the existence of clay minerals does not always mean to reduce the reservoir quality, it may be good phenomenon to indicate good reservoir quality, e.g., coats of chlorite on sand grains can preserve reservoir quality because they prevent quartz cementation (Heald and Larese, 1974; Bloch et al., 2002; Taylor et al., 2004). Sometimes, the higher content zone of kaolinite is indicative of higher porosity. The reason is that porosity is created when the acid dissolves feldspar to produce kaolinite (Jiang et al., 2010). These all show the positive aspect to clay authigenesis.

The secondary porosity development and its relationship with clay minerals evolution has been investigated in many basins (Bloch et al., 2002; Taylor et al., 2004; Jiang et al., 2010). Let's use Liaodong Bay Sub-basin in Bohai Bay Basin in Northeast China as example again. There clearly exist four secondary porosity development zones for the Tertiary strata, whose depth intervals are 1600-1800 m, 2000-2500 m, 2700-2800 m and 3200-3300 m, respectively (Figure 2A). These intervals are named 1 upward to 4 informally. Their corresponding permeability zones have relative higher values (Fig.2B). The secondary and third secondary porosity zones have relatively larger scale. Correlation between porosity, clay minerals and Ro demonstrates that the secondary porosity zones are related to the rapid transformation of the clay minerals and hydrocarbon generation (Ro>0.5%) (Figure 2A, C, D). The relation between zones of secondary porosity and pressure distribution illustrated that No.3 secondary porosity is just right below the top surface of overpressure. This is probably because that the overpressure can retard compaction and avoid the excessive porosity reduction.

2.5. Petroleum emplacement chronology

Petroleum emplacement chronology is one of the frontier research subjects in both petroleum geology and isotope geochronology. Determining the oil or gas emplacement ages has important implications for oil or gas genesis and resource prediction. Typical relative chronology for oil or gas migration, emplacement, and accumulation is established by petrology, basin tectonic evolution, trap formation, and hydrocarbon generation from the source rock (Kelly et al., 2000; Middleton et al., 2000). So far, the illite K-Ar and $^{40}Ar/^{39}Ar$ dating technique hold significant promise in establishing absolute constraints on the emplacement age of oil and gas.

Since the middle of the 1980s, authigenic illite K-Ar dating has been applied to determine the ages of petroleum migration in the North Sea oil fields and Permian gas reservoirs in Northern Germany (Lee et al., 1985; Liewig et al., 1987, 2000; Hamilton et al., 1989). The dating is based on the hypothesis that "illite is commonly the last or one of the latest mineral cements to form prior to hydrocarbon accumulation. Because the displacement of formation water by hydrocarbons will cause silicate diagenesis to cease, K-Ar ages for illite will constrain the timing of this event and also constrain the maximum age of formation of the trap structure" (Hamilton et al. 1989). Wang et al. (1997) investigated oil or gas emplacement ages in the Tarim Basin by this technique.

Recently, illite $^{40}Ar/^{39}Ar$ dating was considered better than traditional K-Ar dating. Among the advantages of $^{40}Ar/^{39}Ar$ dating over traditional K-Ar methods are that stepwise heating can distinguish contributions from authigenic illite and detrital K feldspar by interpreting their gas release characteristics. The K-Ar dating and total fusion $^{40}Ar/^{39}Ar$ dating, however, yield a meaningless mixing age of the authigenic illite and detrital K feldspar, e.g., gradually rising age spectra are obtained by $^{40}Ar/^{39}Ar$ laser stepwise heating of the illite samples from the Tertiary reservoir sandstones in the Huizhou sag, Pearl River Mouth Basin in South China Sea. The youngest ages at the first steps are interpreted as being caused by contributions from authigenic illite, suggesting that the petroleum emplacement occurred after 11 Ma. The high plateau ages in the high-temperature steps that are rather variable between the seven samples are interpreted as being caused by contributions of detrital K- feldspar in the sandstones (Yun et al., 2010).

2.6. Significance for petrophysical property study

The reservoir petrophysics e.g. porosity, permeability, water saturation and hydrocarbon saturation are the most important properties that define and control qualitatively and quantitatively the reservoir performance. The minerals present in the reservoir especially the clay mineral (Moll, 2001) can play the utmost role, which affects both the reservoir capacity and production because the grain size of clay minerals is generally very small and result in very low effective porosity and permeability, thus any presence of clay in a reservoir may have direct consequences on the reservoir properties (Said et al. 2003).

Characteristics of clays that strongly affect their electrical behavior are clay composition, internal structure, the tremendous surface to volume ratios of most clays and the charge imbalance along the surface of clay minerals. All these clay mineral manifestations have an impact upon the interpretative petrophysical parameters by well logging responses. In order to better understand the well logging response for petrophysical analysis, the type of clay minerals must be taken into account in reservoir evaluation, e.g., Potassium presence in the reservoir can increase radiation on Gamma Ray logs. Sometimes, the log response can indicate the hydrocarbon saturation and clay content, e.g., the high resistivity zone of resistivity log corresponds to intervals with low water saturation, a more restricted distribution of diagenetic clay (mainly chlorite) and the low resistivity zone corresponds to intervals with more widely distributed diagenetic clay and variably reduced permeability (Nadeau,2000).

3. TRADITIONAL CLAY MINERAL CHARACTERIZATION METHODS AND THEIR APPLICATIONS FOR CONVENTIONAL OIL AND GAS EXPLORATION

The traditional methods e.g. XRD (X-ray Diffraction), petrographic microscope, XRF (X-ray Fluorescence) and SEM (Scanning Electron Microscopy) have been widely used for clay minerals characterization in conventional siliciclastic and carbonate reservoirs for many years. X-ray diffraction (XRD) is used to provide information on the rock mineral composition and type of clay minerals and their content. Petrographic microscope can identify the reservoir mineralogy composition, pore types, authigenic clays and cements. Figure 5A illustrates that the Tertiary lacustrine turbidite reservoir in Jiyang Sub-basin of Bohai Bay Basin is mainly composed of quart (Q), K-feldspar (Fs) and calcite based on thin section observation. The secondary porosities include the intergranular and intragranular pores caused by dissolution of kaolinite, feldspar and carbonate cement. Scanning electron microscopy (SEM) provides wide range of information about the morphology, mineral composition, distribution and paragenesis of the neoformed authigenic clay minerals, mechanically infiltrated clays, transformational clays and pedogenic mud aggregates.

Figure 5. Photomicrography A illustrates secondary pores resulted from dissolution in feldspar (Fs) and kaolinite (K) and some intergranular secondary pores by carbonate cement dissolution based on thin section observation. The sample is from Tertiary lacustrine turbidite sandstone at the depth of 3012m of Well Niu-110; Photomicrograph B is a Scanning Electron Microscope (SEM) image showing secondary pores from dissolution of kaolinite the sample is from Tertiary turbidite sandstone at the depth of 2985.3m from well Niu-35. Both the two wells are both located in in Jiyang Sub-Basin, Bohai Bay Basin, Northeast China.

Since SEM has a very large depth of field, and can thus yield a three-dimensional image useful for understanding the structure of a sample. It will help understand the clay mineralogy and their effect on the porosity, permeability and other reservoir characteristics. Figure 5B shows secondary pores resulted from dissolution of authigenic kaolinite exist in the similar turbidite sandstone reservoir as that in Figure 5A in Jiyang sub-basin based on SEM observation.

4. NEW TECHNIQUES AND THEIR APPLICATIONS FOR UNCONVENTIONAL OIL AND GAS EXPLORATION

Over the past decade, interest in shale gas and shale oil reservoirs increased due to commercial success of gas-shale plays in North America. In contrast to conventional oil and gas reservoirs (sandstone and carbonate), these new identified reservoirs typically have very fine-grained rock texture (dominant grain size ≤62.5 μm), low porosity (≤10%) and very low permeabilities (in nanodarcy range). These rocks used to be considered as only source rocks with high organic content (≥ 2% weight fraction Total Organic Carbon, TOC), but now they have been found as reservoir rocks through horizontal drilling and hydraulic fracturing. Gas and oil reserves in the tight shales are huge, US Energy Information Administration (EIA) released a major report in 2011 that there exists potential 6,622 trillion cubic feet (Tcf) of gas contained in shales around the world. Despite the commercial importance of shale formations, their physical properties especially porosity, pore-size distributions and clay mineral fabrics are still poorly understood. Porosity measurements in shales are complicated because of the very fine-grained texture, small pore sizes, extremely low permeability, and the strong interaction of water with clay minerals, which are often important component in these rock types. Shales exhibit dual-porosity structure and have a more complex pore-structure than the sandstones and limestones. One of the biggest challenges in estimating oil and gas transport and storage properties of shales has been a lack of understanding of clay type, clay content, free-gas content, porosity and their relationships. Brittleness of shale has an impact on proppant embedment and maintaining hydraulic-fracture connectivity to the wellbore. High clay-rich shales usually have low Young's Modulus and, by extension, low brittleness index and difficult to frac. So clay minerals play key role in shale gas and shale oil exploration.

Since shales are really fine and tight, estimating reservoir quality in gas shale requires a thorough understanding of pore structure and pore connectivity. MicroCT is a proven technique to resolve pore parameters with a resolution in the order of 1 micrometer. NanoCT technology has resolution down to 200nm but even that may not be enough for gas shale. Gas shales are known to contain finely-dispersed porous organic matter within an inorganic matrix. The porosity within the organic phase has pore and pore throat dimensions typically below 100 nanometers and even down to just a few nanometers, so new techniques are required to characterize the clay minerals in shales. The recent new clay mineral characterization methods include but not limited to FTIR (Fourier transform infrared spectroscopy), QEMSCAN (Automated Mineralogy and Petrography), FIB (Focused Ion Beam), EDS (Energy-dispersive X-ray spectroscopy), etc.

The advancements of special analytical techniques have made significant progress in clay mineral imaging, mineral identification and quantifying by using FEI company's QEMSCAN®, The clay mineral identification and quantification based on the QEMSCAN® EDS spectral analysis method is a reliable alternative to conventional methods. Furthermore, the automated SEM-EDS solution approach provides additional textural information, and the resulting mineral maps can be used to differentiate, in the case of conventional sandstone and unconventional tight shale reservoir rocks, pore linings, from granular mineral alteration products, from intergranular cements, and sedimentary laminations. Figure 6 illustrates QEMSCAN® mineral and texture maps of representative shale samples from China. The Silurian shale (Figure 6A) shows no bedding and has relatively low content of quartz (14.5%) and Cambrian shale (Figure 6B) shows bedding and high content of quartz (62%), dominant clay minerals of two samples are illite and no smectite detected, which indicates that among the two samples the Cambrian shale (Figure 6B) is more easier to frac to produce gas since it has high brittle quartz content and no expandable smectite.

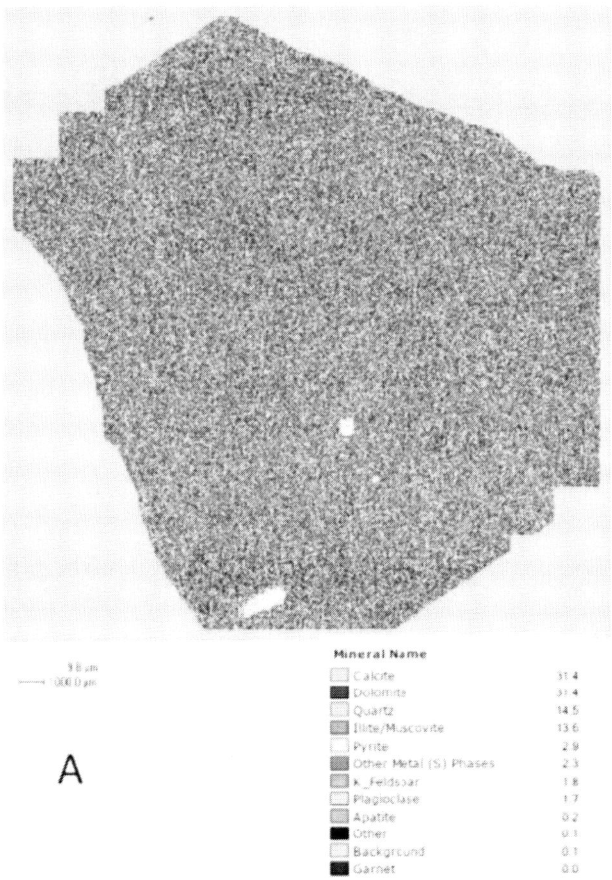

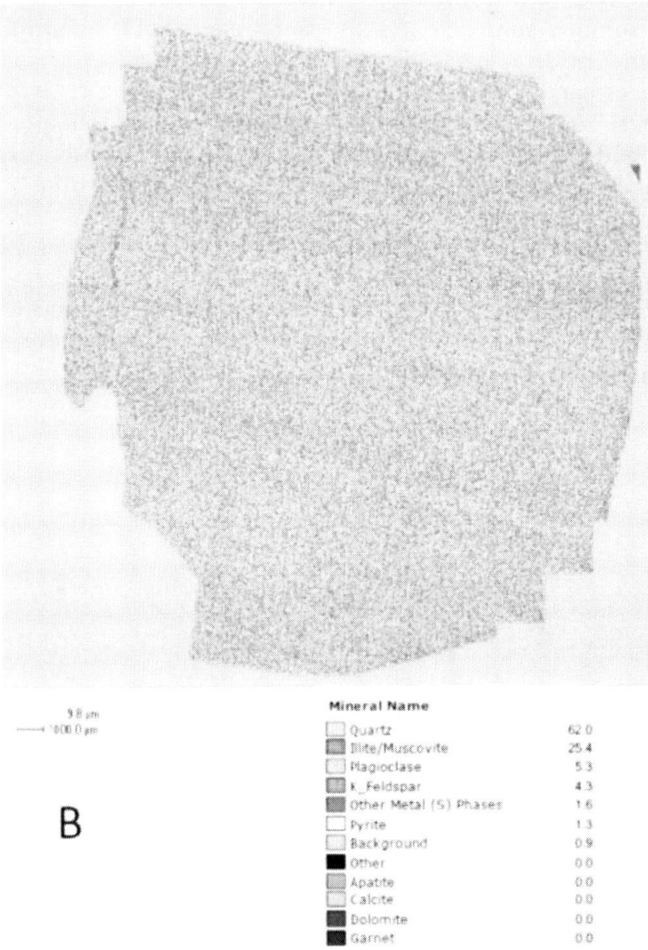

Figure 6. Slices from QEMSCAN showing the rock fabrics and quantitative mineral composition. A is from Silurian shale in Chongqing, China and B is from Cambrian shale in Guizhou, China

The SEM/EDS has also recently been widely used to study shale reservoirs since SEM has a high resolution and EDS's chemical element analysis can help identify mineral composition precisely through the combination of fabric and chemical composition analysis. Figure 7 illustrate this method to identify that a Silurian shale sample from Sichuan, China is mainly composed of quartze, albite, dolomite, illite, chlorite and kerogen by using the SEM/EDS method.

For the intuitive visualization of the nano-pore network and rock fabric architecture, FIB/SEM is the only technology so far with nanometer resolution in 3 dimensions to reveal the reservoir architecture of broad ion polished shale. The high resolution of a Scanning Electron Microscope (SEM) combined with the precise cutting capability of a Focused Ion Beam (FIB) enables rendering of 3D reconstructions with resolution of a few nanometers (Lemmens et al., 2011). The FIB is capable of removing a controlled amount of material to create a subsequent 2D section parallel and aligned with the previous one, with inter-section spacing of the order of 10nm, and having resolution of a few nanometers in the section plane. Figure 8 is an example of FIB/SEM slice of a Silurian shale reservoir in Sichuan Basin in China, which renders the intra-organic (kerogen) nano-scale pores and illite presence in the kerogen. These nano-scale pores can store huge amount of gas in the basin. In this way, after careful combination of the subsequent slices, a 3D model with nanometers resolution can be obtained. Figure 9 shows an example of 3D reconstruction for a US Paleozoic shale reservoir (Zhang and Klimentidis, 2011).

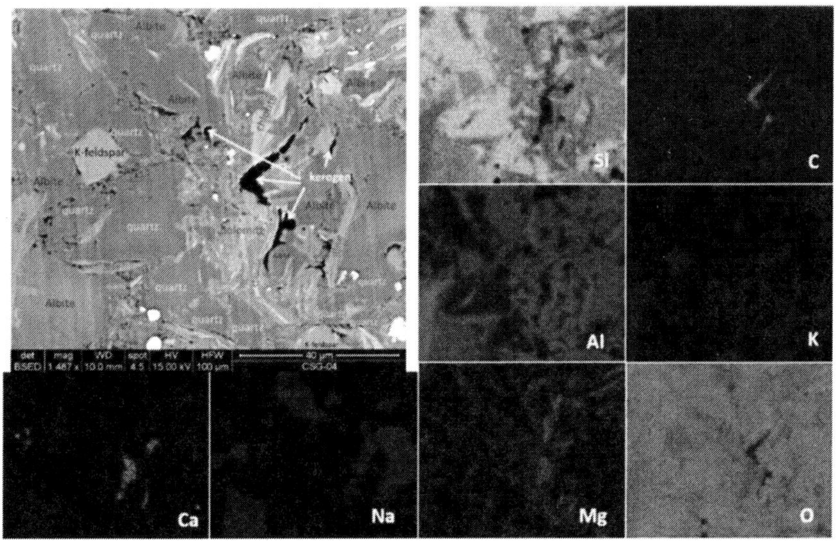

Figure 7. Mineral composition identification based on Scanning Electron Microscope (SEM) and Energy-dispersive X-ray spectroscopy (EDS). The sample is from Silurian shale in Pengshui, Sichuan, China. The upper left and the rest of the slices have the exact same view area under microscope even though their scales are different.

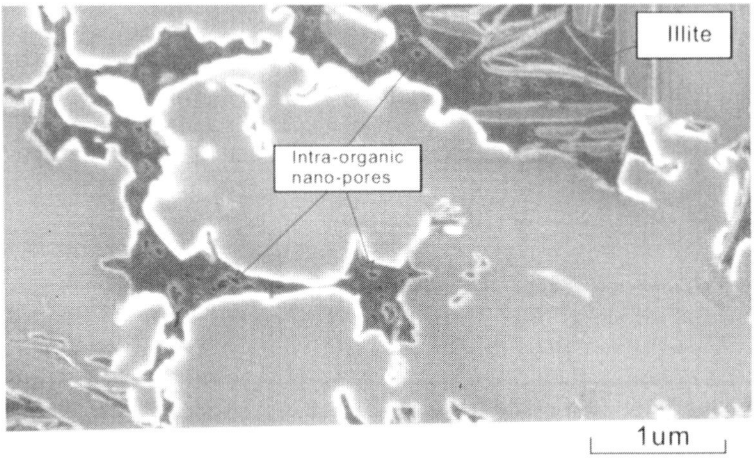

Figure 8. The FIB/SEM slice of Silurian shale reservoir from East Sichuan Basin, China. The sample depth is 2164.8m below surface.

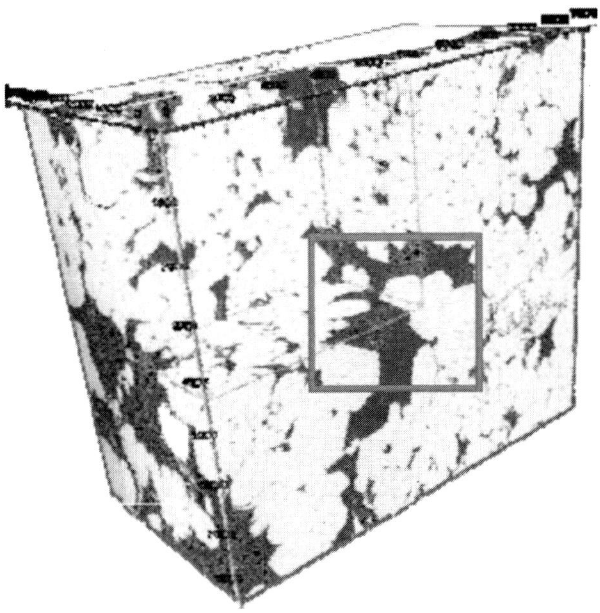

Figure 9. reconstruction of a US Paleozoic shale core sample from a series of FIB/SEM imaging slices (from S. Zhang and R.E. Klimentidis, 2011). The red rectangle area with about 200nm in width shows the nano-pores developed in the kerogen.

5. SUMMARY

The clay minerals are important compositions in source rocks and reservoir rocks that can generate and store oil and gas respectively. The presence of clay minerals strongly influences the physical and chemical properties of conventional sandstone, carbonate and unconventional shale.

Regionally, the clay minerals can be used to interpret and understand such perspectives as the basin evolution on tectonics, sedimentation, burial and thermal history, to infer the sedimentary environment and to correlate strata, etc.

For clay minerals in source rocks, they are important for quality evaluation of the hydrocarbon generation, expulsion and migration. Clay minerals help concentrate organic matter by adsorption and subsequently act as catalyst to generate petroleum. The transformation of montmorillonite to illite and increasing ordering of I/S can indicate the hydrocarbon generation and expulsion events.

For clay minerals in reservoir rocks, their presence has an important impact upon reservoir properties such as porosity and permeability and upon those measured physical data that are used to evaluate reservoir quality. Geologists use clay minerals information to decipher the burial diagenetic process and reveal the pore type and pore evolution. Even though they are usually considered to be detrimental to reservoir quality because they can plug pore throats and can be easily compacted, other diagenetic processes may enhance porosity through the forming of secondary porosity through providing porosity by clay dissolution, creating micropores in clays and coating of chlorite on grains to prevent quartz cementation.

The recent emerging shale oil and gas exploration requires state-of-art imaging and characterization techniques to study the application of clay minerals in the exploration of this unconventional resource. The modern innovative QEMSCAN® and FIB/SEM/EDS have been playing key roles in the identification and quantitative characterization of clay minerals, which help define the best brittle reservoir interval and avoid exploration failure by choosing the compatible drilling and hydraulic fluids.

REFERENCES

1. Bloch, S., R. H. Lander, and L. M. Bonnell, 2002, Anomalously high porosity and permeability in deeply buried sandstone reservoirs: Origin and predictability: AAPG Bulletin, v. 86, p. 301–328.

2. Brooks, B.T., 1952, Evidence of Catalytic Action in Petroleum Formation Ind. Eng. Chem., 44 (11), pp 2570–2577
3. Brown, A., 1997, Porosity variation in carbonates as a function of depth: Mississippian Madison Group, Williston Basin, in J.A. Kupecz, J. Gluyas, and S. Bloch, eds., Reservoir quality prediction in sandstones and carbonates: AAPG Memoir 69, p. 29--46.
4. Bruce, C. H., 1984, Smectite dehydration--Its relation to structural development and hydrocarbon accumulation in northern Gulf of Mexico: Amer. Assoc. Petrol. Geol, Bull. 68, 673-683.
5. Burst, Jr.,J.F.,1959, Post diagenetic clay mineral environmental relationships in the Gulf Coast Eocene. Clays & Clay Minerals, 6, 327-341.
6. Burtner and Warner,1986, relationship between illite/smectite diagenesis and hydrocarbon generation in lower cretaceous Mowry and skull creek shales of the Northern rocky mountain area, Clays and Clay Minerals. Vol. 34, No. 4, 390-402,
7. Drits, V.A., Lindgreen, H., Sakharov, B.A., et al., 2002, Tobelitization of smectite during oil generation in oil-source shales. Application to north sea illite-tobelite-smectitevermiculite. Clays and Clay Minerals, v. 50, no.1, p. 82-98
8. Foscolos, A. E., Powell, T. G., & Gunther, P. R., 1976, The use of clay minerals and inorganic and organic geochemical indicators for evaluating the degree of diagenesis and oil generating potential of shales. Geochirn. Cosrnochirn. Acta, 40, 953-966.
9. Griffin, G.M., 1971, Interpretation of X-ray diffraction data, in Carver, R.E., ed., Procedures in Sedimentary Petrology: New York, Wiley, p. 541–569.
10. Grim, R. E., 1947, Relation of clay mineralogy to origin and recovery of petroleum: Am. Assoc. Petroleum Geologists Bull., v. 31, no. 8, p. 1491-1499.
11. Hamilton, P. J., S. Kelley, and A. E. Fallick, 1989, K-Ar dating of illite in hydrocarbon reservoirs. Clay Minerals, v. 24, p. 215 231
12. Heald, M. T., and R. E., Larese, 1974, Influence of coatings on quartz cementation: Journal of Sedimentary Petrology, v. 44, p. 1269–1274.
13. Hoffman, J. and Hower, J.,1979, Clay mineral assemblages as low grade metamorphic geothermometers: Application to the thrust faulted disturbed belt of Montana: in Aspects of Diagenesis, P. A. Scholle and P. S. Schluger, eds., SEPM Spec. Publ. 26, 55-79.
14. Hower, J., Eslinger, E. V., Hower, M. E., and Perry, E. A., 1976, Mechanism ofburial metamorphism ofargillaceous sediment: Mineralogical and chemical evidence: GeoL Soc. Amer. Bull. 87, 725-737.
15. Jiang, S., Wang, H., Cai, D.,et al., 2010, The secondary porosity and permeability characteristics of Tertiary strata and their origins, Liaodong Bay Basin, China. Energy, Exploration & Exploitation. 28(4): 207-222
16. Jiang, S., Wang, H., Cai, D., et al., 2011, Characteristics of the Tan-Lu Strike-Slip Fault and Its Controls on Hydrocarbon. Accumulation in the Liaodong Bay Sub-Basin, Bohai Bay Basin, China. Advances in Petroleum Exploration and Development. Vol. 2, No. 2, 2011, pp. 1-11
17. Kelly, J., J. Parnell, and H. H. Chen, 2000, Application of fluid inclusions to studies of fractured sandstone reservoirs. Journal of Geochemical Exploration, v. 69, p. 705 709.
18. Lee, M., J. L. Aronson, and S. M. Savin, 1985, K/Ar dating of time of gas emplacement in Rotliegendes sandstone, Netherlands. AAPG Bulletin, v. 69, p. 1381 1385.
19. Legate, C. E. and Johns, W. D.,1964, Gas-chromatographic examination of several systems from clay minerals and organic materials: Beitr. Mineral Petroyraphie. 10, 60-69.
20. Lemmens, H.J., Butcher, A.R. and Botha, P.W.S.K., 2011, FIB/SEM and SEM/EDX: a new dawn for the SEM in the core lab? Petrophysics, 62 (6):452-456
21. Liewig, N., N. Clauer, and F. Sommer, 1987, Rb-Sr and K-Ar dating of clay diagenesis in Jurassic sandstone oil reservoir, North Sea. AAPG Bulletin, v. 71, p. 1467 1474.
22. Leiwig, N. and Clauer,N. 2000, K-Ar dating of varied microtextural illite in Permian gas reservoirs, northern Germany. Clay Minerals, 35, 271-281
23. Middleton, D., J. Parnell, P. Carey, and G. Xu, 2000, Reconstruction of fluid migration history in northwest Ireland using fluid inclusion studies. Journal of Geochemical Exploration, v. 69, p. 673 677

24. Moll, W.F., Jr., 2001, Baseline studies of the clay minerals society source clays: Geological origin. Clays and Clay Minerals, V. 49, pp. 374–380.
25. Nadeau,P.H., 2000, The Sleipner Effect: a subtle relationship between the distribution of diagenetic clay,reservoir porosity, permeability, and water saturation. Clay Minerals, 35, 185-200
26. Pettijohn, F.J., 1975, Sedimentary Rocks, 3rd Edition: New York, Harper & Row, 628 p.
27. Pollastro, R.M., 1993, Considerations and applications of the illite/smectite geothermometer in hydrocarbon-bearing rocks of Miocene to Mississippian age. Clay and Clay Minerals, Vol.41, No.2, 119-133.
28. Powers, M.C.,1957, Adjustment of clays to chemical change and the concept of the equivalence level: Clays & Clay Minerals, Proceedings of the Sixth National Conference, 309-326.
29. Powers, M. C. 1967, Fluid-release mechanisms in compacting marine mudrocks and their importance in oil exploration. Am. Assoc. Petrol. Geol. Bull. 51: 1240, 54
30. Said, A., Abdallah, M. and Alaa E., 2003, Application of well logs to assess the effect of clay minerals on the petrophysical parameters of Lower Cretaceous Reservoirs, North Sinai, Egypt. EGS Journal, V. 1, No. 1, pp. 117-127.
31. Schoonmaker, J., Machenzie, F.T. and Speed, R.C., 1986, Tectinic implications of illite smectite diagenesis, Barbados Accretionary Prism. Clays and Clay Minerals. 34 (4):465-472
32. Taylor, T. R., R. Stancliffe, C. I. Macaulay, and L. A. Hathon, 2004, High temperature quartz cementation and the timing of hydrocarbon accumulation in the Jurassic Norphlet Sandstone, offshore Gulf of Mexico, U.S.A., in J. M. Cubit, W. A. England, and S. Larter, eds., Understanding petroleum reservoirs: Towards an integrated reservoir engineering and geochemical approach: Geological Society (London) Special Publication 237, p. 257–278.
33. Tissot, B.P. and Welte, D.H., 1984, Petroleum; Natural gas; Geochemical prospecting; Geology; Prospecting, Springer-Verlag (Berlin and New York), 699P
34. Waples, D. W., 1980, Time and temperature in petroleum formation: Application of Lopatin's method to petroleum exploration. Amer. Assoc. Petrol. Geol. Bull. 64, 916-926.
35. Wang, F. Y., P. He, S. C. Zhang, M. J. Zhao, and J. J. Lei, 1997, The K-Ar isotopic dating of authigenic illites and timing of hydrocarbon fluid emplacement in sandstone reservoir: Geological Review, v. 43, p. 540 546.
36. Weaver, C.E., 1960, Possible uses of clay minerals in search for oil. Bull. Am. Assoc. Petrol. Geologists 44, 1505-1518.
37. Webb,J.E.,1974, Relation of Oil Migration to Secondary Clay Cementation, Cretaceous Sandstone, Wyoming AAPG Bulletin, v. 58, p. 2245-2249
38. Yariv, S., 1976, Organophilic pores as proposed primary migration media for hydrocarbons in argillaceous rocks. Clay Sci. 5: 19-29
39. Young, A., McIver, R. D., 1977, Distribution of hydrocarbons between oils and associated fine-grained sedimentary rocksPhysical chemistry applied to petroleum geology. Am. Assoc. Petrol. Geol. Bull. 61: 14 07-36
40. Yun J B, Shi H S, Zhu J Z, et al., 2010, Dating petroleum emplacement by illite 40Ar/39Ar laser stepwise heating, AAPG Bull, 94: 759-771
41. Zeng, J. and Yu, C.,2006, Hydrocarbon migration along the Shengbei fault zone in the Bohai Bay basin, China: The evidence from geochemistry and fluid inclusions. Journal of Geochemical Exploration, 89(1-3), 455-459.
42. Zhang, S. and Klimentidis, R.E., 2011, porosity and permeability analysis on Nanoscale fibsem 3d imaging of shale rock, Society of Core Analysis Meeting, SCA2011-30, 1-12

CHAPTER 10

Synthesis of Ultramarine from Reservoir Silts

Yin-Hsiu Hsiao, Yun-Hwei Shen and Dah-Tong Ray *

Department of Resources Engineering, National Cheng Kung University, 1 Da-Hsueh Road, Tainan 70101, Taiwan R.O.C.

ABSTRACT

Ultramarine blue was synthesized by using reservoir silts as a major raw material to replace traditional kaolin clay. The reservoir silts were pretreated to collect the fines, in which the main mineral composition is illite clay. The fine part was mixed with sodium carbonate, sulfur, and activated carbon, and calcined at 800°C for 8 h. The products were examined by X-ray powder diffraction (XRD) and a scanning electron microscope (SEM) to identify the mineral phases and particle morphology. The color characteristics were measured according to the CIELab system. Brilliant blue ultramarine powders were successfully synthesized with a chroma of 52.4, which is competitive with the commercial ultramarine.

Keywords: reservoir silts; ultramarine; synthesis

1. INTRODUCTION

The annual sedimentation rate of the world's reservoirs averages about 1%, but it is much higher in Asia [1,2]. Due to climatic, geographic and geologic situations, the reservoir sedimentation problem is very serious in southern Taiwan, R.O.C. All the streams have steep slopes in their catchment areas.

The rainfall intensity and runoff flow are strong during the summer typhoon season. Even worse, major irrigation and city water supply reservoirs are located in Pliocene and Pleistocene shale and mudstone areas. The accumulated sediment volume in these reservoirs accounts for about one third or more of the reservoir designed capacity [3]. To extend the operation life of these reservoirs, desiltation is essential. Reservoir desiltation involves three major processes, namely the removal, deposition and reuse of the silt. The reuse of the silt is the most important one because improper handling of dredged silts will interrupt the removal operation.

The major mineral components of reservoir sediments in southern Taiwan are sand (mostly quartz) and clay (illite and chlorite) [4]. Since the hardness of quartz is greater than that of the clay minerals, these two types of minerals are concentrated in different size ranges, i.e. sand in the coarse-grained range and clay minerals in the fine-grained range. Using sedimentation technique in mineral processing, it is easy to separate out clay minerals. From the point of view of chemical composition, after the removal of quartz, reservoir silts can be regarded as a natural aluminosilicate mineral source.

Ultramarine with its deep vivid blue is a valuable inorganic pigment. Natural ultramarine is derived from grinding lapis lazuli, which is a rock composed of lazurite with small amounts of calcite, pyroxene, etc. Most of today's ultramarine is synthetically produced based upon processes invented by Guimet in 1826 and Gmelin in 1828 [5].

Ultramarine or lazurite has the same crystal structure of sodalite,$Na_8[(Al,Si)_6O_{24}]C_{l2}$, which is a feldspathoid group mineral with a zeolite structure. In ultramarine, the chloride ions are replaced by the sulfur radical anions S_3^- and S_2^-. The chemical formula of ultramarine thus is $M_8[(Al,Si)_6O_{24}]X_n$, where M is an extra-framework cation, mainly Na^+; X is an extra-framework anion, such as S_2^-, S_3^-, etc., and n may vary from 2 to 4 [6,7,8,9]. The anions and cations balance the charge on the ordered aluminosilicate framework. The close-packed cuboctahedra $(Al_3Si_3O_{12})^{3-}$ called β cages organize the sodalite (SOD) structure (Figure 1a) [10].The chromophores of ultramarine are polysulfide radical anions encapsulated inside β cages (Figure 1b) [7]. Theoretical maximum insertion of chromophores is one S_x^- per β cage (x = 2–4). The polysulfide radicals S_2^-, S_3^- and S_4/S_4^- are responsible for the yellow, blue and red colors, respectively [7,10,11,12,13,14,15]. The cause of color is due to the absorption band at about 400 nm (S_2^-), 600 nm (S_3^-) and 520 nm (S_4/S_4^-), resulted from the transition of electrons among the molecular orbitals of these polysulfide radical anions from ground state to low-lying excited electronic states [13,14,15,16,17,18].

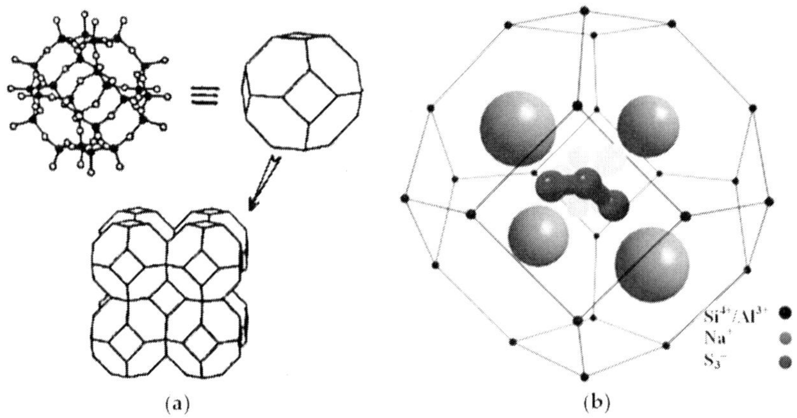

Figure 1. Schematic illustration of (a) sodalite (SOD) structure composed of β cages [10] and (b) blue chromophore S_3^- encapsulated in β cages and surrounded by four sodium cations [7].

The ratio of Si to Al determines the amount of chromophores encapsulated inside β cages. When fewer Al ions are present in the framework, the required number of sodium ions is reduced for lattice stabilization, and more sodium ions are free to act as counter-ions for the polysulfide. The polysulfide content can be increased and thereby improve color quality [19,20]. Within limits set by stability considerations, the proportions of the chemical constituents can vary. Some empirical formulas, like$Na_{6.9}Al_{5.6}Si_{6.4}O_{24}S_{4.2}$ [19], are reported.

Traditionally, the raw material used in the synthesis of ultramarine is kaolin clay. The industrial process of ultramarine synthesis involves the dehydration of kaolin at about 550 °C. The dehydrated clay is mixed with sulfur, sodium carbonate and a reducing agent, e.g., coke or active carbon. Heating of the mixture produces a synthesis reaction [21]. In the reaction, the formation of ultramarine blue proceeds in two stages: first a reduction stage followed by an oxidation stage. In the reduction stage, the mixture is heated to about 750 °C, and held at that temperature for a given period of time. The container for the mixture is covered so that a reducing atmosphere can be maintained within it. When the holding time is reached, the temperature starts to descend, and the reaction enters the oxidation stage. The oxidation stage is achieved by either keeping the mixture at about 450 °C for a fixed time or simply slowing the cooling rate. During this stage, S_2^- ions are oxidized to S_3^- ions to produce a vivid blue hue [20,21].

Besides kaolin, other aluminosilicate sources, including pure chemicals, natural or synthetic minerals and wastes have been used. Chemicals, such as sodium silicate and tetramethylammonium hydroxide [22]; silicon tetrachloride and aluminum chloride [8]; silica, diatomaceous earth and sodium aluminate [23]; natural or synthetic minerals, like zeolite A, X, Y [24,25,26], cancrinite [27], hydroxycancrinite [28]; wastes, such as fly ash [20] and kaolin mine tailings [29], were reported to be technologically feasible.

In a previous study, the reservoir silts were first treated hydrothermally to transform them to hydroxycancrinite [28]; then the hydroxycancrinite was transformed to ultramarine blue via a solid-state reaction process. In this study, reservoir silts were used directly to synthesize ultramarine. The reaction conditions, such as raw materials ratio, calcination temperature and mineral composition of silts on the product phase and color properties, were investigated.

2. EXPERIMENTAL

2.1. Materials

The reservoir silt samples were taken from the city water supply of Nan-Hua Reservoir that is located in Tainan County. The silts were screened through a 200 mesh sieve (75 μm), to remove the non-mineral constituents, such as humus, etc., and then the material was designated as the "raw" silt. The raw silts were further classified via hydraulic sedimentation to separate out the <10 μm, <5 μm, and <2 μm parts.

The size distribution of the raw silts is shown in Figure 2. The distribution fits quite well that of the Gates-Gaudin-Schuhmann distribution function with the distribution modulus of 0.77 and d50 of 5.7 μm. From the distribution curve, it can be determined that the <10 μm, <5 μm and <2 μm parts account for 62, 47, and 33 wt % of the raw silts, respectively.

The XRD patterns of the raw and classified silt samples are shown in Figure 3. The diffraction peaks show that the minerals comprising the silt are quartz, illite, clinochlore and albite. As the size of the silts is decreased, the peaks of quartz and albite decrease rapidly, whereas those of illite and clinochlore are more prominent. The change in mineral composition shows that size classification can separate quartz and albite, which are unusable in the synthesis reaction, from clay minerals effectively.

The chemical compositions of all silt samples determined by XRF are listed in Table 1. With the removal of quartz from classified silts, the molar ratio of Si/Al decreases from 2.77 to 1.42.

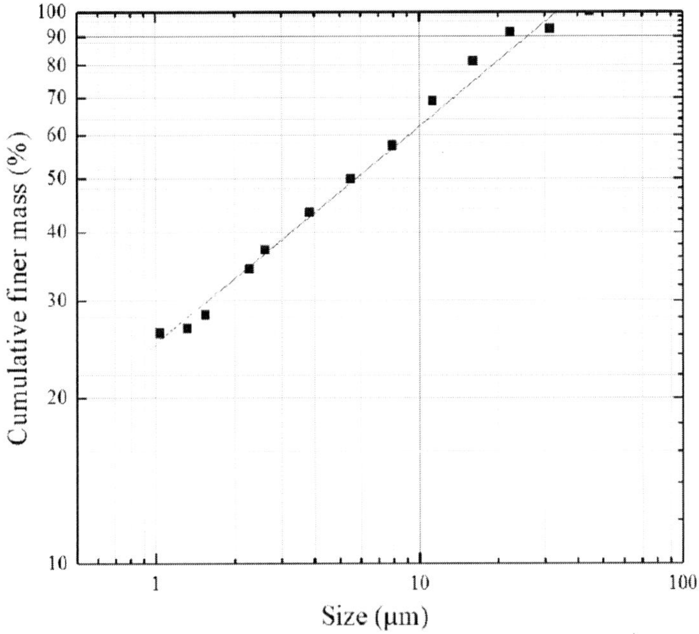

Figure 2. Particle size distribution of raw silts.

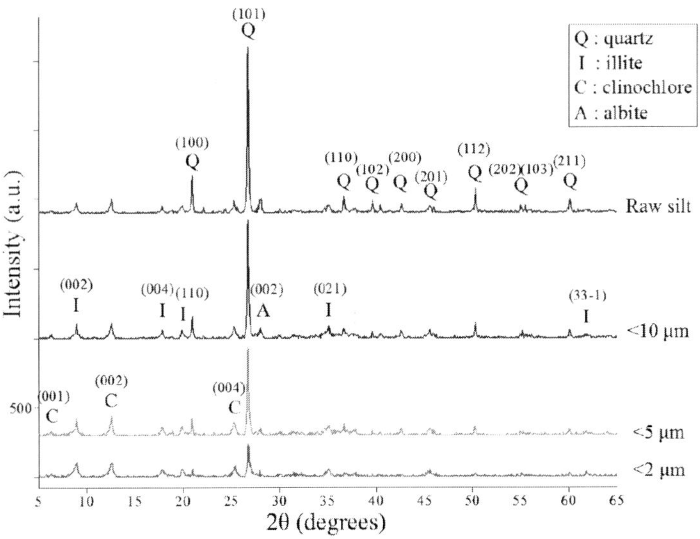

Figure 3. XRD patterns of raw and classified silts.

Table 1. Chemical analysis of raw and classified silts.

Components	Raw	<10 μm	<5 μm	<2 μm
Na_2O	a	a	0.62	1.15
MgO	2.03	2.23	2.48	2.75
K_2O	3.33	3.93	4.33	4.81
CaO	1.48	1.29	1.21	1.22
Fe_2O_3	6.11	7.34	7.97	8.29
Al_2O_3	19.61	22.25	24.74	27.74
SiO_2	63.97	60.86	56.9	51.48
Si/Al	2.77	2.32	1.95	1.58

a: Below the detection limit 0.44%.

The SEM images of the raw and classified silts are shown in Figure 4. It is clear that with decreasing particle sizes, the major mineral images gradually vary from massive, which represents quartz, to sheet-like, which represents clay minerals.

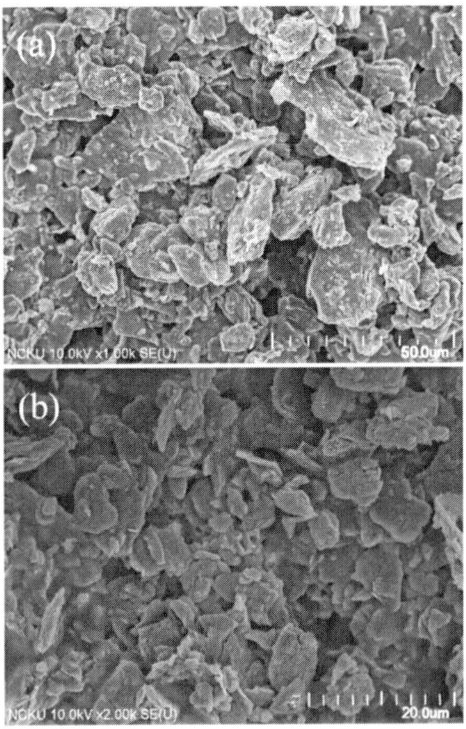

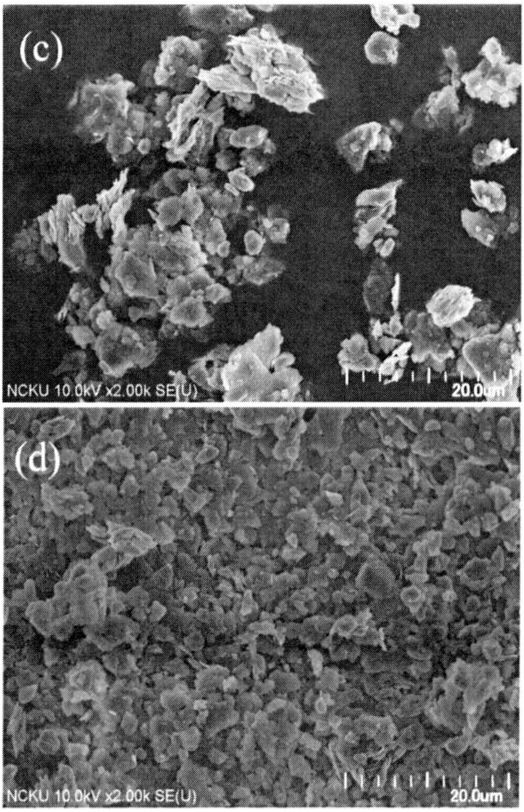

Figure 4. SEM micrographs of (a) raw; (b) <10 μm; (c) <5 μm and (d) <2 μm silts.

2.2. Procedures

The silts were fully mixed with sulfur (purified, Katayama Chemical Co. Ltd., Osaka, Japan) and active carbon (100–400 mesh, Sigma-Aldrich Chemie GmbH, Munich, Germany). An aliquot of sodium carbonate (reagent grade, Fisher Scientific, Leicestershire, UK) solution of assigned concentrations was slowly added to the powder and was hand-kneaded to balls and then dried in air. It was reported [29,30] that in the reaction, sulfur and sodium carbonate are reacted as:

$$3Na_2CO_3 + 1.25S_8 \leftrightarrow 2Na_2S_4 + Na_2S_2O_3 + 3CO_2 \quad (1)$$

From the Na_2CO_3/S_8molar ratio in the above equation, the Na_2CO_3/S_8 weight ratio is approximately equivalent to 1. This ratio was held constant in this study. The dried balls were placed in an alumina crucible covered with a lid and calcined at 700, 800 and 900 °C for 2, 4, 8 and 12 h.

After completion of the reaction, the crucible was allowed to cool down to room temperature in the furnace. The cooling took about 8 h. The reacted balls were mixed with deionized water, stirred thoroughly and filtered several times to remove all dissolved ions. The solids were dried at 120 °C for 24 h and for further analyses.

2.3. Characterization

The mineral phases of the sediment and products were characterized using an X-ray diffractometer (DX-2600, Fangyuan Co., Dandong, China) with Cu K_α (λ = 1.5418Å) radiation and a scan rate of 0.04°/s. The chemical compositions were analyzed using an X-ray Fluorescence Analyzer (NEX CG EDXRF elemental analyzer, Rigaku Corp., Tokyo, Japan). The particle size distribution was determined using an Andreasen pipette. The morphology and individual particle sizes were examined using a Hitachi 8000 High-Resolution Scanning Electron Microscope (HRSEM, Hitach Ltd., Ibaraki, Japan). The color characteristics of the synthesized products were measured using a sphere spectrophotometer SP-60 (X-Rite Inc., Grand Rapids, MI, USA) employing the CIELab (Commission Internationale de l'Éclairage) color spaces system. The measurements (Standard Illuminant D65/10°) performed were chromaticity or color saturation (C* = 0 for grey color, C* = 100 for pure chromatic components), lightness (L* = 0 for a black sample, L* = 100 for a white sample) and hue (h* = 0°, 90°, 180°, 270° for red, yellow, green and blue colors, respectively).

3. RESULTS AND DISCUSSION

3.1.Effects of Calcination Condition

3.1.1. Reaction Temperature

The <5 μm silts were used in these calcination tests. The mass ratio of (Na_2CO_3+ S_8) and active carbon to silt were maintained at 1.5 and 0.2, respectively. The XRD patterns of the products calcined at 700, 800 and 900 °C for 8 h are shown in Figure 5. For all reaction temperatures, lazurite was the major phase formed. Nepheline was also formed for temperatures below 800 °C. When reaction temperature reached 900 °C, nepheline disappeared.

The color properties of the products are listed in Table 2. When calcination was performed at 800 °C, the chroma of the product achieved optimum. Both lower or higher than this temperature were deleterious. This could be because that lower temperature does not promote the reaction efficiently, whereas higher temperature accelerates the loss of sulfur chromophores.

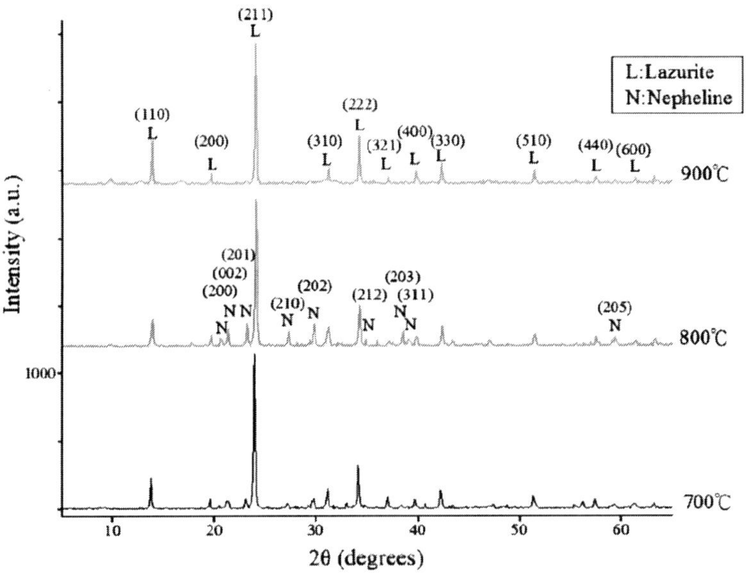

Figure 5. XRD patterns of the products derived from <5 μm silts calcined at 700, 800, and 900 °C for 8 h ((Na_2CO_3 + S_8)/silts=1.5, C/silts=0.2).

Table 2. Color properties of products derived from <5 μm silts calcined at 700, 800, and 900 °C for 8 h ((Na_2CO_3 + S_8)/silts=1.5, C/silts=0.2).

Calcination Temperature	L*	C*	h*	Pantone
700 °C	36.25	40.30	266.8	
800 °C	37.86	51.12	275.5	
900 °C	38.81	28.24	244.2	

The results of Figure 5 and Table 2 shows that the color properties are determined by crystal structure as well as the amount of polysulfide radical anions kept in the β cage. The latter is more important, as can be seen from the XRD patterns of the 800 °C product. Although some nepheline is still present, it represents the best color properties.

3.1.2. Reaction Duration

Figure 6 shows the XRD patterns of the products calcined at 800 °C for 2–12 h. After 2 h of calcination, lazurite formed with a minor amount of nepheline. Lazurite obviously is the stable phase, as it can be seen that all diffraction peaks kept increasing even after 12 h of calcination. Nepheline on the other hand, seemed to be an intermediate phase, and disappeared gradually as calcination proceeded. The color properties of the products are listed in Table 3. The best chroma was obtained at the reaction time of 8 h. It is obvious for long time calcination that S_3^- chromophores may vaporize from the clathrate. According to the above results, 800 °C and 8 h were used as the calcination conditions for the next test.

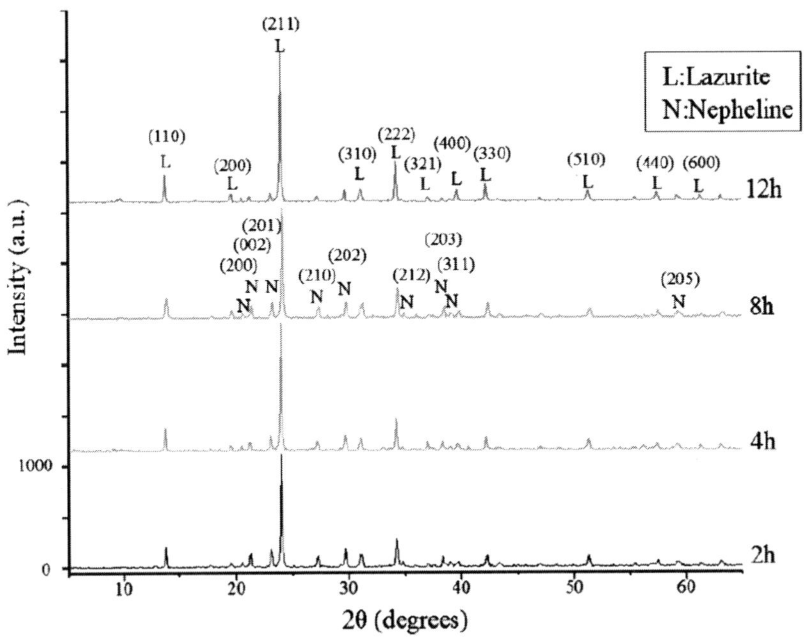

Figure 6. XRD patterns products derived from <5 μm silts calcined at 800 °C for 2–12 h ((Na_2CO_3 + S_8)/silts = 1.5, C/silts=0.2).

Table 3. Color properties of products derived from <5 μm silts calcined at 800 °C for 2–12 h ((Na_2CO_3 + S_8)/silts = 1.5, C/silts = 0.2).

Calcination Time	L*	C*	h*	Pantone
2 h	41.75	37.8	258.1	
4 h	42.27	36.15	256.6	
8 h	37.86	51.12	275.5	
12 h	39.78	33.08	250.8	

3.2.Effects of Raw Materials Ratio

3.2.1. (Na_2CO_3+ S_8)/Silts Ratio

The <5 μm silts were used in this test. There are two concerns about raw material ratios, one is the weight ratio of Na_2CO_3+ S_8 to silts (at Na_2CO_3/S_8 = 1), the other is the active carbon to silts. The mixtures were heated at 800 °C for 8 h. The color properties of the products of all tests are summarized in Table 4.

The effect of (Na_2CO_3+ S_8)/silts ratio on the color properties (chroma, lightness and hue) of the products for reactions with active carbon/silts ratio = 0.2 is shown in Figure 7. With the (Na_2CO_3+ S_8)/silts ratio increased from 1 to 1.5, 2, the chroma increased from 38.9 to 50.2, 53.3, while lightness and hue still remained at the similar levels but much improved to around 30° and 275°, respectively. For the economical reason, the (Na_2CO_3+ S_8)/silts ratio = 1.5 was adopted in the later tests.

3.2.2. Active Carbon/Silts Ratio

The effect of active carbon/silts ratio on the products color properties is shown in Table 5 and Figure 8. For active carbon/silts ratios ≤ 0.2, with the increase of active carbon/silts ratio, the chroma and lightness were improved prominently from 44.5, 45.3 to 50.2 and 41.7, 37.3 to 31.2, respectively. Increasing the ratio to 0.35 and 0.5, the color properties deteriorated. The chroma decreased to 31.9 and 12.9, while the lightness became 29.5 and 46.3. The function of active carbon is to provide a reduced atmosphere in the reaction container to transform elemental sulfur to S_2^-. However, over-dosing of active carbon may deter the formation of S_2^- by its reaction with the sulfur to form carbonyl sulfide (COS). The oxidation stage, in which the yellow chromophore S_2^- is oxidized to the blue chromophore S_3^-, can also be influenced by carbon reacting with oxygen in the crucible to produce carbon monoxide [20]. The hue of the products of active carbon/silts ratio at 0.35 and 0.5 dropped to 258.9° and 157°, inclining to a green (180°) tone, probably due to incomplete oxidation. Thus the proper active carbon/silts ratio is 0.2.

Table 4. Color properties of products derived from <5 μm silts at different ($Na2CO3$+ $S8$)/silts ratio (800 °C, 8h).

$(Na_2CO_3 + S_8)$/Silts	C/Silts	L*	C*	h*	Pantone
1	0.2	28.92	38.89	270.3	
1.5	0.2	31.23	50.23	275.7	
2	0.2	32.25	53.30	279.5	

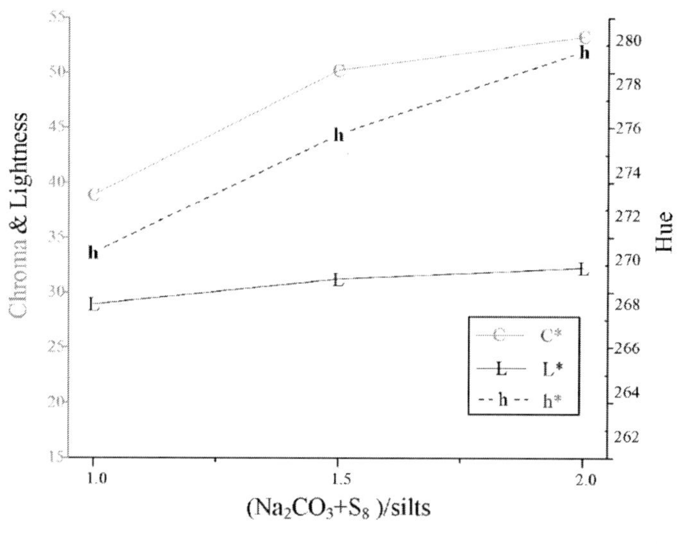

Figure 7. Effects of $(Na_2CO_3 + S_8)$/silt mass ratio on the color properties of the products derived from <5 μm silts (C/silts = 0.2, 800 °C, 8 h).

Table 5. Color properties of products derived from <5 μm silts at different active carbon/silts ratio (800 °C, 8 h).

$(Na_2CO_3 + S_8)$/Silts	C/Silts	L*	C*	h*	Pantone
1.5	0	41.26	40.90	266.3	
1.5	0.075	41.66	44.53	272.9	
1.5	0.15	37.29	45.33	270	
1.5	0.2	37.86	51.12	275.5	
1.5	0.35	29.50	31.86	258.9	
1.5	0.5	46.27	12.92	157	

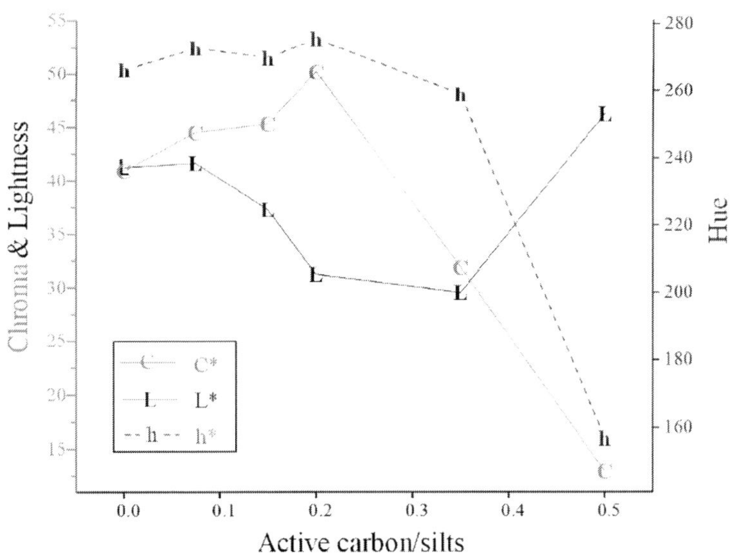

Figure 8. Effects of activated carbon/silt mass ratio on the color properties of the products derived from <5 μm silts ($(Na_2CO_3 + S_8)$/Silts=1.5, 800 °C, 8 h).

3.3. Effects of Mineral Composition of Starting Silts

The results of Table 1 and Figure 4 show that as the silt sizes are fine, more clay minerals are contained in the silts. In this test, the raw and classified silts were used to investigate the effects of mineral composition on the ultramarine synthesis. The reactions were performed at 800 °C for 8 h with (Na_2CO_3+ S_8)/silts ratio and the active carbon/silts ratio as 1.5 and 0.2, respectively.

Figure 9 is the XRD patterns of the products synthesized from different silts. For raw (Si/Al ratio = 2.77), <10 μm (Si/Al ratio = 2.32) and <5 μm (Si/Al ratio = 1.95) silts, both lazurite and nepheline were formed, since the peak heights of lazurite are low, its amount may not be high. For < 2 μm (Si/Al ratio = 1.58) silts, lazurite was the only phase and it was very distinct. Kowalak et al. [25] used different types of zeolite as aluminosilicate sources to synthesize ultramarine. Their results showed that the sodalite structure could be synthesized using zeolite A (Si/Al ratio = 1) and zeolite X (Si/Al ratio = 1–1.5), while it could not using zeolite Y (Si/Al ratio = 1.5–3), mordenite (Si/Al ratio = 5–10) and ZSM-5 (Si/Al ratio ≥10). They concluded that the sodalite structure can be synthesized from aluminosilicate sources with small Si/Al ratios.

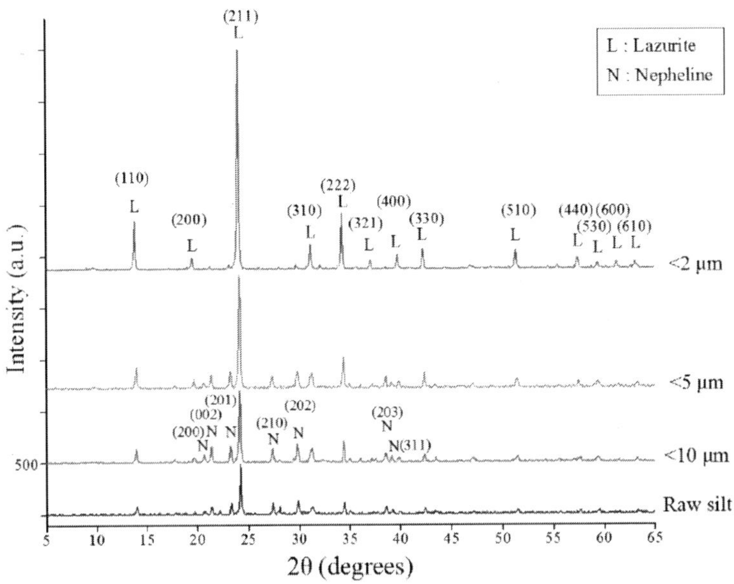

Figure 9. XRD patterns of the products derived from raw, <10 μm, <5 μm and <2 μm silts ((Na_2CO_3 + S_8)/silts = 1.5, C/silts = 0.2, 800 °C, 8 h).

Large Si/Al ratios are unfavorable to the formation of sodalite phase and favor the formation of other undesired phases. From the results of Figure 9, <5 μm silts seem to be on the margin between good and bad raw materials and agree with Kowalak et al.

The SEM images of the products are shown in Figure 10. According to Cork [19,20], the mean particle size of a commercial ultramarine should range from 0.7 to 5.0 μm. It is obvious that the products derived from the classified silts with particle size less than 5 μm and so on meet the requirement.

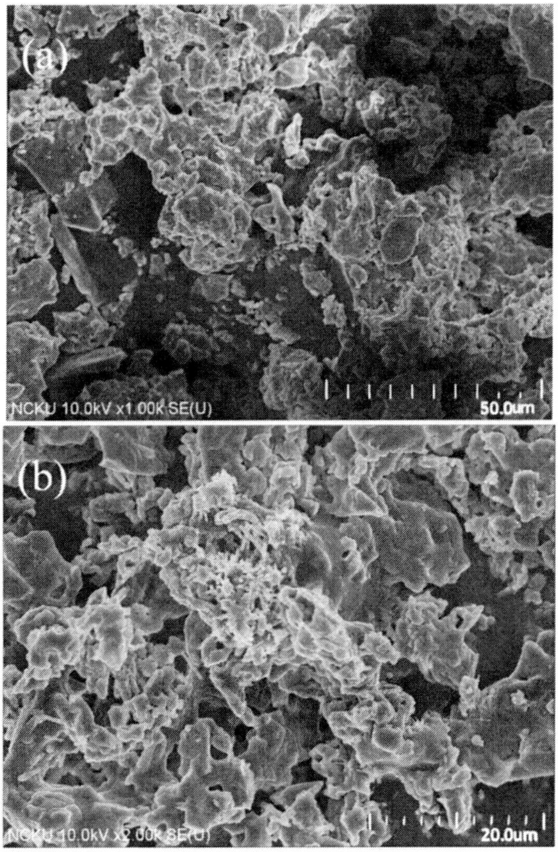

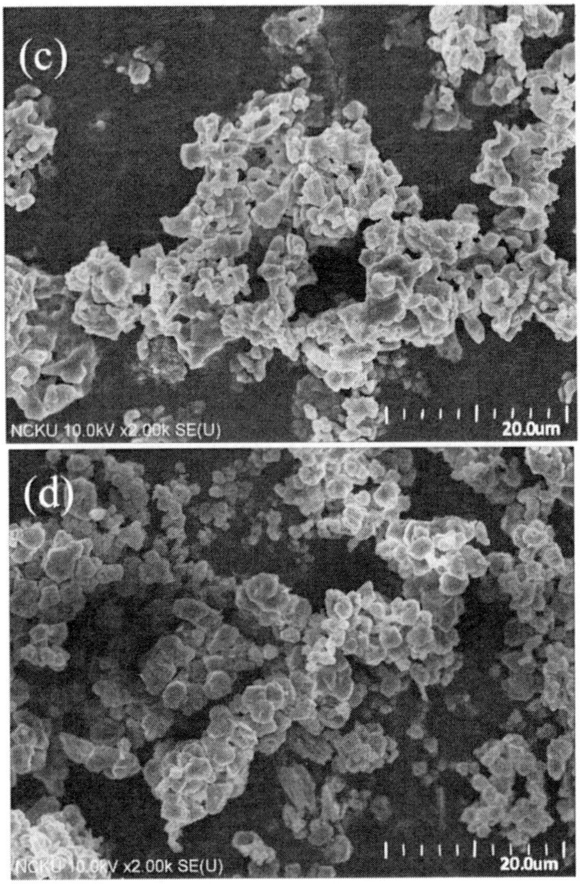

Figure 10. SEM micrographs of products derived from different silt samples:(a) raw; (b) <10 μm; (c) <5 μm and (d) <2 μm silts ((Na_2CO_3 + S_8)/silts = 1.5, C/silts = 0.2, 800 °C, 8 h).

The color properties of the products from the four starting materials are listed in Table 6. It shows that as the particle size becomes fine, the color properties make great progress, and when the particle size of the silts is <5 μm, the chroma of the products can achieve over 50 with lightness below 40 and hue around 270°. From the results of Table 1, the Si/Al ratio of <5 μm silts is 1.95, therefore either from the criterion of the sodalite phase formation or the color performance, a requirement of the silt size for the synthesis of ultramarine is passing 5 μm. In Table 6, the color properties of additional three ultramarine samples, a commercial painting pigment (Sennelier l'Aquarelle), a natural lapis lazuli bead, and a recent report of ultramarine synthesized from kaolin clay [31], are also listed. It is obvious that the ultramarine of this study is competitive.

Table 6. Color performance of products derived from raw and classified silts ((Na_2CO_3 + S_8)/silts = 1.5, C/silts = 0.2, 800 °C, 8 h) and three other natural and synthetic ultramarine samples.

Particle Size	L*	C*	h*	Pantone
Raw	30.19	22.40	253.3	
<10 μm	34.68	35.94	266.4	
<5 μm	37.86	51.12	275.5	
<2 μm	38.48	52.38	271.2	
Sennelier l'Aquarelle 315 serie 2	24.04	61.52	299.6	
Lapis lazuli bead	26	23.84	282.7	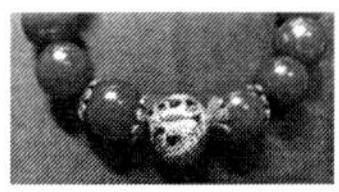
Ultramarine synthesized from kaolin [31]	33.7	53.7	290.1	

4. CONCLUSIONS

This study provides a possibility to synthesize ultramarine blue from reservoir silts, thereby offering a new utilization of the silts. To remove the unusable quartz and keep the useful clays, so as to decrease the Si/Al ratio of the silts to less than 2, the raw silts must be classified to obtain the fine part. In this study the particle sizes must be smaller than 5 μm. Ultramarine can be synthesized with the conditions: Na_2CO_3/S_8 wt. ratio = 1; (Na_2CO_3 + S_8)/silts ratio of 1.5; active carbon/silts ratio of 0.2 and calcination at 800 °C for 8 h. Under the optimal conditions, the ultramarine synthesized from <2 μm silts has the following color properties: chroma 52.4; lightness 38.5 and hue 271.2°.

ACKNOWLEDGMENT

The Administration Bureau of Nan-Hua Reservoir is greatly appreciated for the supply of the silt samples.

AUTHOR CONTRIBUTIONS

Dah-Tong Ray conceived and designed the experiments; Yin-Hsiu Hsiao performed the experiments; Yun-Hwei Shen analyzed the data; Yin-Hsiu Hsiao and Dah-Tong Ray wrote the paper.

REFERENCES

1. Chaudhry, M.A.; Rehman, H.-U. Worldwide experience of sediment flushing through reservoirs. MURJET 2012, 31, 395–408.
2. International Commission on Large Dams Introduction. Sedimentation and Sustainable Use of Reservoirs and River Systems; Stellenbosch University: Stellenbosch, South Africa, 2009; pp. 17–19.
3. Sinotech Engineering Consultants LTD. The aquatic environment in Taiwan. In The Climate Change Knowledge Database and Data Integration Platform (4/4); Water Resources Agency, Ministry of Economic Affairs: Taipai, Taiwan, ROC, 2013; pp. 2–8.
4. Wu, M.L. 2009. The Study on the Clay of Reservoir in Taiwan. Master's Thesis, NCKU, Tainan, Taiwan, 2009.
5. Kendrick, E.; Dann, S.E.; Hellgardt, K.; Weller, M.T. The effect of differents precursors on the synthesis of ultramarine blue using a modified test furnace. Stud. Surf. Sci. Catal. 2004, 154, 3059–3066.
6. Hassan, I.; Peterson, R.C.; Grundy, H.D. The structure of lazurite, ideally $Na_6Ca_2(A1_6Si_6O_{24})S_2$, a member of the sodalite group. Acta Cryst. 1985, C41, 827–832.
7. Climent-Pascual, E.; Sáez-Puche, R.; Gómez-Herrero, A.; Romero de Paz, J. Cluster ordering in synthetic ultramarine pigments. Microporous Mesoporous Mater. 2008, 116, 344–351.
8. Prener, J.S.; Ward, R. The preparation of ultramarines. J. Am. Chem. Soc. 1950, 72, 2780–2781.
9. Tarling, S.E.; Barnes, P. The structure and Si, Al distribution of the ultramarines. Acta Cryst. 1988, B44, 128–135.
10. Reinen, D.; Lindner, G.-G. The nature of the chalcogen colour centres in ultramarine-type solids. Chem. Soc. Rev. 1999, 28, 75–84.
11. Fabian, J.; Komiha, N.; Linguerri, R.; Rosmus, P. The absorption wavelengths of sulfur chromophors of ultramarines calculated by time-dependent density functional theory. J. Mol. Struc. Theochem 2006, 801, 63–69.
12. Clark, R.J.H.; Cobbold, D.G. Characterization of sulfur radical anions in solutions of alkali polysulfides in dimethylformamide and hexamethylphosphoramide and in the solid state in ultramarine blue, green, and red. Inorg. Chem. 1978, 17, 3169–3174.
13. Goslar, J.; Lijewski, S.; Hoffmann, S.K.; Jankowska, A.; Kowalak, S. Structure and dynamics of S3− radicals in ultramarine-type pigment based on zeolite A: Electron spin resonance and electron spin echo studies. J. Chem. Phys. 2009, 130, 204504.

14. Heinemann, C.; Koch, W.; Lindner, G.-G.; Reinen, D. Electronic spectrum of S_2^-, the electron affinity of S2, and the binding energies of neutral and anionic S3 clusters. Phys. Rev. 1995, 52, 1024.
15. Cotton, F.A.; Harmon, J.B.; Hedges, R.M. Calculation of the ground state electronic structures and electronic spectra of di-and trisulfide radical anions by the scattered wave-SCF-Xα method. J. Am. Chem. Soc. 1976, 98, 1417–1424.
16. Orna, M.V. Colorant usage from antiquity to the Perkin Era. In The Chemical History of Color; Springer: Heidelberg/Berlin, Germany, 2013; pp. 47–78.
17. Chivers, T. Ubiquitous trisulfur radical ion S_3^-. Nature 1974, 252, 32–33.
18. Chivers, T.; Elder, P.J.W. Ubiquitous trisulfur radical anion: Fundamentals and applications in materials science, electrochemistry, analytical chemistry and geochemistry. Chem. Soc. Rev. 2013, 42, 5996–6005.
19. Cork, W.B. Ultramarine pigments. In Pigments, Inorganic; Wiley-VCH: Weinheim, Germany, 1993; pp. 87–92.
20. Landman, A.A.; De Waal, D. Fly ash as a potential starting reagent for the synthesis of ultramarine blue. Mater. Res. Bull. 2004, 39, 655–667.
21. Gobeltz, N.; Demortier, A.; Lelieur, J.P.; Duhayon, C. Encapsulation of the chromophores into the sodalite structure during the synthesis of the blue ultramarine pigment. J. Chem. Soc. Faraday Trans. 1998, 94, 2257–2260.
22. Arieli, D.; Vaughan, D.E.W.; Goldfarb, D. New Synthesis and insight into the structure of blue ultramarine pigments. J. Am. Chem. Soc. 2004, 126, 5776–5788.
23. Kumins, C.A.; Tuckahoe, N.Y. Process for Making Ultramarine. U.S. Patent 2544694, 13 March 1951.
24. Lelieur, J.-P.V.; Duhayon, C.L.; Hautecoeur, N.S.; Demortier, A.M.; Lede, B.J.; Coopman, P.J.; Leghie, P. Ultramarine Pigment Synthesis Process. U.S. Patent 7632347 B2, 15 December 2009.
25. Kowalak, S.; Pawłowska, M.; Miluśka, M.; Stróżyk, M.; Kania, J.; Przystajko, W. Synthesis of ultramarine from synthetic molecular sieves. Coll. Surf. A Physicochem. Eng. Asp. 1995, 101, 179–185.
26. Kowalak, S.; Jankowska, A.; Łączkowska, S. Preparation of various color ultramarine from zeolite A under environment-friendly conditions. Catal. Today 2004, 90, 167–172.
27. Kowalak, S.; Jankowska, A.; Zeidler, S. Ultramarine analogs synthesized from cancrinite. Microporous Mesoporous Mater. 2006, 93, 111–118.
28. Chen, K.T.; Ray, D.T. A study on the synthesis of zeolites and ultramarine pigment from reservoir sediments. Taiwan Min. Ind. 2013, 65, 19–27.
29. De Menezes, R.A.; Da Paz, S.P.A.; Angélica, R.S.; De Freitas Neves, R.; Castella Pergher, S.B. Color and shade parameters of ultramarine zeolitic pigments synthesized from kaolin waste. Mater. Res. 2014, 17, 23–27.
30. Gobeltz, N.; Demortier, A.; Lelieur, J.P. Identification of the products of the reaction between sulfur and sodium carbonate. Inorg. Chem. 1998, 37, 136–138.
31. Sancho, J.P.; Restrepo, O.J.; Garcia, P.; Ayala, J.; Fernandez, B.; Verdeja, L.F. Ultramarine blue from Asturian "hard" kaolins. Appl. Clay Sci. 2008, 41, 133–142.

CHAPTER 11

Visualization of trace-element zoning in fluorapatite using BSE and CL imaging, and EPMA and μPIXE/μPIGE mapping

Katarzyna Gros[1], · Ewa Słaby[1], · Hans-Jürgen Förster[2], · Przemysław P. Michalak[3], · Frans Munnik[4], · Jens Götze[3], · Dieter Rhede[2]

1 Institute of Geological Sciences, Polish Academy of Sciences, Twarda 51/55, 00-818 Warsaw, Poland

2 Helmholtz-Zentrum Potsdam, Deutsches GeoForschungsZentrum, Telegrafenberg, 14473 Potsdam, Germany

3 Technische Universität Bergakademie Freiberg, 09596 Freiberg, Germany

4 Helmholtz-Zentrum Dresden-Rossendorf, 01314 Dresden, Germany

ABSTRACT

In this paper, zonation patterns of trace elements in fluorapatite are discussed that were visualized using four analytical techniques, namely back-scattered electrons (BSE) and cathodoluminescence (CL) imaging, electron probe micro-analysis (EPMA), and micro-proton-induced X-ray/gamma ray emission (μPIXE/μPIGE) mapping. Each method demonstrates the in-grain compositional variations in a slightly different way. Both BSE and CL provide qualitative data, and the internal textures are displayed in most detail. Additionally, CL points to specific elements enriched in certain growth zones. Qualitative EPMA maps show detailed zonation patterns for specific elements (with high spatial resolution), which are in general correspondence with the patterns observed in BSE and CL images. The μPIXE/μPIGE maps are fully quantitative and the detection limits are relatively low compared to EPMA mapping. In present spot measurements μPIXE demonstrates lower detection limits than EPMA, however, the latter could be considerably improved by extending the acquisition times. There is no significant overlap of REE (rare earth elements) peaks in the acquired μPIXE energy spectra, however, when multiple REEs are present with sufficiently high concentrations, peak deconvolution may pose some difficulties. Spatial resolution of μPIXE/μPIGE images is not sufficiently high to reflect minor textural features, which also result from the greater interaction depth of the proton beam. However, major growth zones are distinguishable. Even though each method has their advantages and limitations, when applied together, they provide an almost complete characterization of compositional variability in trace-element-bearing minerals.

Keywords: Trace-element zoning, Fluorapatite, BSE, CL, EPMA, μPIXE/μPIGE

1. INTRODUCTION

Trace elements, including the rare-earth-elements (REEs), are important petrogenetic indicators and provide useful information about a wide range of processes taking place in magmatic–hydrothermal as well as metamorphic and sedimentary environments. During the last two decades, focus has been placed on their behaviour during magmatic differentiation and post-magmatic fluid-related processes, which can be inferred from whole-rock and mineral chemistry (e.g. Allègre et al. 1977; Sha and Chappell 1999; Belousova et al. 2002; Deering and Bachmann 2010). Detailed investigation of trace-element contents in minerals and their spatial variations within single grains provide insight into the dynamics of magmatic differentiation and chemical heterogeneities within the melt from which these minerals crystallized (e.g. Dempster et al. 2003; Słaby et al. 2007a, b; Streck 2008; McLeod et al. 2011; Churikova et al. 2013). Therefore, a high-resolution visualization of compositional zoning in accessory minerals, the major carriers of the REEs, is necessary for fully revealing their growth history.

Several analytical methods for textural and compositional imaging, such as backscattered electrons (BSE) and cathodoluminescence (CL) imaging and electron probe micro-analysis (EPMA) are routinely applied in petrogenetical and geochemical studies (e.g. Ginibre et al. 2002; Reed 2005; Götze et al. 2013 and references therein; North-Valencia et al. 2014; Wiedenbeck et al. 2014). Although all of those permit visualizing of growth textures in minerals, the information given by each of them is unique due to different physical principles each method is based on. Backscattered electrons images primarily demonstrate variations in the relative electron density (Z contrast). The contrast is displayed using a grey-scale, where brighter tones indicate a higher mean atomic number, for instance corresponding to relative enrichment of fluorapatite in heavy elements, such as REEs (e.g. Lloyd 1987). However, the intensity of the BSE signal may also depend on crystal orientation if differently oriented crystals of one mineral species are considered (crystal orientation contrast; e.g. Lloyd 1987; Prior et al. 1999), structural variations, for instance among polymorphs of the same chemical compound (phase contrast; e.g. Mosenfelder 2000; Nasdala et al. 2007) and structural defects such as dislocations or micro-porosity (electron channelling contrast; e.g. Kaneko et al. 2005; Nasdala et al. 2007, 2009). Cathodoluminescence imaging reveals the presence of specific structural defects, e.g. vacancies or impurities in the form of trace elements. These cause electron excitation during interaction with an electron beam and induce luminescence of varying energy (Götze et al. 2013). Therefore, the growth zoning revealed by CL usually enables one to distinguish among domains of different origin. However, elevated contents of some ions (e.g. Fe^{2+} or As^{5+}) in the mineral structure can result in significant quenching of the luminescence intensity, which may limit the amount of information provided by CL (e.g. Filippelli and Delaney 1993; Perseil et al. 2000; Kempe and Götze 2002). The compositional zoning can be further demonstrated by elemental-distribution maps, e.g. obtained by EPMA. These maps can be produced for each element present above its detection limit and with a lateral resolution comparable to those of BSE and CL images (e.g. North-Valencia et al. 2014). However, longer acquisition times and, hence, longer sample exposure to the electron beam may inhibit mapping of certain elements, e.g. Na or F. Measurement of F in some minerals, for instance fluorapatite, is problematic owing to diffusion of F under the electron-beam irradiation promoted by specific crystallographic orientations (e.g. Stormer et al. 1993; Stock et al. 2015). Mobility of Na during EPMA measurement has been documented in glasses and some minerals (e.g. Nielsen and Sigurdsson 1981; Spray and Rae 1995; Morgan and London 1996). However, applying shorter acquisition times and lower accelerating voltage (12–15 kV) can: 1) significantly minimize the diffusion and mobility and 2) improve the sensitivity of both F and Na. For trace elements, e.g. REEs, significantly longer acquisition times are required.

Less conventional X-ray based methods that are being applied for imaging zonation patterns are micro-particle-induced X-ray emission (μPIXE) and micro-particle-induced gamma ray emission (μPIGE) (Reimold et al. 1999; Habermann et al. 2000; Ryan 2000, 2011; Vagelli et al. 2003; Dekov et al. 2013; Li et al. 2013; Fuchs et al. 2014). In these techniques, an ion beam (usually protons) is used to induce characteristic X-rays (for elements with $Z > 11$) or gamma rays (for light elements with $Z < 16$). Advantages of these methods over the more conventional X-ray-based methods include generally lower detection limits in mapping, an improved ability to measure light elements (e.g. Na and F), and no need for calibration standards for concentration calculations.

The present study aims at comparing the usability of the aforementioned methods in visualization of compositional zoning in small minerals and evaluating the differences in quality of the obtained results, generally applying standard "routine" measuring conditions. The criteria that we discuss include spatial resolution of the obtained images, detection limits, interaction volume, separation of the relevant element peaks and overall data-acquisition time. As an example we use REE-bearing fluorapatite ($Ca_5(PO_4)_3[F,Cl,OH]$), which is often zoned and contains variable amounts of REEs. Fluorapatite is a challenging mineral in terms of chemical analysis because it commonly contains elements of widely differing atomic number. Analytical problems are due to 1) the coexistence of light and heavy elements which usually require different measurement conditions, and 2) energy lines which may overlap with each other, e.g. K-lines of light elements are located in the same energy range as L- and M-lines of heavy elements. The advantages and limitations of each technique are discussed based on the differences in information obtained by each analytical method.

2. SAMPLES AND ANALYTICAL METHODS

Investigated fluorapatite grains come from the Karkonosze granitoid batholith forming the central part of the Karkonosze – Izera Massif (Western Sudetes, SE Poland, NE Czech Republic). The pluton is composed of many different types of rocks covering a wide range of silica content (49–78 wt%) and is formed mostly by mixing of mantle-derived lamprophyric and crust-derived granitic magmas (e.g. Słaby et al. 2007a, b; Słaby and Martin 2005, 2008). Rock samples used in the present study include three granodiorites – RUD1, RUD2, FOJ, one porphyritic granite – MIL, and equigranular granite – EQU. Fluorapatite crystals are usually 20–

200 μm in size and demonstrate complex zonation patterns, indicating significant trace-element variations during crystal growth (Przywóski 2006; Słaby 2006; Lisowiec et al. 2015). Grains were separated from the crushed and sieved rock by a combination of magnetic and heavy liquid (LST Fastfloat, Polytungstates Europe, Chippenham, UK) separation methods and subsequently prepared as polished grain mounts embedded in epoxy.

Back-scattered electrons and CL images as well as EPMA and μPIXE/μPIGE elemental distribution maps were collected for five fluorapatite grains - EQU-1ap, RUD1-5ap, RUD1-4ap, MIL-4ap, RUD2-12ap. Spot measurements were performed for an additional fluorapatite grain separate (FOJ-5ap) by EPMA and μPIXE. BSE images were acquired using a FEI Quanta 650 MLA-FEG scanning electron microscope operating at 25 kV accelerating voltage and 1 nA electron beam current (Table 1).

Cathodoluminescence direct images were obtained using a "hot cathode" CL microscope HC1-LM (cf. Neuser et al. 1995). Such a system is characterized by a defocused electron beam, therefore the luminescence image is obtained from the entire irradiated area. The system operated at 14 kV accelerating voltage and a beam current of 0.2 mA (current density of about 10 μA/mm^2) (Table 1). Thin-sections were placed upside-down and irradiated from below (the emitted luminescence is observed through the entire thin-section). Luminescence images were captured "on-line" during CL operations using a Peltier cooled digital video-camera (OLYMPUS DP72). Cathodoluminescence spectra in the wavelength range 390 to 1000 nm were recorded with an Acton Research SP-2356 digital triple-grating spectrograph with a Princeton Spec-10 CCD detector that was attached to the CL microscope by a silica-glass fiber guide. Spectra were measured under uniform conditions (wavelength calibration by a Hg-halogen lamp, spot width 30 μm, measuring time 1 s).

Qualitative elemental distribution maps were collected using a field-emission electron probe micro-analyzer (JEOL JXA-8500 F). The analytical conditions for elemental mapping included an acceleration voltage of 20 kV, a beam current of 20 nA, and a beam size of 1 μm (see Table 1). Depending on the size of the respective fluorapatite grain and actual chemical composition, dwell times per pixel were set to 400–2000 ms. Both maps and single-spot analyses were performed in the WDS (wavelength-dispersive spectrometry) mode (using five spectrometers, containing LDE1, TAP, LIF, PET and PETH crystals). This was done because the main focus of the present study are minor and trace elements. The analytical conditions for quantitative spot analysis included an acceleration voltage of 20 kV, a beam current of 10 nA, and a beam diameter of 10 μm (to minimize diffusion of Na and F) (Table 1). Reference materials, X-ray lines, WDS crystals, counting times, and detection limits for all elements are quoted in Table 2. The background counting times were always set to one half of the respective peak counting time. The CITZAF routine in the JEOL software, which is based on the $\Phi(\rho Z)$ method (Armstrong 1995), was used for data processing.

The principles of µPIXE are given in Johansson et al. (1995) and Wang and Nastasi (2010) and are briefly reviewed here. It is a relatively non-destructive, standard-free method based on the interaction between an accelerated high-energy (MeV) ion beam (usually protons) and sample atoms, resulting in their ionization and emission of characteristic X-rays. The high mass of ions, even protons, compared to electrons reduces the primary bremsstrahlung and thus increases the sensitivity compared to a method based on X-ray emission, such as EPMA. The X-ray radiation is usually collected by an energy-dispersive detector. Depending on the type of studied material and measurement conditions, the method is capable of routine quantitative analysis of elements from Na to U, with limits of detection in the range of 10 to 100 ppm. Simultaneously to µPIXE, µPIGE can be applied to detect gamma rays produced by nuclear reactions between the ions (mostly protons or deuterium) and sample atoms, by connecting an additional detector to the experimental chamber. Because the protons have to overcome the Coulomb barrier of the sample atoms' nuclei, mainly light elements ($Z < 16$), such as Li, Be, B, F, Na, and Al, are detectable by this method, which therefore is complementary to µPIXE.

The µPIXE/µPIGE measurements were performed at a microbeamline set-up connected to a HVEE 3 MV Tandetron Accelerator. The beamline is kept under high vacuum and is several meters long, along which the beam is guided and focused on the sample holder in the sample chamber. Operating conditions were: proton beam, 3.5 MeV beam energy, 1 nA beam current and 5×5 µm^2 beam size (Table 1). A Silicon Drift Detector (SDD) collimated to 80 mm^2 area was used for µPIXE with an energy resolution of 145 eV (for Mn-Kα) and an HPGe detector with a relative efficiency of 76 % and an energy resolution of 2 keV for 1332 keV gamma rays was applied for µPIGE. Mylar absorbers of varying thickness (50, 130, 230 µm) were used in front of the µPIXE detector in order to attenuate the Ca and P K-lines intensities, which are major elements of fluorapatite. These extra layers reduce the transmission for low energy X-rays and, therefore, decrease the detection efficiency for low Z elements. The HPGe detector is also mounted outside the sample chamber, separated with a 1-cm glass window and approximately 1 cm of air from the sample. This only reduces the transmission for the lowest energy gamma rays (between 100 and 200 keV). X-ray lines and detection limits for all elements are given in Table 2. Total acquisition time for one elemental-distribution map was 3–4 h. Spot measurements were performed similar to the mapping, but a smaller area (ca. 3×3 pixels) was selected for spectrum acquisition. The Gupix software (Campbell et al. 2000) was used to reduce the analytical data.

Table 1 Compilation of analytical parameters of the four techniques

Method	Accelerating voltage	Beam current	Beam diameter/ spot width
BSE	25 kV	1 nA	-
CL	14 kV	0.2 mA	30 μm
EPMA	20 kV	20 nA	1 μm (mapping) 10 μm (spot analysis)
μPIXE/μPIGE	3.5 MeV	1 nA	8×5 μm^2

3. RESULTS

3.1 BSE and CL imaging

The BSE and CL images (see top rows in Figs. 1, 2, and 3) display in detail the zonation patterns within fluorapatite crystals. Differences in BSE intensity allow us to distinguish among growth stages within the grains and their specific characteristics. Large contrasts imply major compositional variations between adjacent zones (e.g. top rows in Figs. 1a-b, 2b). Automatic adjustment of the BSE contrast does not allow the distinction of thin zones several micrometers in size if the compositional contrast is too small. In this situation, CL is the superior technique as it is usually more sensitive to growth zones in fluorapatite (e.g. MacRae et al. 2012). Here, zones as narrow as 1–2 μm are distinguishable. However, by using manually enhanced contrast settings, the level of details in BSE images can match the quality shown by CL images. Apart from irregular or primary growth zoning, additional features visible on BSE and CL images include micro-inclusions of other minerals (e.g. top rows in Figs. 1b, 2b), cracks (seen slightly better on BSE images, e.g. top rows in Figs. 1a, 2b) and dissolution boundaries (recognizable in both types of images, e.g. top rows in Figs. 2a, 3). Cathodoluminescence of fluorapatite is basically activated by two ions or groups of ions: Mn^{2+} -yellow luminescence and REE^{3+} - violet/blue luminescence (Götze et al. 2001; Kempe and Götze 2002; Götze 2012). Even slight variations in concentration of these elements at the ppm level cause a variation in the peak intensity, producing a change in the overall luminescence and, therefore, makes it possible to distinguish growth domains. Such zones may be indiscernible in BSE images (e.g. the two innermost zones in grain RUD1-4ap, top-left image in Fig. 2a), but visible using CL because of different amounts of REE and/or Mn in the fluorapatite structure. (top-right image in Fig. 2a).

Table 2 Analytical parameters of spot measurements performed by EPMA and μPIXE

Element	X-ray line EPMA	WDS crystal	Counting time [s]	Calibrant	Detection limit EPMA [ppm]	X-ray line μPIXE	Detection limit μPIXE [ppm]
F	Kα	LDE1	20	CaF_2	470	-	-
Na	Kα	TAP	10	Albite	270	-	-
Ca	Kα	PETH	10	Apatite	90	Kα	66
Mn	Kα	LiF	20	Rhodonite	200	Kα	17
Fe	Kα	LiF	20	Hematite	190	Kα	16
As	-	-	-	-	-	Kα	4
Sr	Lα	TAP	20	Celestite	320	Kα	4
Y	Lα	TAP	50	YPO_4	180	Kα	10
Zr	-	-	-	-	-	Kα	14
La	Lα	LiF	50	$LaPO_4$	470	Lα	86
Ce	Lα	LiF	50	$CePO_4$	410	Lα	67
Nd	Lβ	LiF	50	$NdPO_4$	480	Lα	121
Tm	Lβ	LiF	50	$TmPO_4$	-	Lα	34
Yb	Lα	LiF	50	$YbPO_4$	260	Lα	44
Hf	-	-	-	-	-	Lα	22
Th	Mα	PETH	50	Th-metal	110	Lα	12
U	Mβ	PETH	50	U-metal	160	Lα	14

In addition, CL wavelength spectra were collected from both zones (yellow and violet) of the grain RUD1-5ap (Fig. 4). The Mn^{2+}-activated band is present only in the yellow-zone located at around 580 nm (dashed line in Fig. 4). It has one of the highest intensities (next to Dy^{3+}), which is expressed by this dominating yellow luminescence. Bands activated by REE^{3+} are also present, they are however more pronounced in the violet-zone spectrum (full line in Fig. 4). These include Dy^{3+} bands at around 480 and 580 nm, Sm^{3+} bands at around 600, 640, 700 and 800 nm, and Nd^{3+} bands at around 870–900 nm Additionally, a very faint Tb^{3+} peak can be recognized at 550 nm.

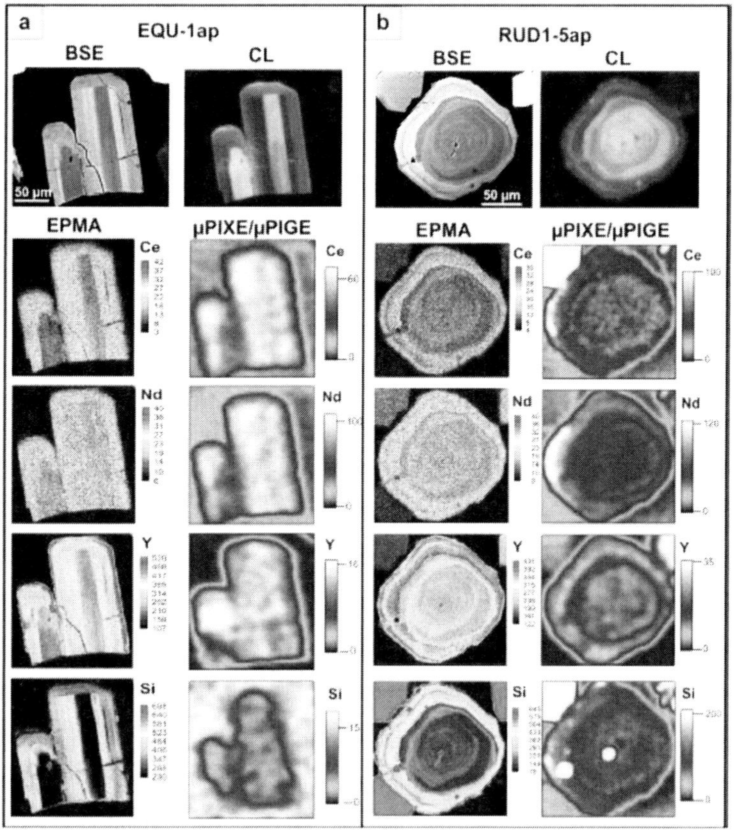

Figure. 1 Backscattered electrons (BSE; top-left), cathodoluminescence (CL; top-right) images, EPMA (bottom-left column) and µPIXE/µPIGE (bottom-right column) elemental distribution maps of two fluorapatite grains: **a** – EQU-1ap, **b** – RUD1-5ap. Ce, Nd and Y were measured by µPIXE , Si by µPIGE. Color scale in EPMA and µPIXE/µPIGE maps represents the intensity of collected X-rays [counts]

A wide band near ~400 nm probably represents "intrinsic" electron defects related to oxygen in the PO_4 groups (Götze 2012). Violet-blue luminescence demonstrated by this particular zone is probably a result of a combination of all present REE^{3+}, intrinsic electron defects and the absence of the Mn^{2+}-activated band.

3.2 EPMA imaging

The EPMA elemental distribution maps of fluorapatite accurately reflect the compositional variations in single grains and reproduce the zonation patterns seen in the BSE and CL images (bottom-left column in Figs. 1, 2, and 3). The pixel size of 1 × 1 μm^2 was sufficiently small to document even the thinnest growth zones (e.g. bottom-left column in Figs. 1b, 2a). All mapped elements are usually positively correlated. Zones enriched in all of them are also those which display violet luminescence (Nd is one of the CL activators, however, the luminescence is the combined result of a greater spectrum of REEs). Differences in the spatial resolution of the EPMA maps can be observed between: 1) individual elements – Ce and Nd exhibit lower level of details compared to Si and Y due to lower concentrations and, thus, the intensity of the X-ray signal; and 2) individual fluorapatite grains, mostly due to different acquisition times (for grain RUD2-12ap, Fig. 3 - 6–7 h, for remaining grains - 16–17 h). The EPMA maps also enable rapid localization of mineral inclusions (grain RUD1-5ap, bottom-left column in Fig. 1b). In addition to the observation, that the zonation seen in BSE images is perfectly mirrored, features such as resorption boundaries are also recorded in the EPMA maps. It is also useful to examine correlations between different elements which may help in the petrogenetic interpretation. For instance, in most grains (Figs. 1b, 2a-b, 3), Ce is positively correlated with Y. However, the grain EQU-1ap (Fig. 1a) contains a rim, in which both lanthanides are antipathetically correlated. Similarly, Y and Si are negatively correlated in this domain, whereas they behaved sympathetically in the core and inner zones. Several grains RUD1-4ap, MIL-4ap and RUD2-12ap - contain small elongated domains or show sharp lines in EPMA Si-maps (bottom-left images in Figs. 2a-b, 3, respectively) showing elevated intensities of the Si signal (red colour). These are not structural but most likely due to colloidal silica not removed from the thin-section surface during preparation.

3.3 μPIXE/μPIGE imaging

The μPIXE (and μPIGE in case of Si) elemental distribution maps show internal zoning of fluorapatite grains, however the smallest achievable beam size of 5 μm (leading to pixel size 5 × 5 μm^2) is too large to resolve the textural patterns in sufficient detail (bottom-right column in Figs. 1, 2, and 3). For instance, only two distinct zones are discernible in grains RUD1-5ap (bottom-right column in Fig. 1b) and RUD1-4ap (bottom-right column in Fig. 2a), whereas the thin subzones well visible in BSE and CL images and EPMA maps are not displayed. Mineral inclusions appear as round spots of higher intensity (bottom-right column in Fig. 1b).

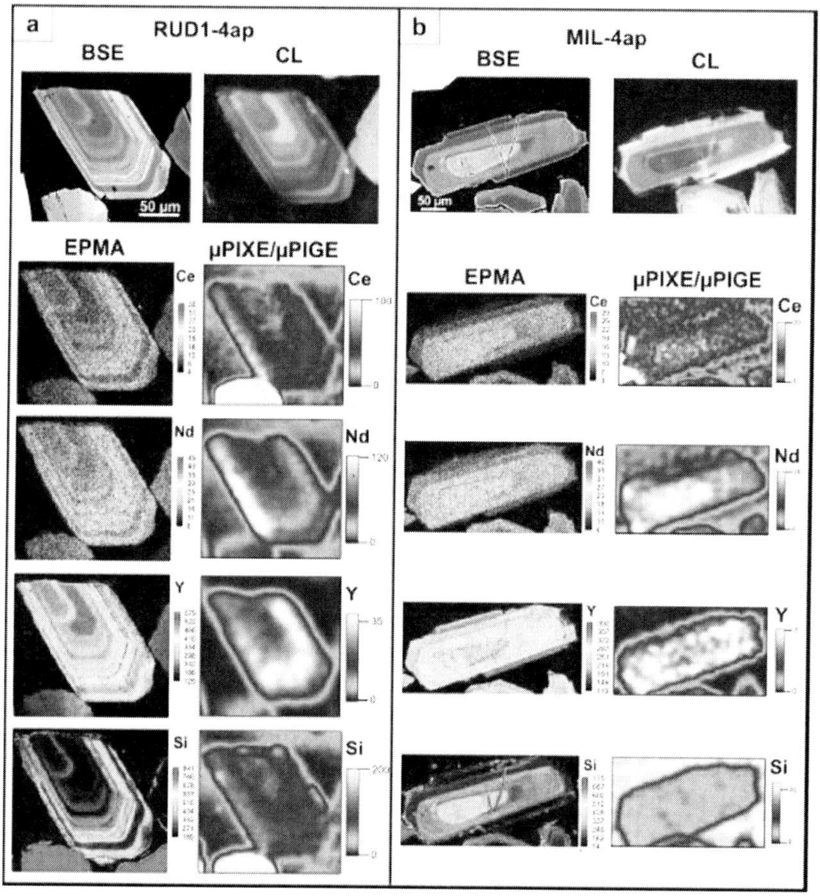

Figure. 2 Backscattered electrons (BSE; top-left), cathodoluminescence (CL; top-right) images, EPMA (bottom-left column) and µPIXE/µPIGE (bottom-right column) elemental distribution maps of two fluorapatite grains: **a** – RUD1-4ap, **b** – MIL-4ap. Ce, Nd and Y were measured by µPIXE , Si by µPIGE. Color scale in EPMA and µPIXE/µPIGE maps represents the intensity of collected X-rays [counts]

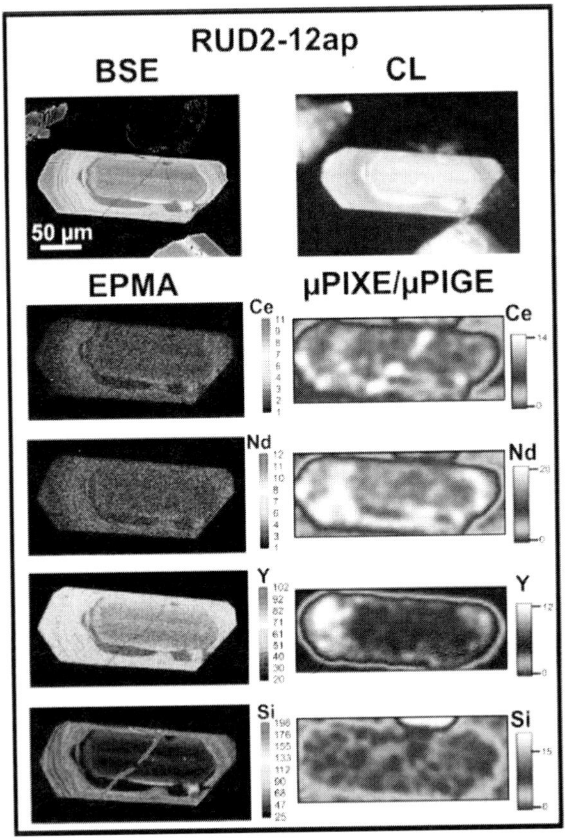

Figure. 3 Backscattered electrons (BSE; top-left), cathodoluminescence (CL; top-right) images, EPMA (bottom-left column) and μPIXE/μPIGE (bottom-right column) elemental distribution maps of one fluorapatite grain – RUD2-12ap. Ce, Nd and Y were measured by μPIXE , Si by μPIGE. Color scale in EPMA and μPIXE/μPIGE maps represents the intensity of collected X-rays [counts]

Not all features observed in the μPIGE maps represent real compositional variations in fluorapatite. The bright yellow colour in Si maps (bottom-right columns in Figs. 1a, 2b) is derived from the silica-rich glass (on which the grains were mounted). High-energy gamma rays are emitted from much greater depths then the low-energy Si-K X-rays, hence, the high intensity of the Si signal from glass detected by μPIGE. Another artefact is the bright colour on one side of the grains (e.g. on the left side of RUD1-5ap, bottom-right column in Fig. 1b). It simply relates to the fixed position of the detector which, located on that side, collects more generated X-rays than from other parts of the grain.

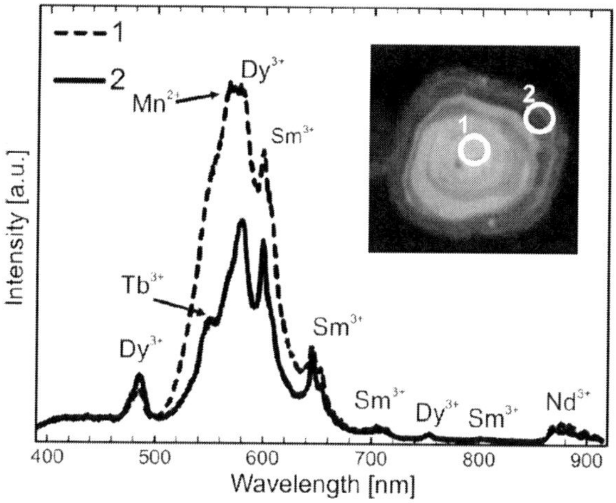

Figure. 4 Cathodoluminescence wavelength spectra of two zones from fluorapatite grain RUD1-5ap. Dashed line represents the yellow, Mn^{2+} activated luminescence (core), whereas the full line demonstrates REE^{3+} emission bands of the violet rim zone

For each grain we show four maps with the most visible zonation patterns (bottom-right columns in Figs. 1, 2, and 3). Elements detected by μPIXE include Ca, Mn, Fe, As, Sr, Y, La, Ce, Nd, Tm, Yb, Th and U (in addition, Zr and Hf are occasionally present), whereas μPIGE energy spectra show peaks related to F, Na, Si and P. Three spot measurements were performed by both μPIXE and EPMA in one grain (FOJ-5ap, Fig. 5) to compare the obtained concentrations and limits of detection. The corresponding spots were located as close to each other as possible. The energy spectra (μPIXE and μPIGE) of spot #3 are given in Fig. 6. Although Si is usually present in the μPIGE spectra of the fluorapatite grains studied, the spectrum of the spot measurement (Fig. 6b) does not contain any significant Si peak. This is probably due to low concentrations of this element in fluorapatite and the large absorption of the Si gamma rays from glass. The quantitative results of μPIXE and EPMA are given in Table 3. For most elements, the μPIXE/EPMA ratios are close to 1. Occasionally, μPIXE measurements yielded contents twice as low as those determined by EPMA.

In addition to Si, four μPIGE maps for F and Na were collected for two selected grains (Fig. 7a – RUD1-5ap, b – RUD1-4ap) in order to verify if any compositional zoning can be detected. There is no apparent zonation in case of fluorine and only very subtle hints of zoning in Na maps. As EPMA mapping of Na and F pose difficulties due to long exposure of the sample to the electron beam and, hence, high mobility of these elements, single spot EPMA measurements were performed for grain RUD1-5ap (Fig. 7). The BSE-bright rim is indeed enriched in Na compared to BSE-darker zones by ~300 ppm (Table 4). However, it may not be a sufficiently large concentration contrast to be discriminated as a separate zone. In case of fluorine, there are no significant variations between the measured spots (Table 4).

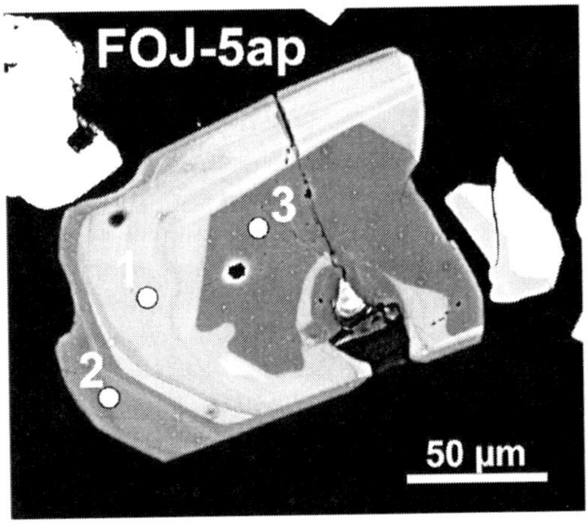

Figure. 5 BSE image of one fluorapatite grain – FOJ-5ap, in which EPMA and μPIXE spot measurements (1,2 and 3) were performed

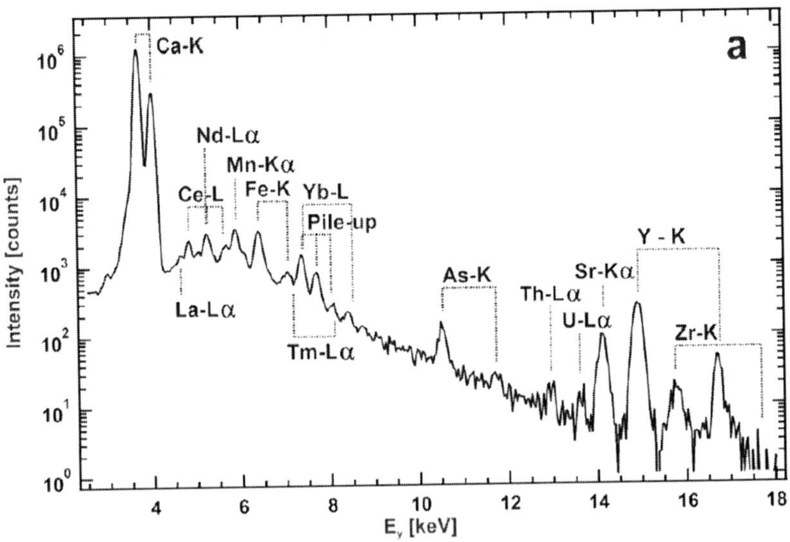

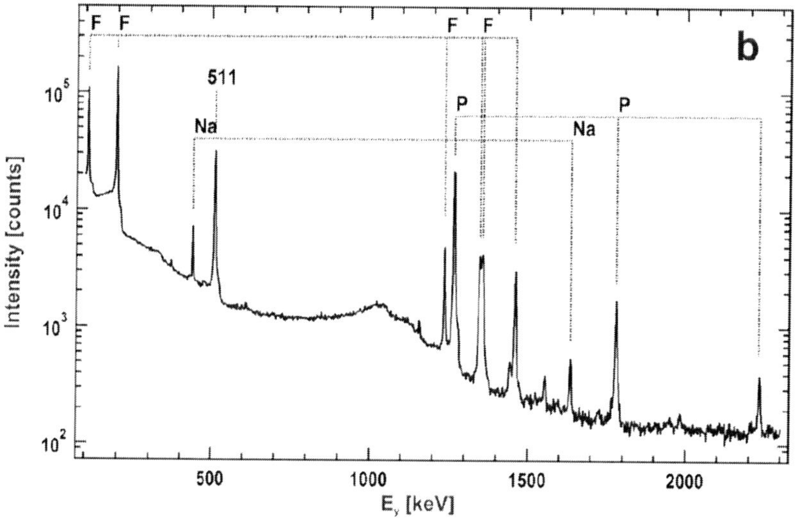

Figure. 6 μPIXE (**a**) and μPIGE (**b**) spectra with marked X-ray/gamma ray lines of all detected elements. Line marked 511 in the PIGE spectrum is caused by positron-electron annihilation both in the background and in the sample. The source of the 1550 keV emission is unknown

4. DISCUSSION

The textural and compositional information obtained by each of the applied methods differ in terms of several important criteria. The first one is the character of the acquired data. Both BSE and CL imaging provide merely qualitative data (however, in case of CL, quantitative analysis of elements giving rise to luminescence is generally available, based on the intensity of emission bands in CL spectra). The contrast in BSE intensity allows the observation of growth zoning and reveals periods of enrichment and depletion of elements during crystal growth. In this case more detailed information is given by CL imaging. Analysis of CL spectra gives evidence of REE enrichment in BSE-bright and violet zones due to increased intensity of REE^{3+}-activated bands. Specific emissions and resulting luminescence of fluorapatite domains may already give hints on its origin. According to present knowledge, yellow luminescence activated by Mn^{2+} is typical of fluorapatite crystallizing from silicic melts (and related hydrothermal fluids), whereas fluorapatite originating from alkaline or carbonatitic melts show violet CL due to an enrichment in REE^{3+} (Kempe and Götze 2002).

Table 3 Results of µPIXE and EPMA analyses placed in three growth zones in fluorapatite grain FOJ-5ap

Element	FOJ-5ap								
	1* µPIXE	1 EPMA	µPIXE/EPMA ratio	2 µPIXE	2 EPMA	µPIXE/EPMA ratio	3 µPIXE	3 EPMA	µPIXE/EPMA ratio
Ca[wt%]	39.43	38.74	1.02	39.71	39.25	1.012	39.73	39.72	1.00
Mn [ppm]	367	b.d.***		379	448	0.84	354	b.d.	
Fe [ppm]	221	564	0.39	300	295	1.02	172	282	0.61
As [ppm]	b.d.	n.a.		23	n.a.		b.d.	n.a.	
Sr [ppm]	108	b.d.		85	b.d.		136	b.d.	
Y [ppm]	2180	2753	0.79	357	817	0.44	712	19	36.84
Zr [ppm]	b.d.	n.a.		b.d.	n.a.		28	n.a.	
La [ppm]	1569	b.d.		557	1040	0.54	291	b.d.	
Ce [ppm]	4176	4430	0.94	1343	2435	0.55	880	768	1.15
Nd [ppm]	2480	4428	0.56	589	863	0.68	567	889	0.64
Tm [ppm]	192	n.a.		52	n.a.		44	n.a.	
Yb [ppm]	244	448	0.55	b.d.	602	0.08	74	655	0.11
Hf [ppm]	53	n.a.		b.d.	n.a.		b.d.	n.a.	
Th [ppm]	64	b.d.		b.d.	b.d.		b.d.	b.d.	
U [ppm]	45	b.d.		23	192	0.12	b.d.	b.d.	

* 1,2 and 3 represent spots marked in Fig. 7

*** b.d. - below detection limit, n.a. - not analyzed

All concentrations measured by µPIXE are normalized to 100 %

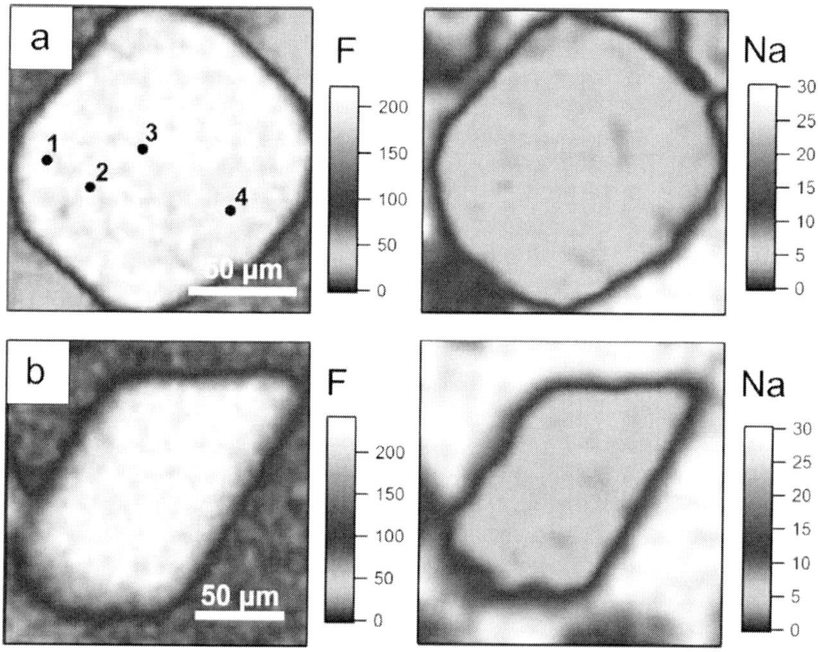

Figure. 7 Elemental distribution maps of Na and F for two fluorapatite grains obtained by μPIGE; **a** – RUD1-5ap, **b** – RUD1-4ap. Color scale represents the intensity of collected gamma rays [counts]. The points with numbers denote the additional EMPA spot measurements presented in Table 2

Table 4 EPMA analyses for fluorapatite grain RUD1-5ap

Grain-point	Na [ppm]	F [wt%]
RUD1-5ap-1	1205	3.68
RUD1-5ap-2	927	3.65
RUD1-5ap-3	957	3.61
RUD1-5ap-4	786	3.65

Another advantage of CL imaging is the possibility to distinguish zones of different luminescence (e.g. yellow and greenish core, top-right image in Fig. 2a), which, however, display identical BSE intensity (top-left image in

Fig. 2a). The presence of specific CL emissions is also helpful in selecting elements for quantitative analysis.

While BSE and CL usually provide qualitative information on compositional zoning in fluorapatite, EPMA and μPIXE/μPIGE are capable of both qualitative and quantitative analysis. The two latter techniques, however, differ in terms of element-detection method. The WDS system used in EPMA is based on selection of characteristic X-rays (with specific wavelengths) by a diffracting crystal. Such mode of detection is sequential, which extends significantly the time of chemical analysis. On the other hand, an EDS system coupled with our μPIXE set-up utilizes energy dispersion and, consequently, all elements are detected simultaneously. Such a method of detection is more rapid, but suffers from the major problem of peak overlapping. Analysing a specific X-ray line in WDS mode is in such case more advantageous due to: 1) better energy resolution, 2) the possibility of choosing a different line instead and 3) numerous corrections for most common pairs of overlapping peaks. The EDS spectra are composed of all detected element peaks, therefore, peak overlapping is an omnipresent problem. Due to similar energies of the REE X-ray lines, the distinction of REE peaks is difficult. In the obtained μPIXE spectra, only La, Ce, Nd, Tm and Yb could be distinguished, whereas elements such as Pr and Sm, most likely also present in the fluorapatite structure, failed to be discriminated On the other hand, μPIXE highlights the presence of elements not regarded as "standard" fluorapatite constituents in EPMA measurements, e.g. As, Zr and Hf (Fig. 6a). In μPIXE, peak overlapping also causes difficulties if both light and heavy elements are being analysed by EDS. K-lines of light elements are located in the same energy range as L-lines of heavy elements. If the concentrations of these heavy elements are sufficiently high (more than several hundreds of ppm), the cross-section of their relevant peaks (i.e. the probability of the specific electron transitions) may be large, thus preventing the recognition of the light elements, or vice versa. Nevertheless, for the present measurements, the problems with overlap in the μPIXE spectra are limited because there are no interfering K-lines, e.g. from Ti, and the REE that are found are relatively far apart, leading to at most double overlap. For the analysis of point spectra using Gupix, the contribution of the overlap to the statistical uncertainty is calculated (Maxwell et al. 1989). It is less than a few percent in most cases, e.g. the statistical uncertainty increases from 1.9 to 3.3 % for the Nd-Lα X-ray line for sample FOJ-5ap. For elements with $Z < 16$ (F, Na, Si and P in fluorapatite), the peak overlapping may be partially overcome by using an additional μPIGE detector. However, even among gamma ray emission lines, peaks with identical energies exist if the nuclear reaction products are the same. The identification and quantification has to be done using several lines that have different relative intensities and cross-sections. For instance, the most intense line for Si at 1779 keV is at the same position as the second most intense line for P. If the Si concentration and, hence, the peak is sufficiently high, then this line may be assigned to Si.

This applies to the maps in Figs. 1, 2, and 3 where the Si signal derives from the silica glass. The deconvolution between Si and P is further supported by the presence of other gamma ray lines. In the case of the spot measurement, however, the intensity ratio between all P-peaks indicates that the peak at 1779 keV corresponds most probably exclusively to P (Fig. 6b). For the two grains EQU-1ap and MIL-4ap (bottom-right columns in Figs. 1a, 2b), the Si map shows most probably no contribution of P, in which case the fluorapatite would be more P-rich than the surrounding silica glass. The grain RUD2-12ap (bottom-right column in Fig. 3), however, may show some contribution of P, as the contrast between the silica glass and the grain is not that significant. In case of both μPIXE and μPIGE, a careful analysis of all peaks present in the spectra and their ratios, as well as a basic knowledge on the possible composition of the studied fluorapatite, are crucial in qualitative and quantitative analysis of mineral composition. Taking into account the analytical conditions applied in both EPMA and μPIXE/μPIGE, EPMA of Si and P concentrations should provide lower detection limits (due to no peak overlapping in either EDS or WDS spectra) than μPIGE.

The EPMA-generated elemental distribution maps discussed in this study are qualitative. Transformation of X-ray intensities into concentrations is principally possible, but would have required longer acquisition times (additional background measurements in each spot). In contrast, μPIXE/μPIGE maps consist of multiple point measurements, each with their relevant energy spectra. Therefore, element concentrations can be readily calculated for every spot.

Unlike the presented maps, the spot measurements performed by EPMA and μPIXE allow a quantitative comparison of both techniques. Our results show that the μPIXE spot compared to EPMA measurements provide lower detection limits (by one or even two orders of magnitude, Table 2) for the given dwell times employed in this analysis. When similar acquisition times are applied in both techniques, lower detection limits are expected to be achieved by μPIXE. This is due to the lower peak-to-background ratio, resulting from lower Bremsstrahlung generated by ions compared to electrons (Wang and Nastasi 2010). In such a case, the only shortcoming is the potential overlapping of REE peaks in the μPIXE energy spectra, which may result in increased detection limits (however, as discussed earlier, this may not always be a problem, as evidenced by our results). On the other hand, electron probe microanalysis is capable of achieving similarly low (or even lower) detection limits as our μPIXE measurements (due to significantly better peak resolution in the energy in WDS spectra). This would, however, require significantly longer measuring times (especially for trace elements), far exceeding those applied in μPIXE. In the case of mapping, similar acquisition times in EPMA and μPIXE would also result in lower detection limits for the latter technique, as the counting time of a single spot during EPMA mapping is very short. The comparison of the acquisition times for a qualitative map performed by EPMA (16–17 h) and quantitative map produced by μPIXE (3–4 h) evidently demonstrates the advantage of μPIXE.

The discrepancies in element concentrations between μPIXE and EPMA (Table 3) may also be related to different acquisition times and peak-overlap resolution of each method. However also the interaction volume (mainly the depth) from which the characteristic X-rays are generated, has to be taken into consideration. In EPMA spot measurements, the interaction depth for the applied analytical conditions (20 kV, 10 μm beam diameter) in fluorapatite is ca. 2.5 μm for the majority of measured elements, including La, Ce, Nd, Y (a density of 3.2 $g/cm^3$for fluorapatite has been assumed), based on Monte Carlo simulations. In μPIXE spot measurements, the beam diameter is slightly smaller (8×5 μm^2), which decreases the width of the interaction volume, however, the depth is significantly larger compared to EPMA. For μPIXE, the analysis depths can be calculated for each element by Gupix and are listed in Table 5. The analysis depth depends on two factors: the cross-section for ionisation, which decreases with depth, and the absorption of X-rays which increases with depth. Therefore, the analysis depths can be determined only for a fixed percentage of the total yield of detected X-rays (here it is 67 %, 90 % and 95 %). The analysis depth of μPIGE is in general much higher because the absorption of the gamma rays in the sample can be neglected and it depends only on the reaction cross-section as a function of depth. The analysis depths for F and Na are presented in Table 5 for gamma rays of 197 keV and 440 keV energy, respectively. The width of the μPIXE interaction volume depends exclusively on the beam size and remains constant from the sample surface until near the end of the proton range (unlike the electron based techniques where the interaction volume is pear-shaped and the diameter increases with depth). The analysis depths shown in Table 5 indicate that the information given by μPIXE and μPIGE measurements comes from greater depths compared to EPMA (especially in case of F, Na, Sr, Y, Zr, Th and U), which may hinder the analysis of heterogeneous samples, e.g. strongly zoned fluorapatite. Such discrepancy in interaction volume may be also evidenced by different concentrations of the same element in spot measurements performed by EPMA and μPIXE (Table 3).

The last important feature worth discussing is the spatial resolution of the obtained images, important for textural and compositional studies of zoned minerals. The spatial resolution of an image depends on the beam size and the interaction volume, from which the back-scattered electrons, visible light or X-rays/gamma rays are emitted. Both BSE and CL images distinctly display the most detailed growth textures, where even the thinnest zones are visualized. The interaction volume for back-scattered electrons is relatively small, resulting in high lateral resolution. Moreover, better depth resolution is achieved for low-energy elements, from which the electrons are emitted only in the near-surface part of the specimen (due to absorption in greater depths). In cathodoluminescence (coupled with an optical microscope), the electron beam is defocused and, thus, a large part of the surface is excited. The CL signal is generated over a wide range of depths: from the sample surface to several μm below (Götze and Kempe 2008). Compared to characteristic X-rays, the interaction volume of CL is greater, however, short wavelengths of visible light ensure high spatial resolution. Moreover, the optical resolution of the microscope in use may further limit the spatial resolution of a CL image. If, however, CL is coupled with EPMA, a focused electron beam ensures a smaller analytical spot and, hence, decreased interaction volume from which both characteristic X-rays and CL signal are generated. In this case, spatial resolution of CL imaging is similar to that of EPMA

The level of details seen in EPMA distribution maps indicates that with a sufficiently small pixel size (1×1 μm^2 in this study), the method is capable of reproducing the same spatial resolution as BSE and CL images. With a beam diameter of 1 μm, the interaction volume is ca. 2 μm × 2 μm in size, therefore no significant overlap of the adjacent spots is present.

The relatively low spatial resolution of the maps obtained by our μPIXE/μPIGE set-up can be explained by two facts: 1) the high-energy proton beam is inherently more difficult to focus compared to an electron beam, and 2) the set-up used in this study is not yet optimised for small beam sizes. A smaller beam size and, hence, better lateral resolution could be achieved by further modernisation of the set-up, with an emphasis on improving the focussing properties. State-of-the-art ion micro-beamlines have beam spots smaller than 1 μm (Ryan 2011). However, in addition to features which limit the lateral resolution, the overall spatial resolution of the μPIXE maps is also influenced by the analysis depth. The variation of the analysis depth can lead to differences in elemental distributions obtained by K and L X-ray lines or L and M X-ray lines of the same element if the sample is inhomogeneous in depth (e.g. Munnik et al. 1999). In case of very thin zones (several μm in width, e.g. in grain RUD1-4ap, Fig. 2a) the X-ray signal generated by μPIXE can partially come from the adjacent growth zone. In such case, these zones may not be sufficiently visible on μPIXE maps.

Table 5 The analysis depths for three fractions of the total yield of detected X-ray quanta, given for all elements measured by μPIXE and two elements measured by μPIGE

Element	X-ray/Gamma line	Depth (μm) for given fractional yield		
		67 %	90 %	95 %
F	γ	32	50	53
Na	γ	43	56	60
Ca	K	7	15	19
Mn	K	8	16	21
Fe	K	9	19	25
As	K	21	39	47
Sr	K	25	44	52
Y	K	25	44	52
Zr	K	26	45	52
La	Lα	4	9	12
Ce	Lα	5	10	13
Nd	Lα	6	12	15
Tm	Lα	12	24	31
Yb	Lα	13	25	32
Hf	Lα	14	28	36
W	Lα	16	31	39
Th	Lα	24	43	51
U	Lα	25	44	52

5. CONCLUSIONS

Each technique applied in this study visualized the trace-element zoning within fluorapatite single-crystals in a different way, in terms of the character of the data (qualitative vs. quantitative), method of detection (WDS vs. EDS) and quality (detection limits, spatial resolution). The following conclusions can be drawn from the acquired results:

1) For the separation of REE X-ray lines, EPMA-WDS is a well suitable technique whereas extensive overlapping of peaks may pose problems in the application of μPIXE/μPIGE-EDS techniques.

2) The EPMA element distribution maps produced in this study give only a qualitative image of compositional variations in fluorapatite. μPIXE/μPIGE maps are quantitative, meaning that element concentrations may be calculated for each spot represented by an energy spectrum.

3) The μPIXE mapping (including current set-up) is capable of achieving low detection limits more rapidly (especially for trace elements which is crucial in terms of accessory minerals), compared to EPMA mapping, unless complex trace element peak overlaps are present.

4) The highest spatial resolution is demonstrated by BSE and CL images which enable a detailed investigation of fluorapatite textures. Element mapping by EPMA is able to reproduce nearly the same level of resolution with sufficiently small pixel size of 1×1 μm^2. The beam size generated by the applied μPIXE set-up was not sufficiently small to resolve minor textural features.

5) Interaction volume of each of the applied methods (CL, EPMA, μPIXE/μPIGE) can influence the detection limits, element concentrations and spatial resolution of elemental distribution maps if the analysed sample is inhomogeneous in depth (especially in case of μPIXE).

All studied techniques are complementary and the application of all of them produces a full characterization of compositional variations in minerals on a micro-scale. Imaging in BSE and CL modes is most suitable for primary textural investigations. Major compositional variations can be visualized rapidly by μPIXE and μPIGE, especially by quantitative mapping of trace elements due to low detection limits and the lack of a need for reference materials. The current μPIXE/μPIGE set-up, however, does not provide sufficiently high spatial resolution, especially when applied to minerals with complex growth patterns. Based on the results, a smaller amount of material can be further selected for more time-consuming EPMA-WDS imaging, which ensures high spatial resolution and good peak separation.

ACKNOWLEDGEMENTS

We thank the crew operating the HZDR Tandem accelerator for their valuable technical assistance with the μPIXE measurements. We also thank Axel Renno (Helmholtz Institute Freiberg for Resource Technology, HZDR) for helpful discussions. Constructive reviews by Colin MacRae, Martin J. Streck, Hans-Peter Schertl, and an anonymous expert are gratefully acknowledged. This work has been supported by the European Community as an Integrating Activity 'Support of Public and Industrial Research Using Ion Beam Technology (SPIRIT)' under EC contract no. 227012 and the Polish Government NCN grant no. 2011/01/N/ ST10/ 04756.

REFERENCES

1. Allègre CJ, Treuil M, Minster J-F, Minster B, Albarède F (1977) Systematic use of trace elements in igneous processes. Contrib Mineral Petr 60:57–75
2. Armstrong JT (1995) CITZAF: a package of correction programs for the quantitative electron microbeam X-ray-analysis of thick polished materials, thin films, and particles. Microb Anal 4:177–200
3. Belousova EA, Griffin WL, O'Reilly SY, Fisher NI (2002) Apatite as an indicator mineral for mineral exploration: trace-element compositions and their relationship to host rock type. J Geochem Explor 76:45–69
4. Campbell JL, Hopman TL, Maxwell JA, Nejedly Z (2000) The Guelph PIXE software package III: Alternative proton database. Nucl Instrum Meth B 170:193–204
5. Churikova TG, Ivanov BV, Eichelberger J, Wörner G, Browne B, Izbekov P (2013) Major and trace element zoning in plagioclase from Kizimen Volcano (Kamchatka): Insights into magma-chamber processes. J Volcanol Seismol 7:112–130
6. Deering CD, Bachmann O (2010) Trace element indicators of crystal accumulation in silicic igneous rocks. Earth Planet Sc Lett 297:324–331
7. Dekov VM, Kamenov GD, Abrasheva MD, Capaccioni B, Munnik F (2013) Mineralogical and geochemical investigation of seafloor massive sulfides from Panarea Platform (Aeolian Arc, Tyrrhenian Sea). Chem Geol 335:136–148
8. Dempster TJ, Jolivet M, Tubrett MN, Braithwaite CJR (2003) Magmatic zoning in apatite: a monitor of porosity and permeability change in granites. Contrib Mineral Petr 145:568–577
9. Filippelli GM, Delaney ML (1993) The effects of manganese(II) and iron(II) on the cathodoluminescence signal in synthetic apatite. J Sed Petrol 63:167–173
10. Fuchs S, Przybylowicz WJ, Williams-Jones AE (2014) Elemental imaging of organic matter and associated metals in ore deposits using micro PIXE and micro-EBS. Nucl Instrum Meth B 318(A):157–162
11. Ginibre C, Kronz A, Wörner G (2002) High-resolution quantitative imaging of plagioclase composition using accumulated backscattered electron images: new constraints on oscillatory zoning. Contrib Mineral Petrol 142:436–448
12. Götze J (2012) Application of cathodoluminescence microscopy and spectroscopy in geosciences. Microsc Microanal 18:1270–1284
13. Götze J, Kempe U (2008) A comparison of optical microscope- and scanning electron microscope-based cathodoluminescence (CL) imaging and spectroscopy applied to geosciences. Mineral Mag 72:909924
14. Götze J, Heimann RB, Hildebrandt H, Gburek U (2001) Microstructural investigation into calcium phosphate biomaterials by spatially resolved cathodoluminescence. Materialwiss Werkst 32:130–136
15. Götze J, Schertl HP, Neuser RD, Kempe U, Hanchar JM (2013) Optical microscope-cathodoluminescence (OM–CL) imaging as a powerful tool to reveal internal textures of minerals. Miner Petrol 107:373–392

16. Habermann D, Götte T, Meijer J, Stephan A, Richter DK, Niklas JR (2000) High resolution rare-earth elements analyses of natural apatite and its application in geo-sciences: Combined micro-PIXE, quantitative CL spectroscopy and electron spin resonance analyses. Nucl Instrum Meth B 161–163:846–851
17. Johansson SAE, Campbell JL, Malmqvist KG (1995) Particle-induced X-ray emission spectrometry (PIXE). Wiley–Interscience, New York
18. Kaneko Y, Fukui K, Hashimoto S (2005) Electron channeling contrast imaging of dislocation structures in fatigued austenitic stainless steels. Mater Sci Eng A 400–401:413–417
19. Kempe U, Götze J (2002) Cathodoluminescence (CL) behaviour and crystal chemistry of apatite from rare-metal deposits. Mineral Mag 66:135–156
20. Li X, Han T, Zhu X, Li Y, Zheng Y, Shen H (2013) A mineralogical application of micro-PIXE technique: The Ni-Mo-PGE polymetallic layer of black shales in Zunyi region, South China. Nucl Instrum Meth B 308:1–5
21. Lisowiec K, Słaby E, Förster HJ, Götze J (2015) Polytopic Vector Analysis (PVA) modelling of whole-rock and apatite chemistry from the Karkonosze composite pluton (Poland, Czech Republic). Lithos 230:105–120
22. Lloyd GE (1987) Atomic number and crystallographic contrast images with the SEM: a review of backscattered electron techniques. Mineral Mag 51:3–19
23. MacRae CM, Wilson NC, Torpy A, Dacidson CJ (2012) Hyperspectral CL imaging and analysis extending from UV to NIR. Microsc Microanal 18:1239–1245
24. MacRae CM, Wilson NC, Torpy A (2013) Hyperspectral cathodoluminescence. Miner Petrol 107(3):429–440
25. Maxwell JA, Campbell JL, Teesdale WJ (1989) The Guelph PIXE software package. Nucl Instrum Meth B 43:218–230
26. McLeod GW, Dempster TJ, Faithfull JW (2011) Deciphering magma-mixing processes using zoned titanite from the Ross of Mull granite, Scotland. J Petrol 52:55–82
27. Morgan GB, London D (1996) Optimizing the electron microprobe analysis of hydrous alkali aluminosilicate glasses. Am Mineral 81:1176–1185
28. Mosenfelder JL (2000) Pressure dependence of hydroxyl solubility in coesite. Phys Chem Miner 27:610–617
29. Munnik F, Sjöland KA, Vittiglio G, Ingelbrecht C, Wätjen U (1999) Nuclear microprobe study of metal segregation in quaternary bronze. Nucl Instrum Meth B 158:281–286
30. Nasdala L, Kronz A, Grambole D, Trullenque G (2007) Effects of irradiation damage on the back-scattering of electrons: silicon-implanted silicon. Am Mineral 92:1768–1771
31. Nasdala L, Kronz A, Wirth R, Kennedy A (2009) The phenomenon of deficient electron microprobe totals in radiation-damaged and altered zircon. Geochim Cosmochim Acta 73:1637–1650
32. Neuser RD, Bruhn F, Götze J, Habermann D, Richter DK (1995) Kathodolumineszenz: Methodik und Anwendung. Zbl Geo Pal 1(2):287–306
33. Nielsen CH, Sigurdsson H (1981) Quantitative methods for electron microprobe analysis of sodium in natural and synthetic glasses. Am Mineral 66:547–552
34. North-Valencia SN, Carpenter PK, Jolliff BL, Korotev RL (2014) Using electron-probe microanalysis and quantitative compositional mapping to study lithic clasts in lunar meteorites NWA 2727 and NWA 3170. Microsc Microanal 20:714–715
35. Perseil EA, Blanc P, Ohnenstetter D (2000) As-bearing fluorapatite in manganiferous deposits from St. Marcel-Praborna, Val d'Aosta, Italy. Can Mineral 38:101–117
36. Prior DJ, Boyle AP, Brenker F, Cheadle MC, Day A, Lopez G, Peruzzo L, Potts GJ, Reddy S, Spiess R, Timms NE, Trimby P, Wheeler J, Zetterström L (1999) The application of electron backscatter diffraction and orientation contrast imaging in the SEM to textural problems in rocks. Am Mineral 84:1741–1759
37. Przywóski L (2006) Chemical and CL study on apatite as an indicator of Karkonosze granite petrogenesis. MSc thesis, University of Warsaw
38. Reed SJB (2005) Electron microprobe analysis and scanning electron microscopy in geology. Cambridge University Press

39. Reimold WU, Przybylowicz W, Koeberl C (1999) Micro-PIXE analysis of varied sulfide populations from the Ventersdorp Contact Reef, Witwatersrand basin, South Africa. Nucl Instrum Meth B 158:593–598
40. Ryan CG (2000) Quantitative trace element imaging using PIXE and the nuclear microprobe. Int J Imag Syst Tech 11:219–230
41. Ryan CG (2011) PIXE and the nuclear microprobe: tools for quantitative imaging of complex natural materials. Nucl Instrum Meth B 269:2151–2162
42. Sha LK, Chappell BW (1999) Apatite chemical composition, determined by electron microprobe and laser-ablation inductively coupled plasma mass spectrometry, as a probe into granite petrogenesis. Geochim Cosmochim Acta 63:3861–3881
43. Słaby E (2006) Record of magma differentiation in apatite from Karkonosze hybrids – preliminary results. Miner Soc Poland Spec Pap 29:196–199
44. Słaby E, Martin H (2005) Mechanisms of differentiation of the Karkonosze granite. Miner Soc Poland Spec Pap 26:264–267
45. Słaby E, Martin H (2008) Mafic and Felsic Magma Interaction in Granites: the Hercynian Karkonosze Pluton (Sudetes, Bohemian Massif). J Petrol 49:353–391
46. Słaby E, Galbarczyk-Gąsiorowska L, Seltmann R, Müller A (2007a) Alkali feldspar megacryst growth: geochemical modelling. Miner Petrol 68:1–29
47. Słaby E, Seltmann R, Kober B, Müller A, Galbarczyk-Gąsiorowska L, Jeffries T (2007b) LREE distribution patterns in zoned alkali feldspar megacrysts—implication for parental melt composition. Mineral Mag 71:193–217
48. Spray JG, Rae DA (1995) Quantitative electron-microprobe analysis of alkali silicate glasses: a review and user guide. Can Mineral 33:323–332
49. Stock MJ, Humphreys MCS, Smith VC, Johnson RD, Pyle DM, EIMF (2015) New constraints on electron-beam induced halogen migration in apatite. Am Mineral 100:281–293
50. Stormer JC Jr, Pierson ML, Tacker RC (1993) Variation of F and Cl X-ray intensity due to anisotropic diffusion in apatite during electron microprobe analysis. Am Mineral 78:641–648
51. Streck MJ (2008) Minerals textures and zoning as evidence for open system processes. In: Putirka KD, Tepley FJ (eds) Minerals, inclusions and volcanic processes. Rev Mineral Geochem, vol.69 Mineral Soc Am, Washington DC, 595–622
52. Vagelli G, Borghi A, Cossio R, Mazzoli C, Olmi F (2003) Comparison between major and trace element concentrations in garnet performed by EPMA and micro-PIXE techniques. Spectrochim Acta B 58:699–709
53. Wang Y, Nastasi M (2010) Handbook of modern ion beam materials analysis. Materials Research Society
54. Wiedenbeck M, Bédard LP, Bugoi R, Horan M, Linge K, Merchel S, Morales LFG, Savard D, Souders AK, Sylvester P (2014) GGR biennial critical review: analytical developments since 2012. Geostand Geoanal Res 38(4):467–512

CHAPTER 12

Whole-Rock Geochemistry and Mineralogy of Triassic Montney Formation, Northeastern British Columbia, Western Canada Sedimentary Basin

Edwin I. Egbobawaye

Department of Earth and Atmospheric Sciences, University of Alberta, Edmonton, Canada

ABSTRACT

Inductively Coupled Plasma-Mass Spectrometry (ICP-MS) was used to analyze chemical elements— major, trace and rare earth elements (REE) concentrations, augmented with quantitative X-ray diffraction (XRD) analysis and thin-section petrography for mineralogical characterization of the Triassic Montney Formation in northeastern British Columbia, Western Canada Sedimentary Basin (WCSB). Results from this study indicate that integration of chemical elements with mineralogy shows affinity to the host lithologies. Evidently, chemical elements are the building blocks for minerals, thus, their significances in the interpretation of geological systems are unambiguous. Herein, major elements concentration such as Al, Fe, K, Mg, Ca, Mn in the samples analyzed from the Montney Formation are interpreted as: 1) indication of dolomitization and diagenesis; 2) trace elements—Rb, Th, U, and Cs are related to the organic matter—kerogen in the clay component of the Montney Formation source rock; and 3) transition metals—Sc, V, Co, Cr, Zn show strong affinity with diagenesis in the study interval.

Keywords: Geochemistry, Whole-Rock, Mineralogy, Montney Formation, Thin-Section, Petrography, Trace-Elements, ICP-MS, XRD, Dolomitization, Diagenesis, WCSB

1. INTRODUCTION

The term trace element in geochemistry refers to chemical elements whose concentration is less than 1000 ppm or 0.1% of a rock's composition [1] [2], and concentration greater than the aforementioned values refers to major elements (expressed as weight % oxides, each >0.1%). Elements make up a large portion of the earth's crust, but occur most often as minerals [3]. Elements such as carbon, hydrogen, nitrogen, oxygen, phosphorus, calcium, sulphur, and iron are the principal chemical elements that living organisms utilize in their body tissues for reproduction, energy, and their daily functional activities [4]. The geological significances of these elements are related to being components of the oceans, atmosphere, and as building blocks for minerals, and crustal rocks. As a result, whole-rock element geochemistry is invaluable in understanding the chemical, physical, and geological forces that continually redistributes these elements between living and non-living reservoirs through biogeochemical cycles [4]. Whole-rock chemical element geochemistry is useful in chemostratigraphic correlations, and have been used to decipher the physical and chemical mechanisms, and interpretation of diverse geological systems [5].

The understanding of the variability and affinity of these chemical elements in rocks have led to the identification of chemical signatures in many sedimentary basins of various tectonic regimes [5]. Furthermore, whole-rock element composition provides a means of quantification and calibration of the concentration of chemical elements with mineralogy, thereby giving additional information about host rock lithology and accessory minerals. Whole-rock elements geochemistry can help to decode the interrelationship between components of the oceans, crustal rocks, biogeochemical systems [4] , seawater pH, and atmospheric CO_2 concentrations spanning centuries, or even millions of years [6]. Many paleoclimatic studies have utilized chemical elements such as Aluminum (Al), Manganese (Mn), and Barium (Ba) concentrated in the rock record to unravel paleoclimate and tectonic regimes [5] [7] -[9]. These studies have led to better understanding of the changes that causes chemical weathering in source areas and help to show the variation in climatic changes in ancient rocks on a geological time scales [10] [11]. The application of whole-rock elements geochemistry are numerous in geological interpretation, thus its utility in the study of Triassic Montney Formation.

Stratigraphically, the Montney Formation is the basal stratigraphic unit of Triassic succession in the subsurface of western Canada (Figure 1 and Figure 2). It rests, unconformably upon Belloy Formation—carbo- nate strata of Permian age [12] - [16]. The succession was deposited in a west-facing, arcuate extensional basin on the western margin of Pangaea [16] - [18]. The Montney Formation consists of siltstone, very fine-grained sandstone, bioclastic packstone/grainstone-coquina, in Alberta [15] [16] [19] , interlaminated, interbedded, dolomitic silty-sandstone and rich in organic matter [20]. The intervals composed of fine-grained sandstones within the Montney Formation is the hydrocarbon reservoir. Previous studies of the Montney Formation in Western Canada Sedimentary Basin (WCSB) have mainly focused on sedimentology, regional stratigraphy, biostratigraphy, and ichnology [12] [16] - [18] [21] - [28]. Other workers study the tectonics of WCSB, e.g., [27] [29] -[31] , and oil and gas reservoir of the Montney Formation have been assessed, e.g., [13] [19] [20] [32] - [35]. Despite numerous previous work, however, no work has been published in the journal that utilized chemical elements integrated with mineralogy and petrography to interpret the Montney Formation dolomitization and diagenesis.

This paper evaluate the Triassic Montney Formation dolomitization and diagenesis using chemical elements whole-rock geochemistry coupled with whole-rock mineralogy and thin-section petrography in order to better understand the Montney Formation dolomitization and diagenesis.

2. GEOLOGICAL SETTING

Based on analyses of paleomagnetic data, paleolatitude and paleoclimatic zonation [37] , and fauna record [38] , the Western Canada Sedimentary Basin (WCSB) during the Triassic time was situated at approximately 30°N paleolatitude [37]. The paleoclimate reconstruction suggests that the paleoclimate may have ranged from sub-tropical to temperate [24] [30] [37] [38]. The region has been interpreted to be arid during the Triassic, and was dominated by westerly winds from the west [12] [37] [39].

The WCSB forms a northeast thinning wedge of sedimentary rocks with thickness of more than 6000 meters, which extends southwest from the

Canadian Shield into the Cordilleran foreland thrust belt [24] [40]. The Cordilleran of the WCSB provides the evidence that the origin and development of the basin was associated with tectonic activity attributed to episodes of epeirogenic subsidence; [24] [40] - [42] ; this is interpreted to be post Triassic, especially due to mountain influences [24] [43] and [44] interpreted sediment loading as evidenced by deformed beds, slump structures and small-scale faults as indicators of tectonic influences on the deposition of Triassic successions. Within the Foothills and Rocky Mountain Front Ranges, Triassic rocks were subjected to Jurassic-Cretaceous Columbian and Upper Cretaceous-Lower Tertiary Laramideorogenies, which caused series of imbricate thrust faults and folds in the region [12] [45].

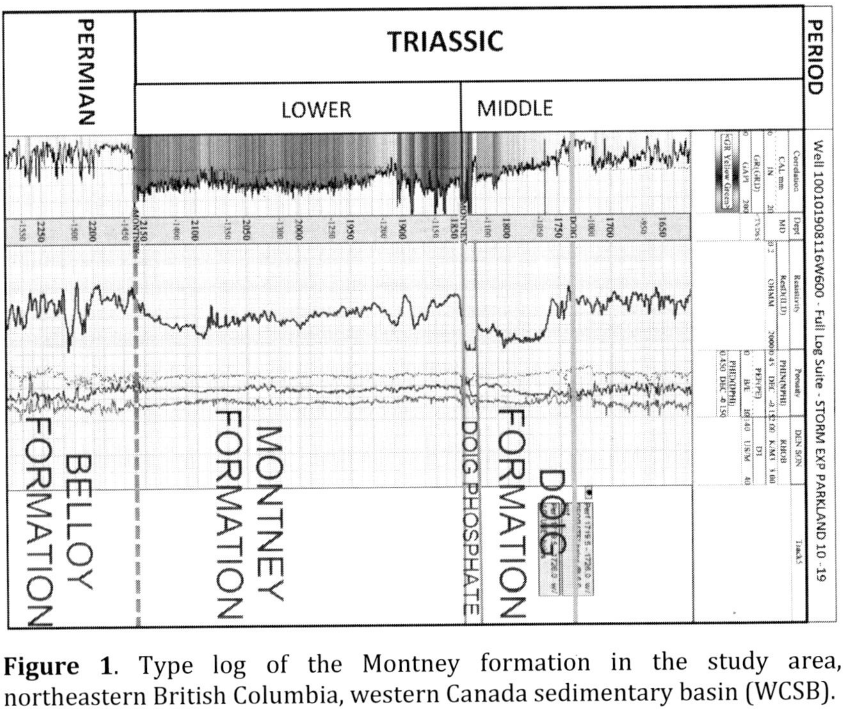

Figure 1. Type log of the Montney formation in the study area, northeastern British Columbia, western Canada sedimentary basin (WCSB).

In Alberta and British Columbia, Triassic sediments were deposited in a central sub-basin known as the Peace River Embayment, which extended eastward from the Panthalassa western ocean onto the North American craton [12]. During the Triassic period, the Peace River Embayment was a low mini basin associated with minor fault block movement caused by a broad downwarp, which resulted in the rejuvenation of structural deformation within the Monias areas of southwest Fort St. John, British Columbia [12] [45].

The Triassic succession thickened westward [12] , and rests unconformably in most areas, upon the Belloy Formation in outcrop of northeastern British Columbia; Carboniferous in parts of northeastern British Columbia and Alberta; and Fantasque Formation in outcrop at Williston [36]. The thickness of the Montney Formation within the Fort St. John area (max: 320 meters thick) in the subsurface (Figure 3), varies in the east along British Columbia/Alberta boarder boundary and to the western portion of British Columbia [20]. The Montney Formation structure map indicates higher paleostructure in the east and low in the western portion of the study area [20]. The structural tilt shows a depositional thinning to the east and north due to erosional removal [12].

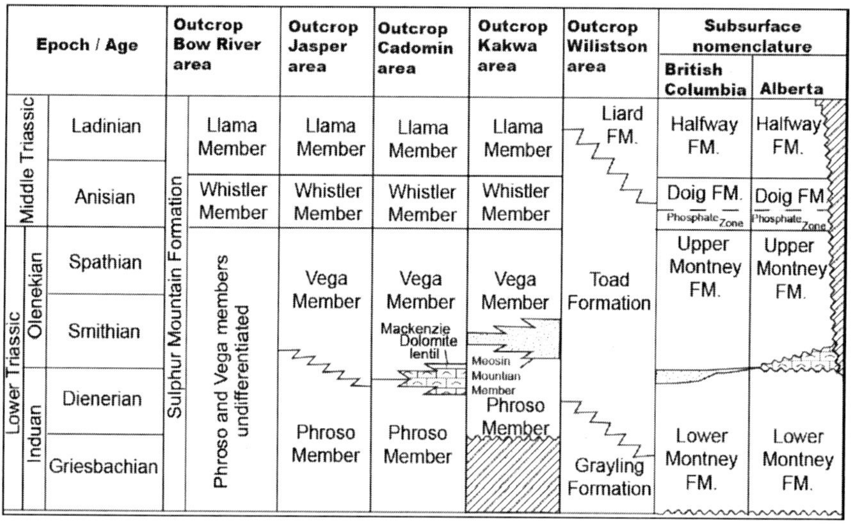

Figure 2. Stratigraphic chart showing lower and middle Triassic deposits and a corrrelation of outcrop with coeval subsurface strata in the western Canada sedimentary basin; modified from [36].

3. METHOD OF STUDY

Rock samples collected from drill cores from five wells in northeastern British Columbia were crushed into powder using pulverizing shatter-box machine at the University of Alberta. Samples were assayed for whole- rock chemical elements geochemical analysis using Inductively Coupled

Plasma?MassSpectrometry (ICP-MS) technique at the University of Alberta Lab. Sample requirement is approximately ±5 ml for ICP-MS measurements. The machine can rapidly determine over 70 elements (Table 1) in <2 ml of sample solution in less than 2 minutes [46]. The functionality of the instruments are based on the principle that by viewing the appropriate region in an Ar plasma tail flame, the atomic and ionic emission lines of analytes can be measured against very low background emissions intensities [46]. The method can show nearly all elements present in a sample (Table 1).

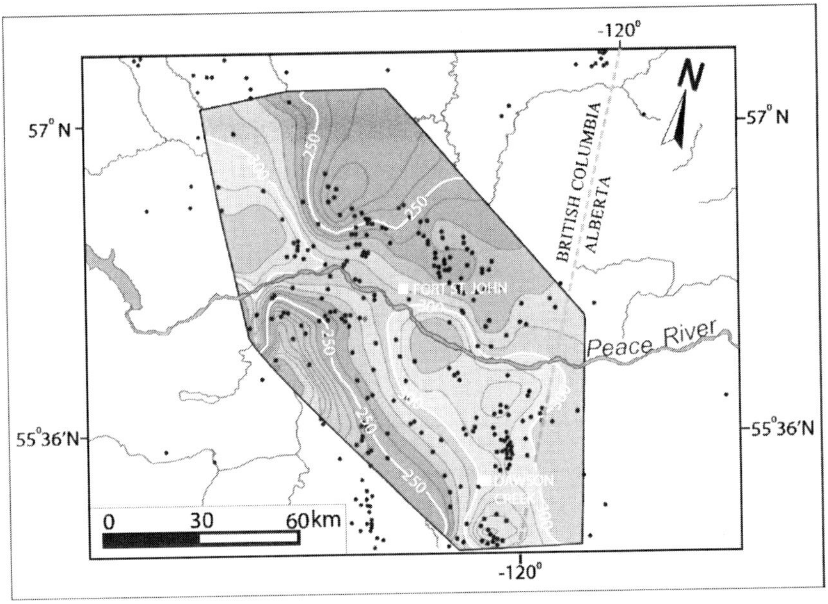

Figure 3. Isopach map of the Montney Formation tight-gas reservoir in the study area showing depositional trends and thickness in northeastern British Columbia, Western Canada Sedimentary Basin (WCSB).

Quantitative X-ray diffraction (XRD) analysis was used to determine whole-rock mineralogy of the Montney Formation (Table 2). XRD is essential in the study of fine-grained sediments because it allows moderately accurate determination of the amount of minerals present in a sample, particularly, magnesium substitution in calcite or dolomite lattices [47] , thus helping to decipher dolomitization and diagenesis.

Thin-sections were examined with petrographic microscope in accordance with the method of thin-section petrography [48] to determine textural characteristics, mineralogical composition, evidence of dolomitizationand diagenesis. Traditional thin-section examination technique was used to determine grain-size, sorting, roundness, and mineralogy (Figure 4).

4. RESULTS

Chemical elements analyzed in the Montney Formation are grouped into three categories, based on their concentration: 1) major rock-forming mineral elements (Al, Ca, Fe, K, Mg, Mn, Na, Si) have the highest concentration (Table 1); 2) trace elements such as Ag, Ba, Cd, Co, Cr, Mo, Ni, Pb, Sn, Sr, V, etc., have low concentration compared to major elements (Table 1 and Figure 5 and Figure 6); and 3) rare earth elements (Figure 7), or Lanthanides, which comprises La, Ce, Pr, Nd, Pm, Sm, Eu, Gd, Tb, Dy, Ho, Er, Tm, Yb, Lu elements (Figure 7).

The integration of chemical elements with mineralogy shows affinity to the host lithologies. Evidently, chemical elements are the building blocks for minerals, thus, their relevance in the interpretations of geological system is unambiguous. Based on data in Table 1, major elements concentrations of Al, Fe, K, Mg, Ca, Mn in the samples analyzed are interpretedas indication of dolomitizationand diagenesis while trace elements—Rb, Th, U, and Cs are related to the organic matter—kerogen in the clay component of the Montney Formation source rock, and transition metals—Sc, V, Co, Cr, Zn show strong affinity with diagenesis in the study interval. The highest level of dolomitization in the Montney Formation occur at the depths of 2042 m and 2055.2 m with magnesium/calcite concentrations (Table 1) in the amount of 42,120 ppm/90,976ppm and 48,152 ppm/98,597ppm respectively.

The concentration of quartz mineral (wt.%) in whole-rock mineralogy data-set is a predictor of calcite and dolomite concentration (Table 2). High quartz concentration yields corresponding low calcite/dolomite concentrations and vice-versa.

4.1. Mineralogy

Whole-rock analysis of bulk mineralogy (Table 2) of the Montney Formation shows that quartz is the dominant mineral by percent volume and abundance, followed by calcite, and then dolomite. Clay total volume in the Montney Formation is relatively high (outside of the study area), but lower than quartz, and in some interval clay is higher than the volume of dolomite and calcite (Table 2). However, within the study area (Figure 2), clay content is relatively low.

Table 1. Whole-rock geochemistry showing chemical element concentrations in Montney Formation.

		Major Elements								Trace Elements								
		Mg	Al	P	K	Ca	Ti	Fe	Mn	Li	Be	B	Co	Cu	Zn	V	Cr
		2	0.2	5	6	31	0.09	3.7	0.03	0.3	0.05	0.03	0.03	0.08	0.1	0.05	2
Well Location	**Depth (m)**	ppm	ppm	ppm	ppm	ppm	ppm	ppm	ppm	ppm	ppm	ppm	ppm	ppm	ppm	ppm	ppm
b-39-H-93-p-9	2042 m	**42,120**	31,116	809	24,111	**90,976**	2795	16,707	297	26.9	51.9	39.1	18.3	43.1	1.2	51.4	<DL
d-39-F-93-p-9	2668.8 m	18,820	33,638	1246	24,251	71,508	3021	14,427	266	15	85.7	64.2	13.7	22.7	1.3	46.4	<DL
d-39-F-93-p-9	2685.4 m	13,338	40,861	2249	30,359	49,140	3668	20,717	253	17.3	77.7	42.5	23.5	89.6	1.4	62.8	<DL
d-39-F-93-p-9	2685.4 m	13,475	41,234	2289	30,453	48,968	3676	20,575	246	17.5	77.9	39	25	111	2	63	<DL
2-19-79-14W6	2048 m	19,780	45,633	1955	36,036	45,915	3877	19,687	253	21	288	26.7	24.1	33.8	1.8	69.3	<DL
2-19-79-14W6	2069.5 m	20,116	36,583	914	29,177	47,584	3397	16,679	332	17.9	530	28.7	19.1	221	1.5	61.2	<DL
2-19-79-14W6	2085 m	11,512	40,016	1493	32,310	36,429	3856	18,929	287	14.4	56.6	39.4	21.2	29.6	1.9	66.1	<DL
7-13-79-15W6	2055.22 m	**48,152**	24,438	1582	20,397	**98,597**	2080	22,999	364	21.8	271	27	30.5	24.6	1.1	70.1	<DL
7-13-79-15W6	2061.3 m	36,224	28,526	688	22,992	79,426	2704	14,057	355	12.7	55.2	40.5	15.2	28.3	1	52.4	<DL
7-13-79-15W6	2084.5 m	26,671	35,233	1717	27,941	61,980	3182	20,491	385	15.7	68.2	27.7	18.6	96.3	1.7	58.9	<DL
7-13-79-15W6	2101.78 m	14,456	39,659	2139	33,885	37,952	3576	26,134	324	12.4	78.5	58.7	29.3	270	2	65.7	<DL
9-29-79-14W6	1973 m	13,408	41,779	938	30,649	50,042	3938	16,527	296	15.6	504	45.1	19.2	82.3	1.6	68.2	<DL
9-29-79-14W6	1973 m	13,545	42,786	1003	31,695	51,007	3832	16,575	295	15.4	494	65.5	16.8	86.1	1.7	66.7	<DL
9-29-79-14W6	1999 m	11,761	36,736	2465	30,541	53,523	3347	15,951	331	11	46.9	28.1	18.1	68.7	1.1	49.9	<DL
GSP-2		**4993**	**61,382**	**1309**	**45,681**	**16,528**	**3521**	**27,936**	**288**		**46.4**	**9.41**	**38**	**121**	**1.3**	**19.8**	**<DL**
Certified value		**5800**	**78,800**	**1300**	**44,800**	**15,000**	**4000**	**34,300**	**320**	**36**	**52**	**7.3**	**43**	**120**	**1.5**	**20**	**-**
% recovery		**86.1**	**77.9**	**100.7**	**102**	**110.2**	**88**	**81.4**	**90**	**88.6**	**89.3**	**128.8**	**88.3**	**100.7**	**87.1**	**99.2**	**-**

	Most dolomitized interval in the Montney Formation based on analyzed samples	**Magnesium/Calcite**

Trace Elements

Well Location	Depth (m)	Ga	Ge	As	Se	Rb	Sr	Y	Zr	Nb	Mo	Ru	Pd	Ag	Cd	Sn	Sb
		0.01	0.02	0.06	0.2	0.04	0.03	0.02	0.09	0.04	0.02	0.01	0.01	0.01	0.06	0.06	0.01
		ppm	ppm	ppm	ppm	ppm	ppm	ppm	ppm	ppm	ppm	ppm	ppm	ppm	ppm	ppm	ppm
b-39-H-93-p-9	2042 m	8.36	1.13	10.9	1.64	77.9	279	23	178	9.22	1.95	0.26	7.86	0.53	0.08	3.39	0.95
d-39-F-93-p-9	2668.8 m	9.16	1.31	8.95	2.9	65	133	24	181	9.19	15.8	0.12	8.04	0.44	0.16	2.77	1.1
d-39-F-93-p-9	2685.4 m	12.2	1.43	17.7	3.53	88	121	28.3	256	11.9	13.8	0.03	11.1	0.59	1.11	3.35	1.17
d-39-F-93-p-9	2685.4 m	12.5	1.42	17.3	2.54	87.9	118	28.3	304	11.9	13.5	0.04	12.9	0.59	1.03	3.01	1.27
2-19-79-14W6	2048 m	14.5	1.6	19.9	3.06	96.8	107	28.4	172	11.7	28	<DL	7.31	0.5	0.34	3.32	3.55
2-19-79-14W6	2069.5 m	10.4	1.73	16.3	7.1	75.6	96.1	28.1	230	10.4	21.4	0.04	9.95	1.18	5.97	3.35	7.23
2-19-79-14W6	2085 m	11.2	1.63	8.01	2	81.3	100	28	281	12.7	4.98	<DL	11.7	0.55	0.28	2.95	0.63
7-13-79-15W6	2055.22 m	7.48	1.21	26	10.2	61.9	128	23.6	178	7.01	141	0.22	7.65	0.82	0.61	2.36	10.5
7-13-79-15W6	2061.3 m	6.83	1.17	7.84	2.71	53.6	111	22.7	195	8.99	28.7	0.23	9.21	0.44	0.35	3.02	0.91
7-13-79-15W6	2084.5 m	9.45	1.36	10.7	1.73	67.9	119	26.7	179	9.54	6.35	0.07	8.28	0.36	1.33	2.81	3.72
7-13-79-15W6	2101.78 m	11.3	1.55	15.6	2.11	79.3	107	29.4	241	11.8	70.1	<DL	11.1	0.79	6.03	3.1	2.16
9-29-79-14W6	1973 m	11.7	1.47	21	7.21	83.4	103	26.7	237	12.5	20.2	<DL	10.9	1.08	2.08	3.53	7.27
9-29-79-14W6	1973 m	12.6	1.41	22	9.26	85.5	103	26.5	230	11.8	21	<DL	10.6	1	2.07	3.23	7.1
9-29-79-14W6	1999 m	9.42	1.3	8.26	3.71	73.3	118	28.5	184	9.85	6.82	<DL	8.54	0.39	1.39	2.71	0.51
GSP-2		**22.3**	**1.92**	**2.14**	**4.54**	**269**	**245**	**24**	**400**	**22.2**	**2.15**	**<DL**	**18**	**0.61**	**0.16**	**7.99**	**0.36**
Certified value		**22**	-	-	-	**245**	**240**	**28**	**550**	**27**	**2.1**	-	-	-	-	-	-
% recovery		**101.2**	-	-	-	**109.9**	**102**	**85.8**	**72.7**	**82.4**	**102.2**	-	-	-	-	-	-

		Rare Earth Elements (REE)										Trace Elements		
		Lu	Hf	Ta	W	Re	Os	Ir	Pt	Th	U	Au	Tl	Pb
		0.04	0.05	0.02	0.08	0.003	0.08	0.04	0.01	0.03	0.01	0.01	0.05	0.03
Well Location	**Depth (m)**	ppm	ppm	ppm	ppm	ppm	ppm	ppm	ppm	ppm	ppm	ppm	ppm	ppm
b-39-H-93-p-9	2042 m	0.34	5.59	1.81	311	0.007	<DL	<DL	5.83	3.67	0.11	<DL	0.78	12.4
d-39-F-93-p-9	2668.8 m	0.36	5.73	1.76	996	0.026	<DL	<DL	4.26	4.89	0.1	<DL	1.88	12.5
d-39-F-93-p-9	2685.4 m	0.43	9.26	1.41	449	0.031	<DL	<DL	7.04	12.1	0.13	<DL	2.43	18.2
2-19-79-14W6	2048 m	0.41	5.32	1.16	236	0.022	<DL	<DL	5.61	6.24	0.09	<DL	2.97	15.6
2-19-79-14W6	2069.5 m	0.41	7.01	1.14	251	0.054	<DL	<DL	6.03	7.45	0.11	<DL	2.41	15.6
2-19-79-14W6	2085 m	0.44	8.61	1.49	490	0.017	<DL	<DL	6.26	5.16	0.13	<DL	1.18	17.1
7-13-79-15W6	2055.22 m	0.32	5.15	0.6	230	0.169	<DL	<DL	5.4	26.9	0.07	<DL	7.29	15.3
7-13-79-15W6	2061.3 m	0.37	6.94	1.69	363	0.007	<DL	<DL	6.87	5.91	0.12	<DL	1.56	8.11
7-13-79-15W6	2084.5 m	0.36	6.08	1.12	223	0.006	<DL	<DL	7.02	5.46	0.09	<DL	0.96	12.7
7-13-79-15W6	2101.78 m	0.41	7.99	1.69	507	0.092	<DL	<DL	6.69	13.1	0.12	<DL	8.2	24.1
9-29-79-14W6	1973 m	0.44	7.99	1.62	423	0.056	<DL	<DL	8.59	7.37	0.12	<DL	3.78	15.1
9-29-79-14W6	1999 m	0.37	6.1	1.05	312	0.021	<DL	<DL	4.32	6.51	0.07	<DL	1.03	15.3
GSP-2		**0.21**	**13.82**	**0.73**	**8.58**	**0.003**	**<DL**	**<DL**	**107**	**2.62**	**0.18**	**<DL**	**1.16**	**42.5**
Certified value		**0.23**	**14**	-	-	-	-	-	**105**	**2.4**	-	-	**1.1**	**42**
% recovery		**91.5**	**98.7**	-	-	-	-	-	**101.6**	**109.1**	-	-	**105.1**	**101.2**

		Rare Earth Elements (REE)															
		Te	Cs	Ba	La	Ce	Pr	Nd	Sm	Eu	Gd	Tb	Dy	Ho	Er	Tm	Yb
		0.02	0.02	0.03	0.03	0.03	0.004	0.03	0.04	0.03	0.03	0.03	0.04	0.02	0.04	0.006	0.05
Well Location	Depth (m)	ppm	ppm	ppm	ppm	ppm	ppm	ppm	ppm	ppm	ppm	ppm	ppm	ppm	ppm	ppm	ppm
b-39-H-93-p-9	2042 m	<DL	3.88	219.63	25.48	42.06	5.96	23.48	4.96	1.14	4.94	0.66	4.11	0.85	2.41	0.35	2.27
d-39-F-93-p-9	2668.8 m	<DL	2.29	354.78	23.45	33.77	5.41	20.78	4.4	1.06	4.46	0.62	4.03	0.83	2.45	0.37	2.34
d-39-F-93-p-9	2685.4 m	<DL	3.7	345.98	30.04	42.3	6.6	25.25	5.29	1.17	5.24	0.74	4.81	1	2.96	0.44	2.81
d-39-F-93-p-9	2685.4 m	<DL	3.65	341.81	31.67	52.23	6.93	26.43	5.47	1.21	5.46	0.73	4.79	0.99	2.96	0.44	2.83
2-19-79-14W6	2048 m	<DL	4.16	380.24	31.14	44.5	6.51	24.91	5.03	1.27	5.16	0.7	4.79	0.99	2.85	0.42	2.8
2-19-79-14W6	2069.5 m	<DL	2.59	330.84	27.21	44.9	6.95	27.1	5.78	1.27	5.63	0.8	4.98	0.99	2.95	0.42	2.72
2-19-79-14W6	2085 m	<DL	2.64	334.07	29.93	43.85	6.62	24.77	5.08	1.12	5.02	0.7	4.75	1	2.96	0.43	2.9
7-13-79-15W6	2055.22 m	<DL	3.36	175.9	22.74	34.06	4.7	17.56	3.54	0.76	3.53	0.5	3.28	0.73	2.2	0.3	2.08
7-13-79-15W6	2061.3 m	<DL	1.99	265.11	24.52	46.27	5.88	22.09	4.48	1.03	4.52	0.64	4	0.83	2.45	0.36	2.33
7-13-79-15W6	2084.5 m	<DL	2.72	535.05	28.08	54.32	7.02	26.63	5.38	1.35	5.71	0.76	4.8	0.96	2.75	0.4	2.5
7-13-79-15W6	2101.78 m	<DL	2.8	339.91	32.6	46.42	6.78	25.18	5.12	1.18	5.26	0.75	4.71	0.99	2.9	0.42	2.77
9-29-79-14W6	1973 m	<DL	3.25	382.29	30.72	55.56	6.88	25.01	4.86	1.13	4.93	0.71	4.71	0.98	2.96	0.42	2.91
9-29-79-14W6	1999 m	<DL	2.32	532.15	28.13	36.39	6.45	25.39	5.21	1.28	5.44	0.74	4.87	1.01	2.84	0.4	2.63
GSP-2		**<DL**	**1.26**	**1190.09**	**186.53**	**407.31**	**54.36**	**198.84**	**26.06**	**2.63**	**15.62**	**1**	**5.67**	**0.96**	**2.64**	**0.28**	**1.75**
Certified value		**-**	**1.2**	**1340**	**180**	**410**	**51**	**200**	**27**	**2.3**	**12**	**-**	**6.1**	**1**	**2.2**	**0.29**	**1.6**
% recovery		**-**	**105.1**	**88.8**	**103.6**	**99.3**	**106.6**	**99.4**	**96.5**	**114.4**	**130.2**	**-**	**92.9**	**95.8**	**120.1**	**97.6**	**109.3**

Table 2. Quantitative X-ray diffraction (XRD) analyses showing whole-rock mineralogy of the Montney formation, British Columbia, western Canada. Data source: B.C. Oil and gas commission.

Depth (meters)	Formation	Quartz	K Feldspar	Plagioclase	Calcite	Dolomite	Pyrite	Marcasite	Apatite	Kerogen	Total Clay
2233.70	Montney	21.2	4.5	6	32.9	20.7	1.8	0	0	5.4	7.6
2236.40	Montney	15.2	5.2	2.1	52.9	9.8	1.1	0	2.8	8.1	2.9
2238.00	Montney	19.8	5.3	5	50	15.8	1.1	0	0	1.2	1.8
2240.80	Montney	19	4.8	3.4	25	20.3	2.5	0	0	17.6	7.3
2242.50	Montney	16.6	2.9	3.5	43.5	5.9	1.1	0	18.6	6.9	1
2245.60	Montney	23	4.5	5.2	36.1	8.2	2	0	6.5	8.7	5.8
2248.10	Montney	30.9	9.1	7.2	14.4	8.9	2.4	0	3.9	15.9	7.3
2251.50	Montney	24.4	5.5	5.4	31.9	7.7	2.2	0	2	11.3	9.6
2252.60	Montney	24	6.9	5.1	28.8	8.7	1.8	0.3	2.2	13.2	8.9
2259.50	Montney	33.6	13	9	10.7	14.7	1.2	0.6	2.5	11.7	3.1
2260.60	Montney	20.7	3.8	9.1	39.5	18.7	0.9	0	0	3.9	3.3
2262.00	Montney	22.9	6.7	9.5	43	10	1.2	0.6	0	1.5	4.7
2262.70	Montney	49	7.9	12.1	9.7	9.5	1.5	0.6	0	0	9.7
2265.20	Montney	20.6	4.7	7.3	53.1	6.2	1.2	0.3	0	2.4	4.4
2273.00	Montney	41.3	7.9	9.7	8.8	15.9	2	0.7	0.9	3.6	9.4
2279.20	Montney	34.6	9	10.9	9.9	17.4	1.6	0.5	0	6.3	9.8
2281.20	Montney	37.5	7.4	10.1	11.8	12.9	1.5	0.5	0	8.1	10.2
2282.40	Montney	39.9	7.8	9.9	10.4	9.8	1.8	0.4	1	9.1	10
2288.40	Montney	38.7	8.5	10.7	10.4	13.5	1.5	0.4	0	5.8	10.4
2294.60	Montney	44.9	7.9	13.6	9	12.6	1.4	0.4	0	0.1	10.2
2299.40	Montney	31.9	6.3	11.4	25.4	10.5	0.9	1	0	4.6	7.9
2317.80	Montney	33.3	7.1	9.9	2.9	33.1	1.1	0.7	0	2.6	9.4
2318.50	Montney	38.3	8.4	12.5	3	20.3	1.3	0.4	0	3.5	12.1
2323.90	Montney	37.8	7.8	12.5	3.4	21.4	1.2	0.4	0	4.5	11
2330.30	Montney	40.8	7.8	13.4	4.8	15.7	2.2	0.8	0	0.1	14.5
2332.80	Montney	31	6.5	8.6	4.7	31.7	1.5	0	0	6.8	9.2
2341.90	Montney	43.2	6.8	12.3	6.2	14.2	1.1	0.5	0	4.8	10.9
2352.39	Montney	22.5	3.5	8	13.7	41.5	1	0	0	3.4	6.3
2354.30	Montney	41	7.8	14.4	4.8	10.4	1.9	0.4	0	2.1	17.1
2355.94	Montney	42.4	7	10.5	5.3	13.4	1.4	0.4	0	5.3	14.3
2360.50	Montney	39.5	8.2	10.1	5.1	15.1	1.6	0.5	0	6.3	13.6
2366.00	Montney	43.4	6.7	13	6.3	9.8	1.6	0.4	0	5.5	13.2
2370.00	Montney	41	8.4	12.4	5.6	11.6	2.9	0.4	0	9.2	8.5
2370.60	Montney	38.5	9.9	11.2	4.9	11.8	3.2	0.4	0	11.4	8.7
2373.00	Montney	42.4	8.2	12.1	4.2	15.3	1.4	0.4	0	4.8	11.2
2377.00	Montney	22	3.5	6.8	27.6	31.6	0.6	0.3	0	3.1	4.4
2380.00	Montney	38.5	8.2	12.7	6.2	11.1	3	0.5	0	10.4	9.4
2383.15	Montney	34.2	6.3	9.2	4.3	27.6	2	0.3	0.3	6.9	8.8
2387.00	Montney	8.3	1.5	3.3	80.8	2.4	1.2	0	0	1	1.5
2390.20	Montney	45.8	6.7	14.6	4.3	11.4	2.1	0.6	0	0	14.5
2392.00	Montney	10.9	1.2	3.7	68.4	7.7	0.5	0.4	0	1.8	5.4

2394.30	Montney	8.1	1.1	3.2	78.2	6.8	0.6	0	0	0.5	1.6
2396.10	Montney	39.5	3.7	10.7	5.9	17.7	2.3	0.5	0	0	19.8
2400.00	Montney	40.4	4.5	13.8	4.6	12	1.9	0.6	0	3.7	18.4
2404.45	Montney	16.4	2.4	8.7	51	6.1	1	0.4	0	4	9.9
2415.70	Montney	31	3.3	7.7	5.2	23	2	0.8	0	3.1	23.8
2419.60	Montney	43	6.7	15.1	4.7	9.2	1.1	0.5	0	2.7	17
2421.80	Montney	41.4	5.5	12.3	3.6	10.3	1.4	1.1	0	4.1	20.4
2427.90	Montney	45.7	7.3	13.7	5.9	5.9	1.4	0.5	0	3.7	15.9
2428.70	Montney	42.3	3.6	13	8.6	6.2	2	0.6	0	4.6	19.1
2433.30	Montney	47.5	5.7	12.2	4.3	4	1.8	0.7	0	4.5	19.3
2435.40	Montney	43.4	5.3	11.8	4.9	6.2	1.7	0.7	0	4.2	21.8
2442.44	Montney	41.7	3.8	9.4	3.6	13.8	1.8	1.2	0	4	20.7
2447.76	Montney	46.2	5.5	10.8	2.2	4.6	2.2	0.6	0	5	22.9
2449.70	Montney	46.3	4	10.9	2.6	6	2.2	0.6	0	4.2	23
2453.90	Montney	42.7	4.6	10.7	3.5	6.4	2.1	0.6	0	4.5	24.8
2470.80	Montney	43.5	4.1	13	3.3	7	1.7	1.2	0	3.8	22.6
2481.70	Montney	43.5	4	9.3	3.5	7	1.8	1.1	0	5.1	24.8
2482.70	Montney	42.5	3.6	9.4	3.5	11	1.5	1.3	0	3.9	23.3
2486.00	Montney	41.6	3.8	10.5	3.6	7.2	1.7	1.3	0	3.7	26.6
2490.20	Montney	40	4.4	11.1	2.9	8	1.6	1.4	0	4.9	25.7
2492.70	Montney	38.7	4.1	9.7	3.1	16.2	1.7	1.1	0	3.2	22.1
2495.15	Montney	40.5	4.2	8.6	3.4	6.5	2.5	1.2	0	4.2	29
2501.65	Montney	44	2.8	8.7	3.7	6.5	1.9	0.8	0	5	26.6
2504.95	Montney	49.8	4.6	10	3.4	5.8	1.5	0.7	0.3	5.2	19.2
2506.15	Montney	25.1	1.2	4.1	2	54.1	0.6	0.6	0	2	10.3
2510.65	Montney	51.7	3.3	7.7	3.3	5.7	2	0.8	0	5.5	20
2512.55	Montney	47.8	4.3	8.4	3.3	5.4	2.2	0.7	0	5.7	22.1
2517.00	Montney	48.9	3.4	5.1	5.6	9	1.9	0.5	0	6.3	19.2
2522.70	Montney	49.1	3.1	4.4	3.9	6.6	2.5	0.8	0	6.7	22.9
2528.10	Montney	50	3.6	5.3	1	4.5	2.9	0.7	0	5.7	26.2
MINIMUM		**8.1**	**1.1**	**2.1**	**1**	**2.4**	**0.5**	**0**	**0**	**0**	**1**
MAXIMUM		**51.7**	**13**	**15.1**	**80.8**	**54.1**	**3.2**	**1.4**	**18.6**	**17.6**	**29**
AVERAGE		**35.7**	**5.5**	**9.4**	**15**	**12.9**	**1.7**	**0.5**	**0.6**	**5.1**	**13.6**

Feldspar, plagioclase, and pyrite constitute significant component of the Montney Formation. Apatite and marcasite only occur as trace (very low by percent volume). Apatite in particular, is related to the Montney/Doig phosphate zone. To exemplify the mineralogical transformation in the Montney Formation, scanning electron microscopy (SEM) and thin-section petrography shows diagenetic modification of Illite clay, and dissolution of organic matter replaced by pyrite (Figure 8 and Figure 9).

Mineralogy is key to characterization of the constituent composition of a rock, particularly, the present of magnesium and calcite are indication of dolomitization and diagenesis in the Montney Formation. Mica is a major component of siliciclastic sediments, and it is particularly more associated with silts and very fine-grained sediments [47]. Thin-section petrographic examination indicates that the dolomite in Montney Formation appears as detrital fragments (Figure 10)—suggesting that the dolomite may have been transported along with other materials, and may not have been formed in-situ. Similar detritus dolomite in siliciclastic have been reported by [49].

The Montney Formation mineralogy suggests that the sediments may have undergone advanced diagenetic alteration as evidenced by transformation of illite clay to polygorskite clay, dissolution of organic matter and replacement by pyrite (Figure 8(b)), authigenic quartz overgrowth and calcite cementation (Figure 11). X-ray diffraction response of illite (clay mineral) is an indication of diagenesis and low-grade metamorphic history of a sedimentary rock [50] -[54]. Kaolinite is transformed to illite during the process of diagenesis, typically at temperatures of 212°F - 230°F (100°C - 110°C), which lies within the oil generating window [55] , and thus confirms diagenesis in the Montney Formation.

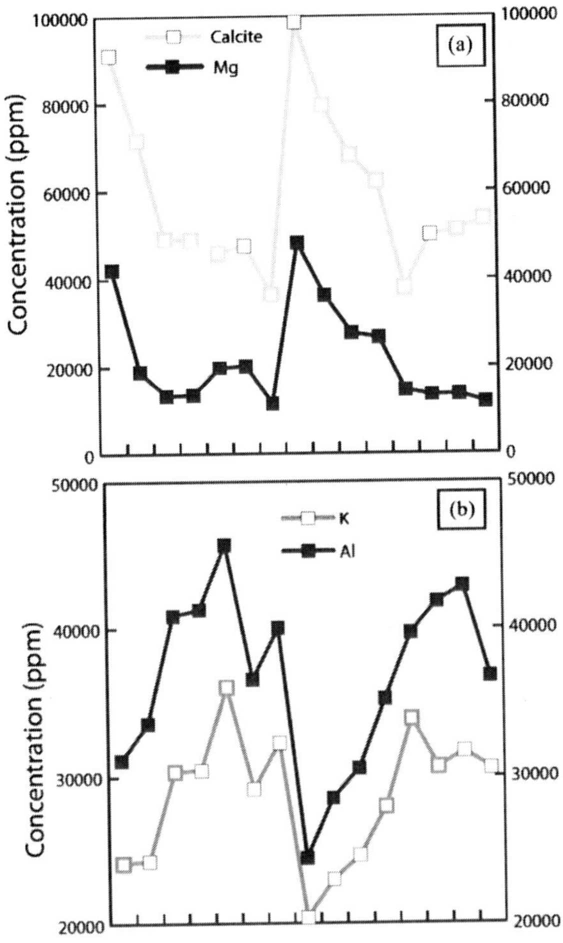

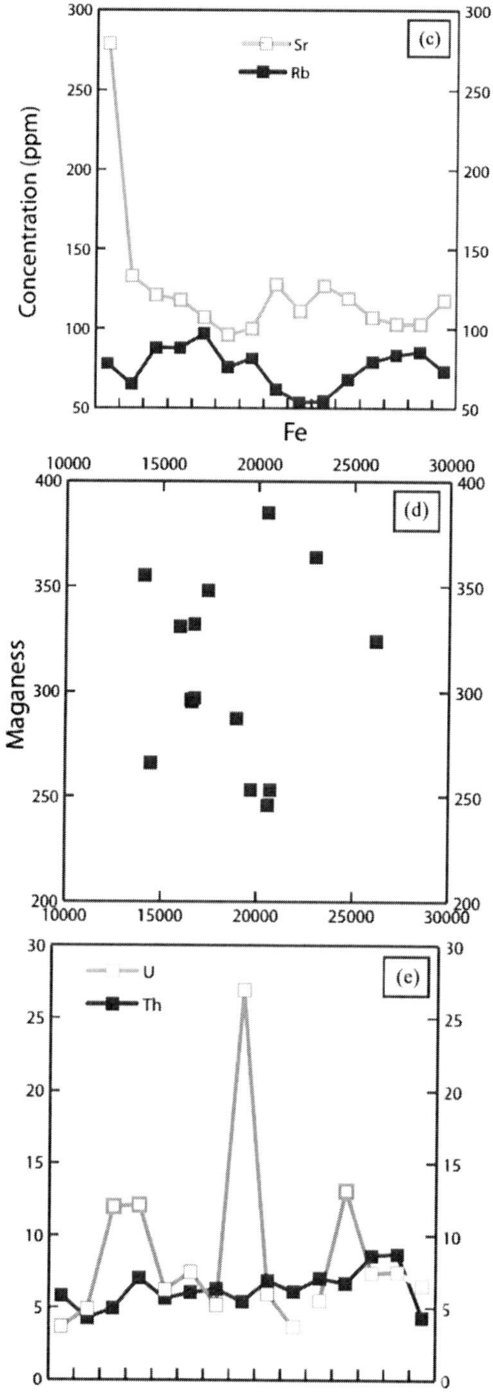
(c)
Sr
Rb
Concentration (ppm)
300
250
200
150
100
50
Fe
10000
15000
20000
25000
30000
(d)
Maganess
400
350
300
250
200
(e)
U
Th
30
25
20
15
10
5
0

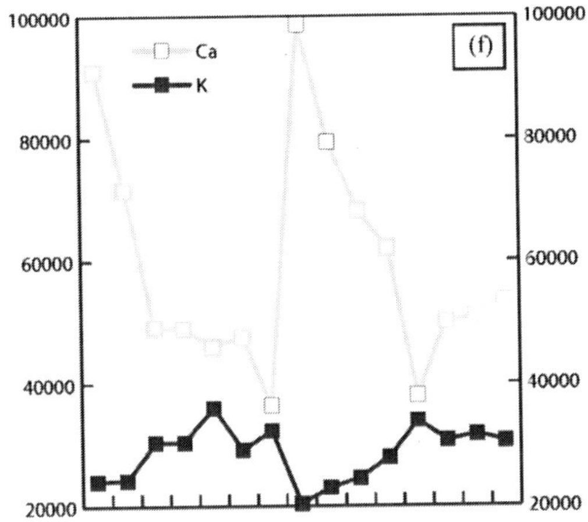

Figure 4. Shows the variation pattern of concentration of chemical elements. (a) Illustrates the composition of dolomite, in which there is higher concentration of magnesium (Mg) relative to calcite (Ca) component. The graph pattern shows good correlation between Mg and Ca. (b) Shows major elements, potassium (K) and aluminum (Al). The concentration of K is very high because of the clay mineral and the organic matter richness of the Montney Formation sediments. The Al concentration is related to the clay mineralogy and partly has affinity to organic matter. (c) Shows alkaline earth metals—Strontium (Sr) and Rubidium (Rb) concentration. (d) Illustrates the concentration of Iron (Fe) and Manganese (Mn). High Fe content and Mn are both related to diagenesis. The concentration of Fe is very high compared to the concentration of Mn. This indicates that Fe has more dominating diagenetic influence in the Montney Formation. (e) Shows the relationship between radioactive elements—Uranium (U) and Thorium (Th). These elements are particularly related to the clay mineralogy and organic matter components of the Montney Formation sediments—source rock kerogen. (f) Shows the relationship between calcite (Ca) and Potassium (K). Evidently, Ca has enormously high concentration due to the stoichiometric co-existence with dolomite.

Pyrite in the Montney Formation is related to post-depositional emplacement [56] caused by the dissolution of organic matter due to diagenesis (Figure 8). Pyrite is an important diagenetic mineral and its occurrence can help define the diagenetic history of sediments [57]. Hence, the occurrence of pyrite in the Montney Formation is a strong evidence of post depositional changes, and interpreted herein to have formed in-situ during diagenesis.

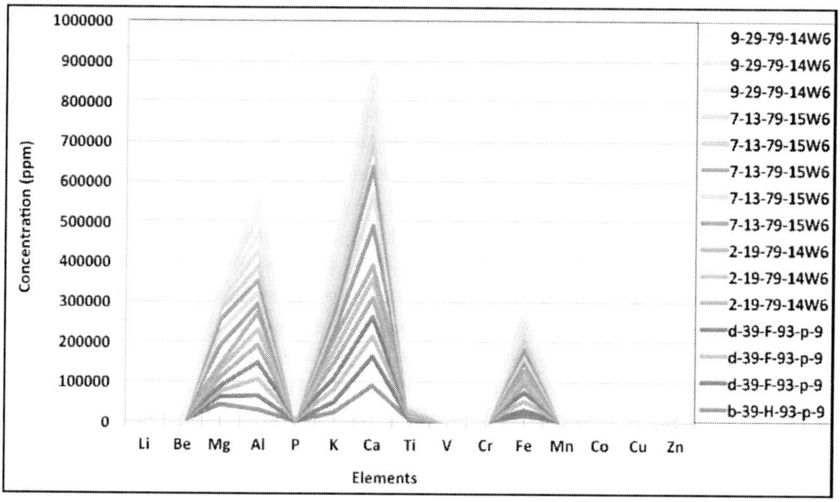

Figure 5. Element concentration showing distinct pattern that reflects the major elements compared to the trace elements with very low concentration.

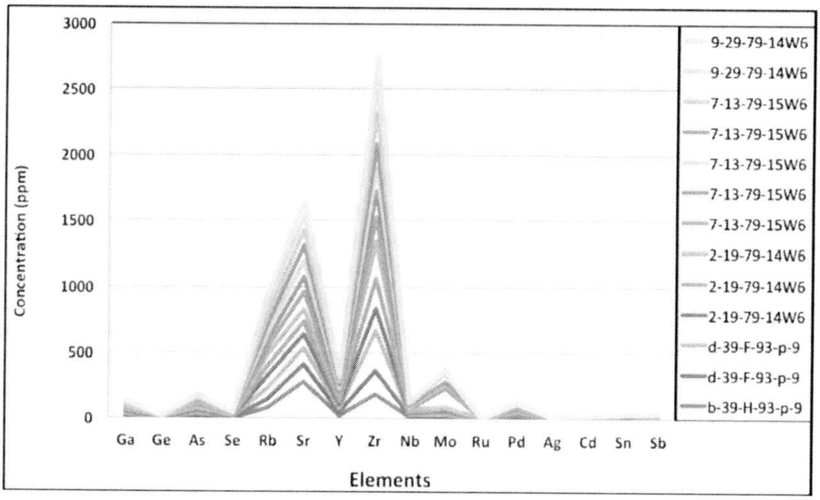

Figure 6. Trace element concentration indicates high values in Rubidium (Rb), Strontium (Sr), and Zirconium (Zr).

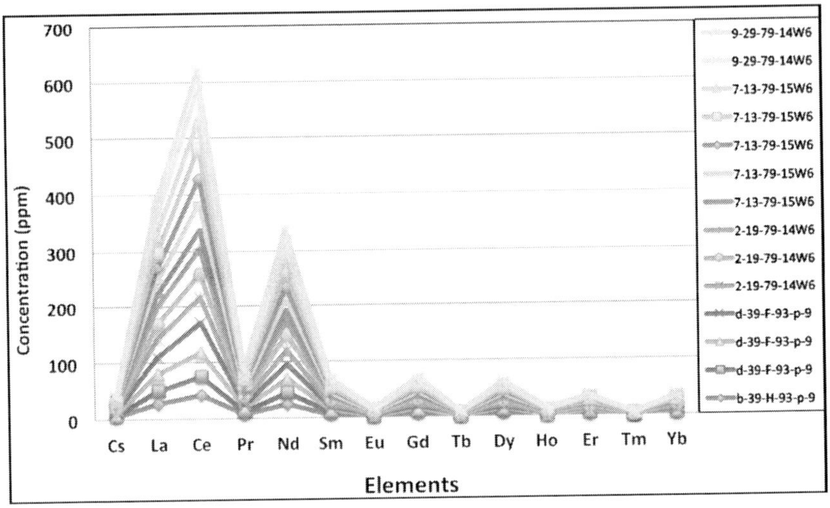

Figure 7. Illustrates the concentration of rare earth elements (REE). The pattern of the graph shows a systematic decrease in their concentration following the pattern in the periodic table with concentration decreasing.

The clay mineral is associated with the organic carbon richness—kerogen component of the Montney Formation source rock. [58] study the relationship between organic matter and shale in modern sediments and found that adsorption of carbon compounds onto clay mineral surfaces played a fundamental role in the burial and preservation of total organic carbon in the sediment. This evidence proved that organic matter are within the smectite inter-layers of clay, which implies that hydrocarbon prone source rock may be closely related to clay mineralogy [58]. The depositional mechanism of silt-sized grains and clay mineral may be related to flocculation of fines due to the affinity of mica mineral as a major component of siliciclastic sediments, which are particularly more associated with silts and very fine-grained sediments [47]. [59] observed that the distance of transport of clay rich sediments affects clay mineral composition, thus they differ from one depositional environment to another (e.g., coastal clay mineral differs from continental slope clay mineral). The observation of [59] shows that chlorite proportion of clay mineral increase from the lower slope to upper slope, corresponding with a decrease in the proportion of smectite. The illite component of the clay generally reaches its maximum on the middle slope [59]. Differential proportion of clay in the different depositional environment may be due to different mineral grain-sizes that characterize different depositional settings, and the different affinities to flocculation effects associated with deposition of hemipelagic [59]. The illitic/palygorskite in the Montney Formation has significant interpretation in relation to petrogenesis. [60] stated that kaolinite tends to flocculate near the river mouths, whereas illite and montmorillonite (smectite) are transported further offshore, thus, supporting a distal depositional environment interpretation for the Montney Formation. [61] - [63] have reported similar depositional environment based on clay mineralogy.

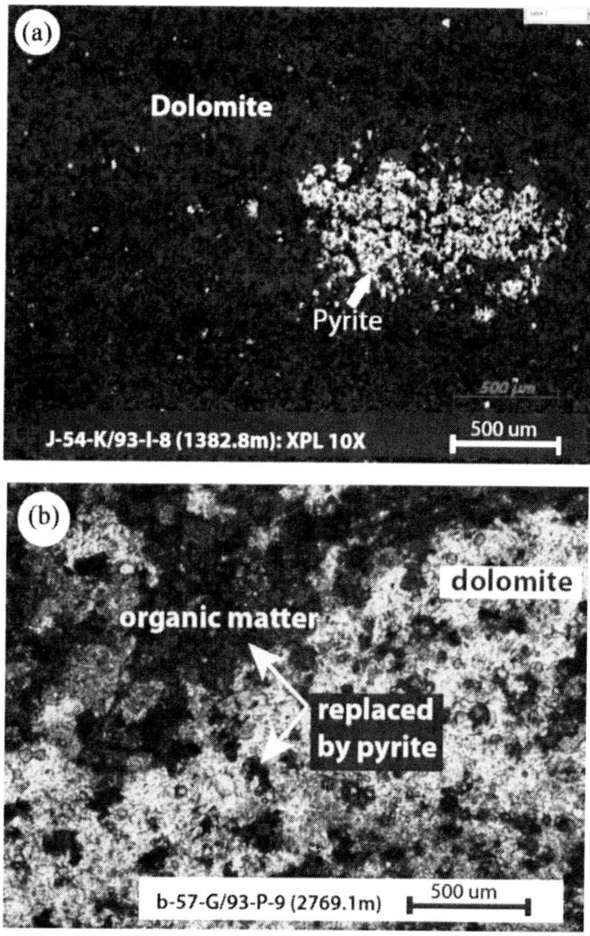

Figure 8. Microphotographs showing thin-section petrography of the Montney Formation illustrating mineralogical composition. (a) Dolomite and pyrite (b) Illustrates pyritized, dolomitized, organic matter rich siltstone. Pyrite replaces organic matter during decomposition of organic matter due to diagenesis.

Figure 9. SEM images showing mineralogy of the Montney Formation. (a) Illustrates a very well formed palygorskite clay—a form of clay mineral growth, formed from the transformation of dickite to Illite, and further transformation of Illite to palygorskite due to severe temperature and pressure in a diagenetic regime. (b) Dolomite and clay mineral.

4.2. Chemical Elements Geochemistry

The major elements, trace elements and Rare Earth Element—REE (Lanthanide) detected in the analyzed samples of the Montney Formation are interpreted based on their concentration, natural abundances in Earth's crust, mode of formation, and affinity with mineralogy. The Montney Formation whole-rock mineralogy (Table 2) as determined by XRD analyses shows average dominance of the minerals: quartz (36.3%), dolomite (15.5%), pyrite (14.1%), calcite (9.4%), K-feldspar (5.6%), marcasite (3.3%), apatite (1.0%), clay (0.7). The abundance by volume of clay minerals (maximum value obtained) indicates that clay is 29% (Table 2).

Detrital minerals such as quartz, feldspar, and clays are a result of the host rock. The Montney Formation host rock is mainly siltstone/mudstone characterized by high concentrations of SiO_2, $CaMg(CO_3)_2$, Al_2O_3, $KALSi_3O_8$, $NaAlSi_3O_8$, $CaCO_3$, Fe_2S_2, TiO_2, K_2O and Na_2O as shown in XRD mineral spectrum. Manganese (Mn) is highly insoluble in oxic environment and is buried as oxyhydroxide particles and coatings of the sediment is mobilized either by reduction of manganese oxyhydroxides [64] , or as a result of solubility equilibria, in which Mn^{+2} then diffuses along a pore-water concentration gradient from anoxic conditions and immobilized by the oxidation of Mn^{+2}resulting in the precipitation of manganese oxyhydroxides [65] - [67]. This process can produce near surface diagenetic manganese contents [68] -[70]. Thus, the present of Mn^{+2} in the Montney Formation indicates association with organic matter in anoxic condition, and the diagenesis associated with the transformation of kerogen during the thermal maturation phase of source rock.

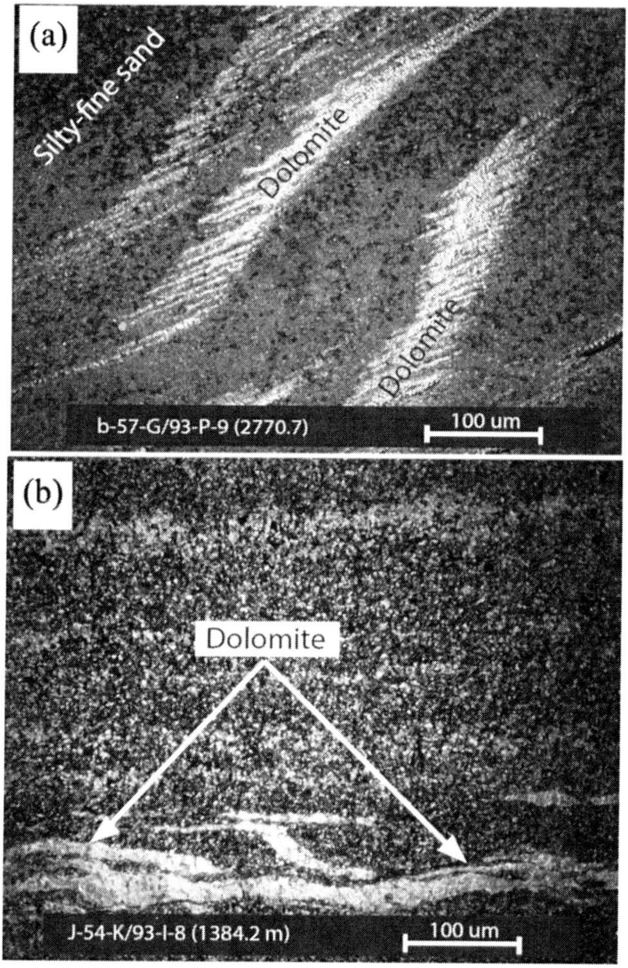

Figure 10. Photomicrographs showing dolomitic silty-sandstone. (a) Shows detrital dolomite resembles ripple. (b) shows detrital dolomite admixed siltstone.

The relatively high amounts of uranium (U), potassium (K), arsenic (As) and barium (Ba) detected in the samples are related to the Montney Formation organic matter richness and high phosphate content in the upper Montney/Doig Formation boundary. These elements (U, K, AS, Ba) are mainly associated with organic matter due to several depositional factors in phosphatic rocks, and are comparable to those detected in other phosphate worldwide [71]. This explains the phosphatic and diagenetic phenomenon in the Upper Montney/Doig formations boundary as evident by the present of apatite mineral in the Montney Formation.

The trace element Gallium (Ga) has chemical affinity with Aluminum (Al) thus its enrichment is attributed to the presence of alluminosilicate minerals [72]. Aluminum make up about 8% of the Earth's solid surface, and it is the third most abundant element in the Earth's crust, second only to oxygen and silicon [73]. The formation of aluminum has been attributed to several mechanisms associated with reducing conditions, and aluminum is usually combined with other elements such as oxygen, silicon or fluorine to form different compounds [73]. Native metals like aluminum can form in reducing oceanic or mud volcanic environments, which are associated with upward migration of basaltic magma [74] , hydrothermal activity [75] , magmatic or metamorphic process [76] , or high temperature hydrocarbon-enriched fluids [77]. However, aluminum can also be formed by exolution of metal-rich fluids in the magma [78] and degassing of magmatic vapors during submarine eruptions, which may lead to reduction of some elements in their metallic state [79]. Other mechanisms for aluminum formation includes: 1) gas condensate from metalliferous fluids together with fairly other dry reducing gasses [74] [75] [80] ; 2) reduction of carbon compounds [81] ; 3) reduction by intratelluric fluids and reaction of aluminum chloride ($AlCl_3$) with H_2 [91]). Aluminum has affinity to clays, and because clay mineral is a component of the Montney Formation, thus, the high content of aluminum in the samples analyzed.

Lithium (Li) concentration is attributed to the presence of mica minerals, whereas, Nickel (Ni), Copper (Cu), Chromium (Cr), and Vanadium (V) enrichments can be attributed to the presence of minerals related to mafic-ultramafic rocks such as vermiculite and chlorite [82] [72]. Copper (Cu) is critical to the behavior of the elements associated with biological processes in the sedimentary cycle [83]. Copper (Cu) is associated with clay mineral fractions, especially those rich in coatings containing organic carbon and manganese oxides [2].

The high amounts of Zirconium (Zr) are related to Hafnium (Hf), and are in general fixed in silicate minerals [72]. Rubidium (Rb) is most likely related to the potassium feldspars contained in the clay component of the Montney Formation. Enrichments in elements such as Barium (Ba), Lead (Pb), Yttrium (Y) was reported by [82] as evidence of diagenetic alterations of the sediments from lagoonal environment. Thus, their presence in the Montney Formation is interpreted as evidence of diagenesis. The element Barium (Ba) is related to biogenic compounds, it is suggested that Barium in the form of Barite was adsorbed by planktonic organisms from seawater and sunk down to the bottom of the sea [71]. The present of Barium (Ba) is associated with the high organic matter richness of the Montney Formation. Barium (Ba) typically migrates and is redeposited along with phosphates

[84]. The Montney and Doig formations boundary have very high phosphate concentration, which contributes organic carbon content.

Iron (Fe) hydroxides seem to be the main sinks for Lead (Pb), Cadmium (Cd), and Zinc (Zn), with less important role by organic matter [85]. Cadmium (Cd) is generally found where there is high concentration of organic matter such as muddy deposit associated with benthic layers [86]. Iron (Fe) and Aluminum (Al) contents provide useful information on the clay, contamination and sedimentation style [87]. Most Iron (Fe) is structurally bound by clay minerals. Fe is also an indication of diagenesis [56].

Boron abundance in carbonates provides a powerful proxy for seawater pH reconstruction [87]. The Boron (B) system in oceanic carbonates is a geochemical tool for paleo-seawater acidity reconstruction and indication of diagenesis [88] [89]. The relatively high concentration of Boron (B) in the Montney Formation sediments implies diagenesis. The work of [90] [91]; and [92] link main sources of oceanic Boron to riverine discharge and ridge hydrothermal fluid expulsion from accretion prisms. The main Boron sinks in the ocean are oceanic crust alteration, adsorption on sediments and co-precipitation in carbonates [90] [93] [94].

The diagenetic behavior of phosphorus (P) in marine sediments is not well documented [64]. However, phosphate is produced during oxidation of organic matter in the sediment by the dissolution of inorganic particulate phosphates, and by the liberation of phosphate associated with iron oxyhydroxides following their reduction to Fe^{+}. [64] stated that it is inferred that phosphate is precipitated in the sediment by unspecified inorganic diagenetic process. Phosphorous does preferentially remain in solution in reducing conditions because under oxic conditions it is bound to iron [95]. Therefore, the behavior of phosphorous can be related to the dissolution and precipitation of iron compounds in response to redox-driven processes [64].

The behaviors of cobalt (Co), copper (Cu) and nickel (Ni) are closely related to manganese during diagenesis [66] [96] [97] [99]. Pore-water data from hemipelagic sediments indicate that both cobalt and nickel reach dissolved maxima in manganese-reduction zone and are precipitated along with manganese-oxidation zone [100].

Cobalt and nickel are progressively depleted in pore-waters of the iron-reduction zone, indicating uptake by the sediment under reducing condition [64]. Pore-water from pelagic and hemipelagic sediments indicate that copper is mobilized from labile solid phases, which is predominantly related to organic matter decomposition during diagenesis [97] -[99]), and reaches a dissolved maximum immediately below the sediment-water interface [64]. In addition, copper is mobilized in the manganese-reduction zone, where it is co-precipitated with manganese in sediments [97]. Trace elements such as La, Ce, Nd, Y, Th, Zr, Hf, Nb, Ti and Sc are mostly used for interpreting provenance and tectonic setting because of their relatively low mobility during sedimentary processes, and their low residence time in seawater [1] [101]. These elements are transported quantitatively into clastic sedimentary rocks during transport of weathered materials, and thus reflect the signature of the parent-source rock [101]. The present of La, Ce, Nd, Y, Th, Zr, Hf, Nb, Ti and Sc elements in the Montney Formation may infer somewhat tectonic stability. Similar interpretation based on the present of La, Ce, Nd, Y, Th, Zr, Hf, Nb, Ti and Sc was documented by [101].

5. DISCUSSIONS

Ever since Dolomieu [102] first introduced the term 'Dolomite' into the geological literature to describe a calcareous rock in the mountain range of northwestern Italy where the rock 'dolomite' was first identified, the calcium magnesium carbonate $CaMg(CO_3)_2$ rock continue to be a subject of vast interest in geology owing to: 1) mode of formation of the mineral dolomite; 2) diagenesis, and implications of dolomitization in reservoir rocks; and 3) its usefulness as paleotemperature indicator, or as paleothermometer.

Dolomite is a rhombohedral carbonate with the ideal formula $CaMg(CO_3)_2$, in which calcium and magnesium occupy preferred sites [103] [104] ; [105] used hydrothermal experiments extrapolated to low temperatures to demonstrate that calcite and dolomite are essentially ideal in composition at 25°C. Thus, any double carbonate crystal of Ca and Mg at 25°C is not essentially pure dolomite, and is either metastable or unstable with respect to calcite [103]. Diagenetic modification of the very fine-grained, silty-sandstone of the Montney Formation may have occurred in stages of progressive oxidation and reduction reactions involving chemical element such as Fe, which manifest in mineral form as pyrite (Figure 8), particularly, during early burial diagenesis, or late stage diagenesis. Similar interpretation was documented by [106]. Mineralogical changes in the form of cementation and mineral replacement involving calcite and dolomite are typical of diagenesis [107] , and are evident in the Montney Formation based on petrographic study and SEM analysis.

Oxidation and reduction reaction mechanisms are crucial in the modification of sediments, shortly after burial, prior to lithification or compaction during which fluids are ejected into the depositional interface [108] [109]. This drives the oxidation and reduction processes involving Fe, sulfur, and carbon [107]. The significant amounts of organic matter (TOC) in the Montney Formation essentially make these elements principal reactants. The carbon compound of the organic matter content is the most rapidly oxidized and consequently contributing energy to drive the Fe into the ferrous state, thereby causing fixation of sulfur as pyrite. Because of the present of organic matter in the Montney sediments, pyrite occurs as scattered "clots" throughout the rock samples—this phenomenon is evident in thin-section petrography (Figure 8). Pyrite is related to post-depositional emplacement [56] caused by the dissolution of organic matter due to diagenesis (Figure 8).

The quantitative analysis of mineralogy (Table 2), chemical elements and their concentration (Table 1) in the Montney Formation, particularly, Ca and Mg show incredibly high concentration owing to mineral precipitation and dolomitization. It is interpreted thus, that calcite may have been precipitated into the interstitial pore space of the intergranular matrix of the very fine-grained silty-sandstone of the Montney Formation as cement by a complex mechanism resulting in the interlocking of grains, welded together by calcite cement. Evidence of grain interlocking is revealed by SEM image showing authigenic quartz overgrowth (Figure 11). It is established through mineralogical composition in this study (Table 2) that the Montney Formation is quartz rich and contains substantial clay minerals as well, including unstable mineral such as feldspar (Table 2). This sort of compositional mixture of quartz, clay and feldspar minerals may have resulted in the decomposition of feldspar along the clay-quartz boundary due to processes involving hydrolysis in the process of diagenesis.

The depositional environment interpreted for the Montney Formation is proximal to distal offshore marine setting [35] [19] [20]. In the marine environment, several weathering and transformation are prevalent. As a result, there exist an exchange of cations, in which the positions of clay minerals are changed, thereby resulting in the substitution of Mg for Ca [103] [105] [110]. In the marine environment, chlorite and illite minerals are formed by fixation of Mg and K in montmorillonite or degraded illite delivered into distal offshore settings due to continental denudation resulting from fluvial processes [110].

The present of illite and palygorskite clay mineral (Figure 9) support the evidence of diagenesis in the Montney Formation formed from ionic solutions in extreme conditions of temperature and pressure. [50] noted that the transformation of clay minerals into illite involves several factors such as preferential flocculation, current sorting of floods (hyperpycnal or hypopycal) intensities, and variations in the composition and concentration of detritus during sedimentation and subsequently enhance clay transformation.

The dolomite in the Montney Formation appears as debris in thin-section petrography (Figure 10). Such allogenic (non in-situ) dolomite may have only played a role (minimal) in the Montney Formation diagenesis, compared to sediments that are mainly composed of biogenic carbonate that are formed completely from aragonite and calcite. Thus, authigenic calcite may have played dominant role in the vast diagenetic phenomenon of the Triassic Montney Formation in northeastern British Columbia based on the evidence of the most dolomitized interval, which have the highest percentage of calcite composition (Table 1)—shaded in black color.

6. CONCLUSIONS

Whole-rock chemical element geochemistry provides a means of understanding the chemical, physical, and geological forces that continually redistributes elements between living and non-living reservoirs through bio-geochemical cycles [4]. The understanding of the variability and affinity of chemical elements and mineralogy in rocks are invaluable, which aid identification and interpretation of chemical elements and mineralogical signatures, thereby serving as a means of quantification, calibration, and correlation in sedimentary basins.

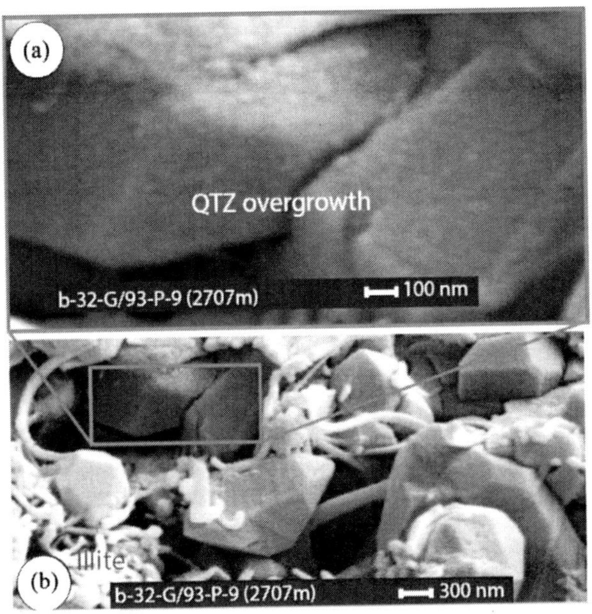

Figure 11. SEM images shows evidence of diagenesis. (a) Insect of Plate B showing authigenic quartz overgrowth due to diagenesis. (b) Shows quartz overgrowth (red rectangle), illite clay and quartz mineral.

Results from this study show that chemical elements such as major, trace and rare earth element (REE) concentrations in the Triassic Montney Formation sediments (very fine-grained, dolomitized silty-sandstone) are related to the host rock mineralogy. The variation patterns of some of the major elements (Al, Fe, K, Mg and Mn) and trace elements (Sc, V, Co, Cr, Zn, Ba, Th, U, Rb and Cs), show strong affinity to the host rock mineralogy. This evidence is supported by quantitative XRD mineralogical analyses (Table 2). Mineralogy is key to characterization of the constituent composition of a rock, hence, mineralogical analysis from this study helped to understand dolomitization and diagenesis of the Montney Formation in northeastern British Columbia, Western Canada Sedimentary Basin (WCSB).

ACKNOWLEDGEMENTS

B.C. Oil and Gas Commission is thanked for providing the Montney Formation well data, lab reports, and for granting the authorization to use the data for publication in a Journal. Special thanks is accorded to Hardy Friedrich, Manager of Communications at the B.C. Oil and Gas Commission, who communicated the authorization to use the data for publication. Darcie Wetland with "Wells-Files" department at B.C. Oil and Gas Commission is thanked for the timely provision of the data and numerous support during data retrieval. The Geoscience B.C is thanked for providing funding for this research. Thanks to GuangCheng Chen for using the ICP-MS machine at the University of Alberta to determine chemical elements in the samples used in this study. Dee-Ann Rollings is thanked for her assistance with SEM image acquisition. The University of Alberta Faculty of Graduate Studies and Research is thanked for providing conducive academic and research environment that enhanced the success of this work. Professors Murray Gingras and J-P Zonneveld are thanked for providing the financial resource for field work and lab analyses. The Editor is thanked for the recommendation that greatly improved the quality of this paper.

REFERENCES

1. Bowen, H.J.M. (1966) Trace Elements in Biochemistry. Academic Press, London, 241 p.
2. Callender, E. (2004) Heavy Metals in the Environment-Historical Trends. In: Lollar, B.S., Ed., Treatise on Geochemistry, Environmental Geochemistry, US Geological Survey, Westerly, 68-105.
3. Cooper, H.K., Duke, M.J.M., Simonetti, A. and Chen, G.C. (2007) Trace Element and Pb Isotope Provenance Analyses of Native Copper in Northwestern North America: Results of a Recent Pilot Study Using INAA, ICP-MS, and LA-MC-ICP-MS. Journal of Archaeological Science, 35, 1732-1747.

4. Summons, R.E. (1993) Biogeochemical Cycles: A Review of fundamental Aspects of Organic Matter Formation, Preservation, and Composition. In: Engel, M.H. and Macko, S.A., Eds., Organic Geochemistry, Plenum Press, New York, 3-21. http://dx.doi.org/10.1007/978-1-4615-2890-6_1
5. Bhatia, M.R. and Crook, K.A.W. (1986) Trace Element Characteristics of Graywackes and Tectonic Setting Discrimination of Sedimentary Basins. Contributions to Mineralogy and Petrology, 92, 181-193. http://dx.doi.org/10.1007/BF00375292
6. Kasemann, S.A., Schmidt, D.N., Bijmaand, J. and Foster, G.L. (2008) In Situ Boron Isotope Analysis in Marine Carbonates and Its Application for Foraminifera and Palaeo-pH. Chemical Geology, 260, 138-147. http://dx.doi.org/10.1016/j.chemgeo.2008.12.015
7. Leinen, M. and Pisias, N. (1984) An Objective Technique for Determining End-Member Compositions and for Partitioning Sediments According to Their Sources. Geochimica et Cosmochimica Acta, 48, 47-62. http://dx.doi.org/10.1016/0016-7037(84)90348-X
8. Dymond, J., Suess, E. and Lyle, M. (1992) Barium in Deep-Sea Sediment: A Geochemical Proxy for Paleoproductivity. Paleoceanography, 7, 163-181. http://dx.doi.org/10.1029/92PA00181
9. Murray, R.W. and Leinen, M. (1993) Chemical Transport to the Seafloor of the Equatorial Pacific across a Latitudinal Transect at 135 W; Tracking Sedimentary Major, Trace, and Rare Earth Element Fluxes at the Equator and the ITCZ. Geochimica et Cosmochimica Acta, 57, 4141-4163. http://dx.doi.org/10.1016/0016-7037(93)90312-K
10. Nesbitt, H.W. and Young, G.M. (1982) Early Proterozoic Climate and Plate Motions Inferred from Major Element Chemistry of Lutites. Nature, 299, 715-717. http://dx.doi.org/10.1038/299715a0
11. Kronberg, B.I., Nesbitt, H.W. and Lam, W.W. (1986) Upper Pleistocene Amazon Deep-Sea Fan Muds Reflect Intense Chemical Weathering of Their Mountainous Source Lands. Chemical Geology, 54, 283-294. http://dx.doi.org/10.1016/0009-2541(86)90143-9
12. Edwards, D.E., Barclay, J.E., Gibson, D.W., Kvill, G.E. and Halton, E. (1994) Triassic Strata of the Western Canada Sedimentary Basin. In: Mossop, G.D. and Shetson, I., Eds., Geological Atlas of the Western Canada Sedimentary Basin, Canadian Society of Petroleum Geologists, 257-275.
13. Podruski, J.A., Barclay, J.E., Hamblin, A.P., Lee, P.J., Osadetz, K.G., Procter, R.M. and Taylor, G.C. (1988) Conventional Oil Resources of Western Canada. Geological Survey of Canada Paper, 149 p.
14. Gibson, D.W. and Edwards, D.E. (1990) An Overview of Triassic Stratigraphy and Depositional Environments in the Rocky Mountain Foothills and Western Interior Plains, Peace River Arch Area. Bulletin of Canadian Petroleum Geology, 38A, 146-158.
15. Zonneveld, J.P., MacNaughton, R.B., Utting, J., Beattty, T.W., Pemberton, S.G. and Henderson, C.M. (2010) Sedimentology and Ichnology of the Lower Triassic Montney Formation in the Pedigree-Ring/Border-Kahntah River Area, Northwestern Alberta and Northeastern British Columbia. Bulletin of Canadian Society of Petroleum Geology, 58, 115-140. http://dx.doi.org/10.2113/gscpgbull.58.2.115
16. Davies, G.R., Moslow, T.F. and Sherwin, M.D. (1997) The Lower Triassic Montney Formation, West-Central Alberta. Bulletin of Canadian Petroleum Geology, 45, 474-505.
17. Orchard, M.J. and Tozer, E.T. (1997) Triassic Conodont Biochronology, Its Calibration with the Ammonoid Standard, and a Biostratigraphic Summary for Western Canada Sedimentary Basin. Bulletin of Canadian Petroleum Geology, 45, 675-692p.
18. Zonneveld, J.P., Gingras, M.K. and Pemberton, S.G. (2001) Trace Fossils Assemblages in a Middle Triassic Mixed siliciclastic-Carbonate Marginal Marine Depositional System, British Columbia. Palaeogeography, Palaeoclimatology, Palaeoecology, 166, 249-276. http://dx.doi.org/10.1016/S0031-0182(00)00212-1
19. Egbobawaye, E.I., Kravchinsky, V.A., Zonneveld, J.-P. and Gingras, M.K. (2011) Magnetostratigraphy Dating and Correlation of the Lower Doig and Upper Montney Formations (Lower Triassic), Northeastern British Columbia, Western Canada. Canadian Society of Petroleum Geologists, CSPG, CSEG, CWLS Convention Abstract, Calgary, Alberta, Canada, 1-5.

20. Egbobawaye, E.I. (2013) Tight Gas Reservoir Characterization in Montney Formation, Northeastern British Columbia, Western Canada. Doctor of Philosophy Thesis, Department of Earth and Atmospheric Sciences, University of Alberta, Edmonton, 431 p.
21. McLean, F.H. and Kindle, E.D. (1950) Geology of Northeast British Columbia. Geological Survey of Canada, Memoir, 259.
22. Hunt, A.D. and Ratcliffe, J.D. (1959) Triassic Stratigraphy, Peace River Area, Alberta and British Columbia, Canada. AAPG Bulletin, 43, 563-589.
23. Armitage, J.H. (1962) Triassic Oil and Gas Occurrences in Northeastern British Columbia, Canada. Journal of the Alberta Society of Petroleum Geologists, 10, 35-56.
24. Gibson, D.W. and Barclay, J.E. (1989) Middle Absaroka Sequence, the Triassic stable Craton. In: Ricketts, B.D., Ed., Western Canada Sedimentary Basin: A Case History, Canadian Society of Petroleum Geologists, 219-231.
25. Gibson, D.W. and Edwards, D.E. (1990) An Overview of Triassic Stratigraphy and Depositional Environments in the Rocky Mountain Foothills and Western Interior Plains, Peace River Arch Area. Bulletin of Canadian Petroleum Geology, 38A, 146-158.
26. Moslow, T.F. and Davies, G.R. (1992) Triassic Reservoirs Facies and Exploration Trends: Western Canada Sedimentary Basin, Canadian Society of Petroleum Geologists. American Association of Petroleum Geologists, Short Course, Calgary, Alberta, 166 p.
27. Henderson, C.M. (1997) Uppermost Permian Conodont and Permian-Triassic Boundary in the Western Canada Sedimentary Basin. Bulletin of Canadian Petroleum Geology, 45, 693-707.
28. Paull, R.K., Paull, R.A. and Laudon, T.S. (1997) Conodont Biostratigraphy of the Lower Triassic Mackenzie Dolomite Lentil, Sulphur Mountain Formation in the Cadomin Area, Alberta. Bulletin of Canadian Petroleum Geology, 45, 708-714.
29. Richards, B.C. (1989) Upper Kaskaskia Sequence: Uppermost Devonian and Lower Carboniferous. In: Ricketts, B.D., Ed., Western Canada Sedimentary Basin—A Case History, Canadian Society of Petroleum Geologists, Special Publication No. 30, Calgary, Alberta, 165-201.
30. Barclay, J.E., Krause, F.F., Campbell, R.I. and Utting, J. (1990) Dynamic Casting and Growth Faulting: Dawson Creek Graben Complex, Carboniferous-Permian Peace River Embayment, Western Canada. Bulletin of Canadian Petroleum Geology, 38A, 115-145.
31. O'Connell, S.C., Dix, G.R. and Barclay, J.E. (1990) Origin, History, and Regional Structural Development of the Peace River Arch, Western Canada. Bulletin of Canadian Petroleum Geology, 38A, 4-24.
32. Walsh, W., Adams, A., Kerr, B. and Korol, J. (2006) "Regional Shale Gas" Potential of the Triassic Doig and Montney Formations, Northeastern British Columbia. Ministry of Energy, Mines and Petroleum Resources, Oil and Gas Division, Resource Development and Geoscience Branch, British Columbia, Petroleum Geology Open File 2006-02, 19 p.
33. Faraj, B., Williams, H., Addison, G., McKinstry, B., Donaleshen, R., Sloan, G., Lee, J., Anderson, T., Leal, R., Anderson, C., Lafleur, C. and Ahlstrom, J. (2002) Shale Potential of Selected Upper Cretaceous, Jurassic, Triassic and Devonian Shale Formations, in the WCSB of Western Canada: Implications for Shale Gas Production. A Report for Gas Technology Institute (GTI), Des Plains, Illinois, USA.
34. Reidiger, C.L., Fowler, M.G., Brooks, P.W. and Snowdon, L.R. (1990) Triassic Oils and Potential Mesozoic Source Rocks, Peace River Arch Area, Western Canada Basin. Organic Geochemistry, 16, 295-305. http://dx.doi.org/10.1016/0146-6380(90)90049-6
35. Egbobawaye, E.I., Zonneveld J.-P. and Gingras, M.K. (2010) Tight Gas Reservoir Evaluation in Montney and Lower Doig Formations, Northeastern British Columbia, Western Canada. American Association of Petroleum Geologists Abstract, 19, 66-67.
36. Orchard, M.J. and Zonneveld, J.-P. (2009) The Lower Triassic Sulphur Mountain Formation in the Wapiti Lake Area: Lithostratigraphy, Conodont Biostratigraphy, and a New Biozonation for the Lower Olenekian (Smithian). Canadian Journal of Earth Sciences, 46, 757-790. http://dx.doi.org/10.1139/E09-051
37. Habicht, J.K.A. (1979) Paleoclimate, Paleomagnetism, and Continental Drift. American Association of Petroleum Geologists Studies in Geology, 9, 1-29.

38. 38. Tozer, E.T. (1982) Marine Triassic Faunas of North America: Their Significance for Assessing Plate and Terrane Movements. Geologische Rundschau, 71, 1077-1104. http://dx.doi.org/10.1007/BF01821119
39. Arnold, K.J. (1994) Origin and Distribution of Eolian Sandstones in the Triassic Charlie Lake Formation, Northeastern British Columbia. Unpublished M.Sc. Thesis, University of Alberta, Edmonton, 320 p.
40. Porter, J.W., Price, R.A. and McCrossan, R.G. (1982) The Western Canada Sedimentary Basin. Philosophical Transactions of the Royal Society of London, A305, 169-182. http://dx.doi.org/10.1098/rsta.1982.0032
41. Price, R.A. (1979) Intracontinental Ductile Crustal Spreading Linking the Fraser River and Northern Rocky Mountain Trench Transform Fault Zones. Geological Society of America, Programs with Abstracts, 11, 499.
42. Monger, J.W.H. and Price, R.A. (1979) Geodynamic Evolution of the Canadian Cordillera—Progress and Problems. Canadian Journal of Earth Sciences, 16, 770-791. http://dx.doi.org/10.1139/e79-069
43. Wittenberg, J. and Moslow, T.F. (1991) Origin and Variability of Over Thickened Sandstone in the Doig Formation, West-Central Alberta. Canadian Society of Petroleum Geologists, Calgary, 146.
44. Cant, D.J. (1986) Hydrocarbon Trapping in the Halfway Formation (Triassic), Wembley Field, Alberta. Bulletin of Canadian Petroleum Geology, 34, 329-338.
45. Berger, Z., Boast, M. and Mushayandebvu, M. (2009) Basement Structures Control on the Development of the Peace River Arch's Montney/Doig Resource Plays. Canadian Society of Petroleum Geologists, 36, 40-45.
46. Jarvis, I. and Jarvis, K.E (1992) Plasma Spectrometry in the Earth Sciences: Techniques and Future Trends. Chemical Geology, 95, 1-33. http://dx.doi.org/10.1016/0009-2541(92)90041-3
47. Tucker, M., Ed. (1988) Techniques in Sedimentology. Blackwell Scientific Publication, 393 p.
48. Sorby, H.C. (1858) On the Microscopic Structure of Crystals, Indicating the Origin of Minerals and Rocks. Quarterly Journal of the Geological Society of London, 14, 453-500 http://dx.doi.org/10.1144/GSL.JGS.1858.014.01-02.44
49. Bone, Y., James, N.P. and Kyser, T.K. (1992) Synsedimentary Detrital Dolomite in Quaternary Cool-Water Carbonate Sediments, Lacepede Shelf, South Australia. Geology, 20, 109-112. http://dx.doi.org/10.1130/0091-7613(1992)020<0109:SDDIQC>2.3.CO;2
50. Weaver, C.E. (1959) The Clay Petrology of Sediments. Clays Mineral, Proc. Natural Conference, Minerals, 6, 154-187.
51. Weaver, C.E. (1960) Possible Uses of Clay Minerals in the Search for Oil. AAPG Bulletin, 44, 1505-1518. http://dx.doi.org/10.1016/b978-0-08-009351-2.50024-6
52. Frey, M. (1970) The Step from Diagenesis to Metamorphism in Pelitic Rocks during Alpine Orogenesis. Sedimentology, 15, 261-279. http://dx.doi.org/10.1111/j.1365-3091.1970.tb02189.x
53. Gill, W.D., Khalaf, F.E. and Massoud, M.S. (1977) Clay Minerals as an Index of the Degree of Metamorphism of the Carbonate and Terrigeneous Rocks in the South Wales Coalfield. Sedimentology, 39, 330-332.
54. Stalder, P.J. (1970) Organic and Inorganic Metamorphism in the Taveyyannaz Sandstone of Swiss Alps and Equivalent Sandstones in France and Italy. Journal of Sedimentary Petrology, 49, 463-482.
55. Burst, J.F. (1969) Diagenesis of Gulf Coast Clayey Sediments and Its Possible Relation to Petroleum Migration. American Association of Petroleum Geologists Bulletin, 53, 73 93.
56. England, G.L., Rasmussen, B., Krapez, B. and Groves, D.I. (2002) Palaeoenvironmental Significance of Rounded Pyrite in Siliciclastic Sequences of the Late Archean Witwatersrand Basin: Oxygen-Deficient Atmosphere or Hydrothermal Alteration? International Association of Sedimentologists, 49, 1133-1156.
57. Hudson, J.D. (1982) Pyrite in Ammonite-Bearing Shales from the Jurassic of England and Germany. International Association of Sedimentologists, 37, 639-667. http://dx.doi.org/10.1111/j.1365-3091.1982.tb00072.x

58. Kennedy, M.J., Pevear, D.R. and Hills, R.J. (2002) Mineral Surface Control of Organic Carbon in Black Shale. Science, 295, 657-660. http://dx.doi.org/10.1126/science.1066611
59. 59. Raitzsch, M., V?lker, D. and Heubeck, C. (2007) Neogene Sedimentary and Mass Wasting Processes on the Continental Margin off South-Central Chile Inferred from Dredging Samples. Marine Geology, 244, 166-183. http://dx.doi.org/10.1016/j.margeo.2007.06.007
60. Einsele, G. (2000) Sedimentary Basins: Evolution, Facies, and Sediment Budget. Springer-Verlag, Berlin, Heidelberg, 791 p. http://dx.doi.org/10.1007/978-3-662-04029-4
61. Carson, B. and Arcaro, N.P. (1983) Control of Clay Mineral Stratigraphy by Selective Transport in Late Pleistocene-Holocene Sediments of Northern Cascadia Basin—Juan de Fuca Abyssal Plain: Implication for Studies of Clay Mineral Provenance. Journal of Sedimentary Petrography, 53, 395-406.
62. Gibbs, R.J. (1985) Settling Velocity, Diameter, and Density for Flocs of Illite, Kaolinite and Montmorillonite. Journal of Sedimentary Petrography, 55, 65-68.
63. Lamy, F., Hebbeln, D. and Wefer, G. (1998) Terrigeneous Sediment Supply along the Chilean Continental Slope: Modern Latitudinal Trends of Texture and Composition. Geologische Rundschau, 87, 477-494. http://dx.doi.org/10.1007/s005310050223
64. Jarvis, I. and Higgs, N. (1987) Trace Element Mobility during Early Diagenesis in Distal Turbidites: Late Quaternary of the Madeira Abyssal Plain, N. Atlantic. Geological Society, London, Special Publications, 31, 179-214. http://dx.doi.org/10.1144/ GSL.SP. 1987.031.01.14
65. Froelich, P.N., Klinkhammer, G.P., Bender, G.R., Luedtke, N.A., Heath, G.R., Cullen, D., Dauphine, P., Hammond, D. and Hartman, B. (1979) Early Oxidation of Organic Matter in Pelagic Sediments of the Eastern Equatorial Atlantic: Suboxic Diagenesis. Geochimica et Cosmochimica Acta, 43, 1075-1090. http://dx.doi.org/10.1016/0016-7037(79)90095-4
66. Klinkhammer, G.P. (1980) Early Diagenesis in Sediments from the Eastern Equatorial Pacific, II. Porewater Metal Results. Earth Planetary Science Letter, 49, 81-101. http://dx.doi.org/10.1016/0012-821X(80)90151-X
67. Burbidge, D.J. and Gieskes, J.M. (1983) A Porewater Solid Phase Diagenetic Model for Manganese in Marine Sediments. American Journal of Science, 283, 29-47. http://dx.doi.org/10.2475/ajs.283.1.29
68. Bonatti, E., Fisher, D.E., Joensuu, O. and Rydell, H.S. (1971) Postdepositional Mobility of Some Transitional Elements, Phosphorous, Uranium and Thorium in Deep Sea Sediments. Geochimica et Cosmochimica Acta, 35, 1815-1819. http://dx.doi.org/10.1016/0016-7037(71)90057-3
69. Sawlan, J.J. and Murray, J.W. (1983) Trace Element Mobilization in the Interstitial Waters of Red Clay and Hemipelagic Marine Sediments. Earth Planetary Science Letter, 64, 213-230. http://dx.doi.org/10.1016/0012-821X(83)90205-4
70. Kodo, D. and Heath, G.R. (1984) Models of Depth Dependent Bioturbation at MANOP Site H in the Eastern Equatorial Pacific. Journal of Geophysical Research, 89, 6567-6570. http://dx.doi.org/10.1029/JC089iC04p06567
71. Stamatiakis, M.G. (2003) Phosphate Deposits of Neogene Age in Greece. Mineralogy, Geochemistry and Genetic Implications. Chemistry, 64, 329-357.
72. Cox, P.A. (1997) The Elements on Earth. Inorganic Chemistry in the Environment. Oxford University Press, Oxford, 284 p.
73. Chen, Z., Huang, C.-Y., Zhao, M.X., Yan, W., Chien, C.-W., Chen, M.H., Yang, H.P., Machiyama, H. and Lin, S. (2010) Characteristics and Possible Origin of Native Aluminum in Cold Seep Sediments from the Northeastern South China. Journal of Asian Earth Sciences, 40, 363-370. http://dx.doi.org/10.1016/j.jseaes.2010.06.006
74. Iyer, S.D., Mascarenhas-Pereira, M.B.L. and Nath, B.N. (2007) Native Aluminum (Spherules and Particles) in the Central Indian Basin Sediments: Implications on the Occurrence of Hydrothermal Events. Marine Geology, 240, 177-184. http://dx.doi.org/10.1016/j.margeo.2007.02.004

75. Dekov, V.M., Mandova, E.D., Dimitrov, K.C. and Rekkalov, K.N. (1995) Native Aluminum in Metalliferous Sediments from the East Pacific Rise Axial Zone (21 S). Marine Geology, 123, 87-103. http://dx.doi.org/10.1016/0025-3227(95)80006-W
76. Howard, K.J. and Fisk, M.R. (1988) Hydrothermal Alumina-Rich Clays and Boehmite on the Gorda Ridge. Geochimica et Cosmochimica Acta, 52, 2269-2279. http://dx.doi.org/10.1016/0016-7037(88)90129-9
77. Novgorodova, M.I. and Mamedov, Y.G. (1996) Native Aluminum from Mud Volcano at the Bulla Island, Caspian Sea. Lithology and Mineral Resources, 31, 301-310.
78. Yang, K. and Scott, S.D. (1996) Possible Contribution of a Metal-Rich Magmatic Fluid to A Sea-Floor Hydrothermal System. Nature, 383, 420-423. http://dx.doi.org/ 10.1038/383420a0
79. Rubin, K. (1997) Degassing of Metals and Metalloids from Erupting Seamount and Mid-Ocean Ridge Volcanoes: Observations and Predictions. Geochimica et Cosmochimica Acta, 61, 3525-3542. http://dx.doi.org/10.1016/S0016-7037(97)00179-8
80. Stolyarov, I.S., Silenko, T.M., Ryabeva, Y.G. and Yurkina, K.V. (1988) Al from the Ukachilkan Tin Deposit, Northeastern Yakutia. International Geological Review, 10, 1136-1140. http://dx.doi.org/10.1080/00206818809466094
81. Deng, Y., Zhang, L., Ou, J. and Zhang, H. (1983) Discovery of Native Aluminum in Fault Zones of Granodiorite. Journal of Guilin Institute of Technology, No. 1, 1-6.
82. Banerjee, D.M. and Saigal, N. (1988) A Study of Proterozoic Phosphorite from Chelima-Pachcherla Area, Kurnool District, Andhra Pradesh, India. Journal of Geological Society of India, 32, 32-39.
83. Leckie, J.O. and Davis, J.A. (1979) Aqueous Environmental Chemistry of Copper. In: Nriagu, J.O., Ed., Copper in the Environment, Wiley, New York, 90-121.
84. Gao, H.Z. (1998) The Biochemical Sedimentary Metallogenic Model of Baritic and Witheritic Deposits in Lower Cambrian in China. Journal of Mineralogy and Petrology, 18, 70-77.
85. Huba, C., Remacle, J., Dubois, D. and Thorez, J. (1983) Factors Affecting the Concentrations of Cadmium, Zinc, Copper and Lead in the Sediments of the Vesdre River (Belgium). Water Research, 17, 1281-1286. http://dx.doi.org/10.1016/0043-1354(83)90253-1
86. Li, Y.-H., Burkhardt, L. and Teraoka, H. (1984) Desorption and Coagulation of Trace Elements during Estuarine Mixing. Geochimica et Cosmochimica Acta, 48, 1879-1884. http://dx.doi.org/10.1016/0016-7037(84)90371-5
87. Paris, G., Bartolini, A., Donnadieu, Y., Beaumont, V. and Gaillardet, J. (2010) Investigating Boron Isotopes in a Middle Jurassic Micritic Sequence: Primary vs. Diagenetic Signal. Chemical Geology, 275, 117-126. http://dx.doi.org/10.1016/j.chemgeo.2010.03.013
88. Vengosh, A., Chivas, A.R., McCulloch, M.T., Tarinsky, A. and Kolodny, Y. (1991) Boron Isotope Geochemistry of Australian Salt Lakes. Geochimica et Cosmochimica Acta, 55 2591-2606. http://dx.doi.org/10.1016/0016-7037(91)90375-F
89. Hemming, N.G. and Hanson, G.N. (1992) Boron Isotopic Composition and Concentration in Modern Marine Carbonates. Geochemica et Cosmochimica Acta, 56, 537-543. http://dx.doi.org/10.1016/0016-7037(92)90151-8
90. Lemarchand, D., Gaillardet, J., Lewin, E. and Allègre, C.J. (2000) The Influence of Rivers on Marine Boron Isotopes and Implications for Reconstructing Past Ocean pH. Nature, 408, 951-954. http://dx.doi.org/10.1038/35050058
91. Joachimski, M.M., Simon, L., van Geldern, R. and Lécuyer, C. (2005) Boron Isotope Geochemistry of Paleozoic Brachiopod Calcite: Implications for a Secular Change in the Boron Isotope Geochemistry of Seawater over the Phaneozoic. Geochimica et Cosmochimica Acta, 69, 4035-4044. http://dx.doi.org/10.1016/j.gca.2004.11.017
92. Simon, L., Lécuyer, C., Maréchal, C. and Coltice, N. (2006) Modelling the Geochemical Cycle of Boron: Implications for the Long-Term δ11B Evolution of Seawater and Oceanic Crust. Chemical Geology, 225, 61-76. http://dx.doi.org/10.1016/j.chemgeo.2005.08.011
93. Spivack, A.J., Palmer, M.R. and Edmond, J.M. (1987) The Sedimentary Cycle of the Boron Isotopes. Geochimica et Cosmochimica Acta, 51, 1939-1949. http://dx.doi.org/10.1016/0016-7037(87)90183-9

94. You, C.F., Spivack, A.J., Gieskes, J.M., Rosenbauer, R. and Bischoff, J.L. (1995) Experimental Study of Boron Geochemistry: Implications for Fluid Processes in Subduction Zones. Geochimica et Cosmochimica Acta, 59, 2435-2442.
95. Berner, R.A. (1973) Phosphate Removal from Sea Water by Adsorption on Volcanogenic Ferric Oxides. Earth and Planetary Science Letters, 18, 77-86. http://dx.doi.org/10.1016/0012-821X(73)90037-X
96. Hem, J.D. (1972) Chemistry and Occurrence of Cadmium and Zinc in Surface Water and Ground Water. Water Resource Research, 8, 661-679. http://dx.doi.org/10.1029/WR008i003p00661
97. Callender, E. and Bowser, C.J. (1980) Manganese and Copper Geochemistry of Interstitial Fluids from Manganese Nodule-Rich Pelagic Sediments of the Northeastern Equatorial Pacific Ocean. American Journal of Science, 280, 1063-1096. http://dx.doi.org/10.2475/ajs.280.10.1063
98. Klinkhammer, G.P. (1980) Early Diagenesis in Sediments from the Eastern Equatorial Pacific. II. Porewater Metal Results. Earth and Planetary Science Letters, 49, 81-101. http://dx.doi.org/10.1016/0012-821X(80)90151-X
99. Sawlan, J.J. and Murray, J.W. (1983) Trace Element Mobilization in the Interstitial Waters of Red Clay and Hemipelagic Marine Sediments. Earth and Planetary Science Letters, 64, 213-230. http://dx.doi.org/10.1016/0012-821X(83)90205-4
100. Heggie, D. and Lewis, T. (1984) Cobalt in Pore Waters of Marine Sediments. Nature, 311, 453-455. http://dx.doi.org/10.1038/311453a0
101. Holland, H.D. (1978) The Chemistry of the Atmosphere and Oceans. Wiley-Interscience, New York.
102. de Dolomieu, D.G. (1791) Sur un de pierrestrés-peueffervescentes avec les acides of phosphorescentes par la collision. Journal de Physique, 39, 3-10.
103. Land, L.S. (1983) The Application of Stable Isotopes to Studies of the Origin of Dolomite and Problems of Diagenesis of Clastic Sediments. In: Arthur, M.A., et al., Eds., Stable Isotope in Sedimentary Geology, Vol. 10, SEPM Short Course, 4.1-4.22.
104. Graf, D.L. and Goldsmith, J.R. (1956) Some Hydrothermal Syntheses of Dolomite and Protodolomite. Journal of Geology, 64, 173-186. http://dx.doi.org/10.1086/626332
105. Goldsmith, J.R. and Heard, C.D. (1961) Subsolidus Phase Relations in the System CaCO3-MgCO3. Journal of Geology, 69, 45-74. http://dx.doi.org/10.1086/626715
106. Degens, E.T. (1967) Diagenesis of Organic Matter. In: Larsen, G. and Chilingar, G.V., Eds., Diagenesis of Sediments, Elsevier Publishing Co., New York, 343-390. http://dx.doi.org/10.1016/S0070-4571(08)70846-X
107. Dapples, E.C. (1967) Diagenesis in Sandstones. In: Larsen, G. and Chilingar, G.V., Eds., Diagenesis of Sediments, Elsevier Publishing Co., New York, 91-125. http://dx.doi.org/10.1016/S0070-4571(08)70842-2
108. Weller, J.M. (1959) Compaction, of Sediments. American Association of Petroleum Geologists, 43, 273-310.
109. Von Engelhardt, W. and Gaida, K.H. (1963) Concentration Changes of Pore Solutions during the Compaction of Clay Sediments. Journal of Sedimentary Petrography, 33, 919-930. http://dx.doi.org/10.1306/74D70F74-2B21-11D7 8648000102C1865D
110. Müller, G. (1967) Diagenesis in Argillaceous Sediments. In: Larsen, G. and Chilingar, G.V., Eds., Diagenesis of Sediments, Elsevier Publishing Co., New York, 128 177. http://dx.doi.org/ 10.1016/S0070-4571(08)70843-4

CHAPTER 13

Resource and Reserve Calculation in Seam-Shaped Mineral Deposits; A New Approach: "The Pentahedral Method"

César Castañón [1,*], Daniel Arias [2], Isidro Diego [1], Agustin Martin-Izard [2] and Yhonny Ruiz [3]

[1] *Department of Mines Research and Exploitation, Oviedo University, C/Independencia 13, E-33004 Oviedo, Spain*

[2] *Department of Geology, Oviedo University, C/Arias de Velasco s/n, E-33005 Oviedo, Spain*

[3] *SolMine SRL, Av/Ramón Castilla 272, Piura, 3P, D1 Peru*

ABSTRACT

In recent years, the introduction of specific software for the evaluation of geological resources and mineral reserves has increased significantly thanks to the use of block models capable of working with large databases and applicable to virtually all types of deposits. It is only in layered, tabular-shaped deposits where the use of block models poses certain drawbacks, not only in terms of visual representation but also during the process of interpolation. Other calculation methods for tabular bodies such as sections, polygons, and triangles work with 2D projections but not with 3D. The "Pentahedral" method described here is undoubtedly an innovative method that allows work to always be conducted in 3D, providing a fairly accurate representation of tabular bodies and making it possible to carry out calculations of resources and reserves using any interpolation method. It is demonstrated with figures and tables of the Carlés mineral deposit, a well-developed exoskarn gold deposit in northwestern Spain (Asturias) where the authors have worked extensively. The pentahedral method takes into account not only geological and stratigraphic data from the model interpolation, but also mining concepts such as "minimum thickness" related to the minimum seam size that can be economically and technically mined, and "overbreak" related to the dilution effect that appears during the mining process due to over-excavation.

Keywords: tabular body; interpolation; pentahedral method; mineral deposit; mineral resources; mineral reserves

1. INTRODUCTION

Once the geological, geochemical, and structural information of a mineral deposit are known, it must be represented in 3D in order to obtain sufficiently accurate estimates of its mineralogical content [1], which will lead to the estimation of both mineral resources and mineral reserves, as is explained in international standards such as those from the Canadian Institute of Mining, Metallurgy and Petroleum (CIM) [2].

According to CIM standards, a Mineral Reserve is the economically mineable part of a Measured or Indicated Mineral Resource demonstrated by at least a Preliminary Feasibility Study. This study must include adequate information on mining, processing, metallurgical, economic, and other relevant factors that demonstrate, at the time of reporting, that economic extraction can be justified. A Mineral Reserve includes diluting materials and allowances for losses that may occur when the material is mined.

Despite the existence of several resource estimation methodologies [3], from traditional to geostatistical ones [4,5,6] that use the block model as a geometrical representation, in the particular case of tabular or seam-shaped deposits these methods do not allow a sufficiently accurate representation or interpolation [7].

Hexahedral block models are widely used to simulate and classify resources. Reference [8] shows a study of 120 recent NI 43-101 mineral deposit technical reports, where kriging was the most elaborate one, although used in only seven of the reports. Other more refined methods, such as the sequential indicator simulation (SIS, a pixel-based model) or indicator kriging, are mainly usable in reservoir simulations. Implicit modeling is starting to be implemented in commercial simulation packages, e.g., Leapfrog Radial Basis Functions, Surpac Dynamic Shells Module or Minesight Implicit Modeler, but is recommended for use as an additional tool to provide initial and quick resource analysis, not as a complete replacement of block models [9].

State-of-the-art software uses algorithms that sequentially divide each block into smaller blocks in order to follow the geological structures (sub-blocking), but this approach does not achieve the level of precision required and results in models with an extremely high number of blocks; see [1]. Additionally, sub-blocking uses the composition of the samples surveyed, but does not make an interpolation using the intersections between the surveys and the mineral structure, considering a cutoff, or a minimal thickness, or even considering lateral dilution due to over-excavation. In this paper, it is the "operational mining dilution" that occurs at the time of mining, as defined in [10], which will be referred to as dilution.

Advances have been made in the modeling of local anisotropy in complex geological structures, see [11], but this method does not include mining parameters such as the minimum thickness values or the dilution.

This paper uses a new method to model tabular deposits, as exemplified by the "Zona M" seam of the northern side of the Carlés deposit, in Spain. This deposit is a gold copper skarn, located in the Rio Narcea Gold Belt [12,13,14] formed at the contact between the Carlés granodiorite and the Nieva limestone of the Rañeces Group [15,16]. It consists of several seam-shaped mineral bodies, some of them with economical grades of Au and Cu, with thicknesses ranging from 1 to 6 m and dips from 40° in the north (55°) direction (eastern areas) to 70° in the north (30°) direction (northern bodies).

In the Carlés deposit there are several seams, each of them continuous. They are separated by faults or are located in different contact areas. As is standard practice in classical block models, bodies or seams divided by faults should be treated as different bodies and studied separately.

RECMIN (from: "REC ursos MIN eros" which translates as "mineral resources" in English) software will be used throughout the research in order to manage data and visualization; see [17]. RecMin is free software used in mineral deposit modeling, developed entirely by the authors over the past 25 years and thoroughly tested in real working conditions at dozens of mining projects.

RecMin code was supplemented with enough algorithms to consider this new calculation method, described further on.

This paper will show how to work with calculation units geometrically different from blocks. The pentahedral method also adds information usually not taken into account in block modeling, such as "lateral dilution" or "mining intercept".

2. RESEARCH OBJECTIVES

The main goal of this research is the evaluation of the seam-shaped mineralized structure using a new methodology that provides a precise representation of morphology and physical and chemical characteristics, and storing this information in a suitable manner in order to proceed with resource estimations. The specific objectives of the method are:

- Three-dimensional (3D) modeling is mandatory;
- It must allow different interpolation algorithms, including geostatistics;
- The calculation detail has to be user-defined;
- Dilution (material below the cutoff) should be considered, if necessary;
- It must allow estimates with a categorization using distances, number of intercepts, or a combination of both parameters;
- It must allow interpolation, taking into account boreholes filtered by distance: neighborhood influence;
- The results will be stored in a Structured Query Language (SQL) database.

3. GEOLOGICAL CHARACTERISTICS AND CONSTRAINTS

The Carlés Cu-Mo-Au ore deposit is located in the Rio Narcea Gold Belt (RNGB), in the Cantabrian zone of the Iberian Massif. It is related to a small postkinematic calcalkaline monzogranite, which intrudes as a cedar tree laccolith into the upper siliciclastic Furada Formation (shales, oolitic ironstones, and, at the top, some carbonate layers of late Silurian age) and the Nieva carbonates (mostly limestones and some dolostones of early

Devonian age), developing skarns bearing copper, molybdenum, and gold of economic interest (Figure 1A).

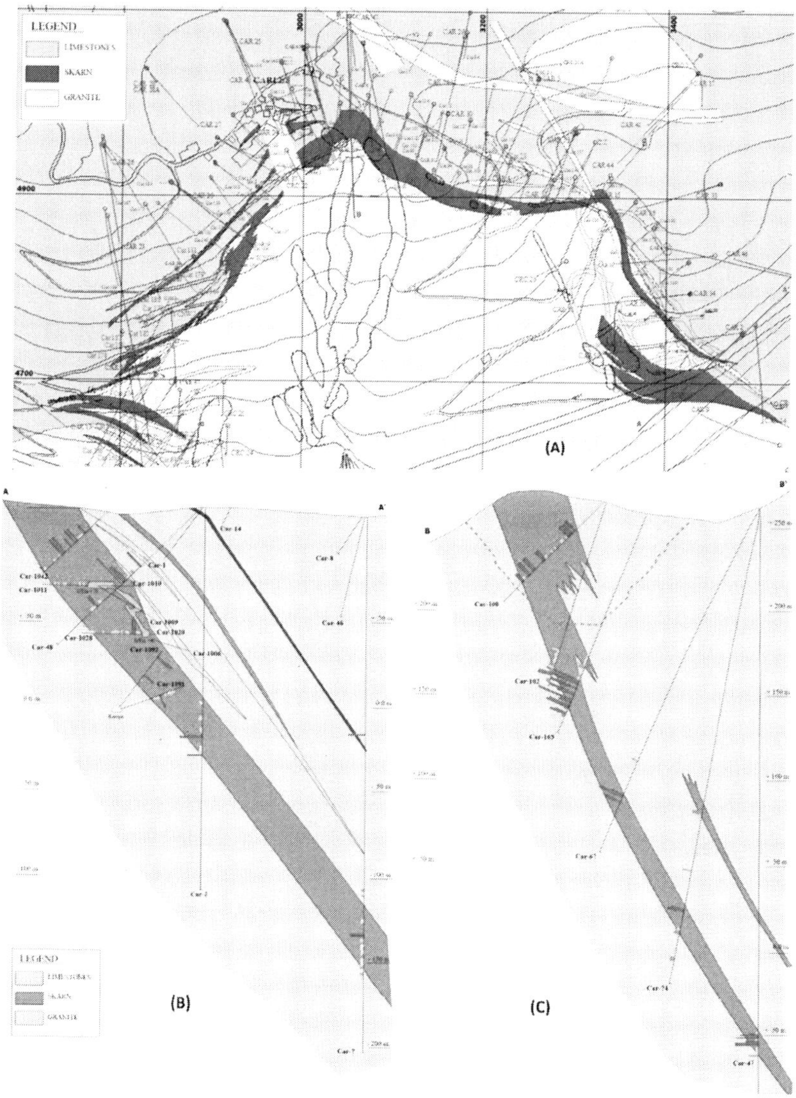

Figure 1. (**A**) Geological map of the Carlés gold-bearing skarn (data provided by Rio Narcea Gold Mines); (**B**) Section of the Carlés deposit (West Carlés); (**C**) Section of the Carlés deposit (North Carlés).

According to the paper by Martin-Izard et al. [16], the Carlés deposit consists mainly of a well-developed exoskarn [18]. The Carlés intrusive is a monzogranite, up to 800 m in diameter, which intrudes along the contact of a Silurian siliciclastic formation and a Devonian carbonate formation. It is a light gray, medium-grained, equigranular igneous rock with plagioclase, biotite, quartz, K-feldspar, and magnesiohornblende. The accessory minerals are zircon, rutile, ilmenite, magnetite, and apatite. The plagioclase (An 30–47) has normal zoning with a common narrow albite-oligoclase (An 8–20) rim. The skarn is at the contact between the monzogranite and the carbonate rock, where a Cu-Au-Mo calcic skarn developed. The mineralized area is divided into three main sectors (east, west and north sectors; See Figure 1), all of which have been studied by Martin-Izard et al. [13]. At the moment the east and north sectors are being exploited (Figure 1A–C), both by open pit and also by underground mining.

The exoskarn is mostly calcic skarn made up of early garnet and pyroxene, and later amphibole, magnetite, and sulfides (Figure 2A,B). The ore is related to the skarn retrogradation and post-skarn veining and faulting.

Figure 2. (**A**) View of the stratigraphic control of the skarn development at the Carlés skarn. The monzogranite intrusion is also controlled by the sedimentary layers; (**B**) Close-up view of stratigraphically controlled Carlés mineralized banded skarn. Bands are dominated by pyroxene-magnetite alternate with garnet-amphibole-rich bands.

The metasomatic phenomena give way to alternating layers, centimeters to meters in thickness, of garnet-dominant and pyroxene-dominant skarn. Retrograde alteration of the garnet-pyroxene skarn to form a pyroxene, amphibole, magnetite and sulfide minerals assemblage is a very important process in the Carlés deposit. Above all, metallic mineralization at Carlés is most important in lenses within amphibole-pyroxene bands.

4. METHODOLOGY

"Zona M" in the northern area of the Carlés deposit is the zone this study will focus on. The calculation procedure was developed in the following phases:

1) Determination of the seam localization from the geological information and survey data.
2) Construction of a 3D-modeled surface at the center of the seam.
3) Generation of an equally-spaced cloud of points in the seam centered surface.
4) Calculation of the intercept values of the surveys crossing the seam.
5) Execution of a geostatistical study of the intercept values.
6) Interpolation of the grade and the thickness of each point of the cloud of points, using the intercept information.
7) Definition of the estimation categories and generation of the calculation units.
8) Database generation and representation.

4.1. Determination of the Seam Localization from the Geological Information and Surveys

For each and every borehole, the intercept interval of the "ZONA M" mineralized body was determined, without considering the grade value, only considering the skarn lithology.

The seam localization has to be broad enough to allow late adjustments as the modeling defines the correct economical limits, from parameters such as cutoff grade, minimal thickness, or lateral dilution. Figure 3 shows graphically the seam limits at a borehole intercept and the associated bar chart displaying grade data.

4.2. Construction of a 3D Modeled Surface at the Center of the Seam

As stated above, once the borehole intervals that intercept the mineralized body have been defined, a surface at the center of this body has to be created, from where the interpolation point mesh will be created. As noted above, the surface does not need to be located exactly, as the software will adjust its final position according to the real values of the borehole intercepts and the body.

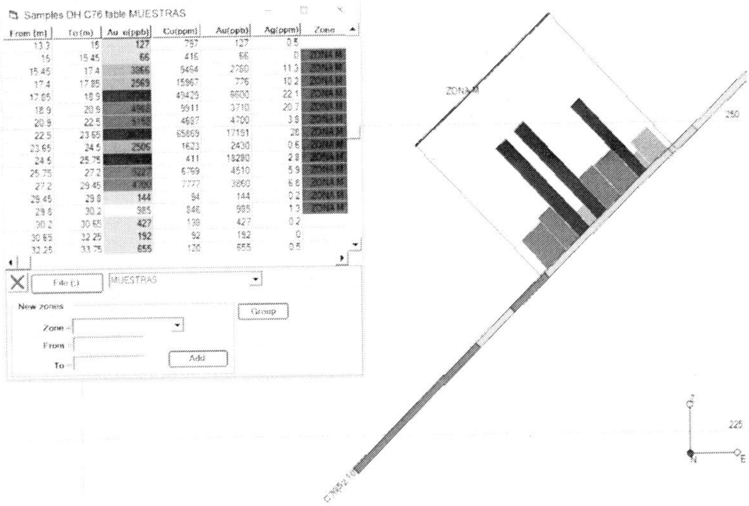

Figure 3. Example of a mineralized area in an intercepting borehole.

The easiest way to proceed is to work section by section (Figure 4). Figure 4A shows a cross section of boreholes depicted in white whereas the intercepts with the body are depicted in red. The green line connects the centers of each red line section. Figure 4B simultaneously shows several sections and their corresponding centerlines, which are triangulated to generate a surface. Figure 4C shows the final calculated surface, which will henceforth be referred to as "T1".

4.3. Generation of an Equally Spaced Cloud of Points in the Seam Centered Surface

In order to have equally spaced interpolation points over the T1 surface, T1 will be remeshed to generate T2, with a point density high enough to reflect geology variation. A high point density will not alter the resource calculations, as interpolation points too close to each other will display the same results as if they were close to the intercepts. This cloud of points will be referred to as NPS from now onwards. The software will automatically cut parallel sections of T1 at a defined distance (5 m in this example, as the borehole density is 25 × 25 and 5 intermediate points between boreholes have been considered) in a selected direction (the *X* axis in this example). Inside each line generated by each cut, points are defined by a certain distance, user-selectable (5 m again in this example) on the perpendicular axis (*Y* in this case). Note that 5 m is a suitable value in the case of the "ZONA M" ore body. This value has to be adjusted according to each ore body size and morphology so as to obtain the necessary accuracy for mining the resource, see [7].

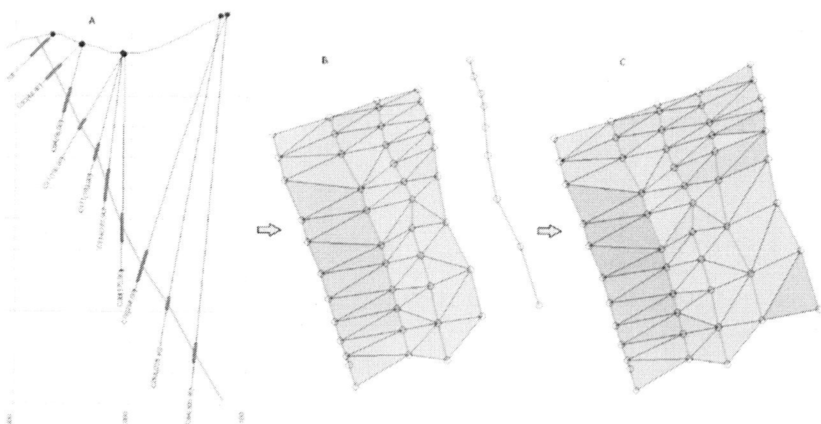

Figure 4. T1 surface generation. (**A**) A cross section of boreholes is depicted in white and the intercepts with the body are depicted in red. The green line connects the centers of each red line section; (**B**) Sections and their corresponding centerlines, which are triangulated to generate a surface. (**C**) Calculated surface ("T1").

Once the evenly spaced cloud of points is generated (Cartesian mesh), the software will create new points at the cross of the diagonals connecting any four points of each generated square (Figure 5C). Figure 5B shows the outcome of this operation. Finally, the algorithm will create triangles that mimic the seam-centered surface. This triangulated surface will be called T2.

All the information is stored in a database (BDS), previously created in the database.

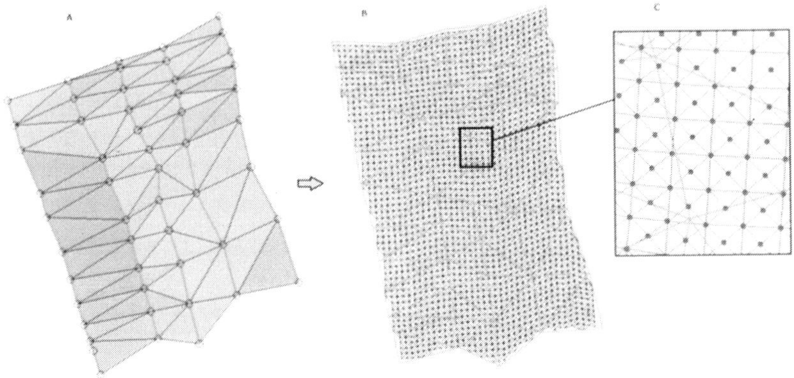

Figure 5. Example of base surface T1 (left (**A**)) converted to calculation Surface T2 (right (**B**)). Detail of meshing points in sections and in crossing diagonals of any generated square (**C**).

4.4. Calculation of the Intercept Values of the Surveys Crossing the Seam

Before the interpolation of grades and thicknesses at each point of the NPS, the economical interval in each borehole (intercept) must be calculated.

Cutoff grade. This is defined as the minimum grade that allows economical exploitation of the ore body, see [19]. In the example that is being used, the "Zona M" ore body of the Carlés deposit hosts three recoverable elements: Au, Cu, and Ag. A new element called "Au_e" element or "Au grade equivalent" can be defined, where an accumulated value of the three elements is stored according to a price-weighting index scheme. In order to simplify the calculation, an average metal recovery value was assumed, understood as the final percentage of each metal that gives a monetary income, deducting the concentration plant recovery, selling costs, concentration transport, penalties, besides other factors.

Minimum thickness. A minimum value of seam thickness must be considered in order to reflect the economic feasibility of the exploitation. It depends on the rock type, the seam dip and the mining method (room and pillar, sublevel stopping, cut and fill). A thickness of 2 m has been considered in this example, according to the mining machinery size that can be used in this deposit mining.

Mining overbreak. A lateral dilution has been defined due to unwanted results of blasting and hole deviations. From our experience with this deposit, 50 cm of overbreak (overexcavation, 25 cm each side) is defined, a value that can be user-selectable

Maximum thickness of waste rock. A usual situation that is found when the intercept of the borehole and the ore body is calculated is the appearance of inclusions, intervals of waste rock inside the ore body and, more commonly, mineralization below the cutoff grade value. The treatment for this type of situation is achieved by defining a maximum waste thickness value whereby, if the waste thickness value exceeds the maximum, it will not be mined (it will remain as a pillar). If the waste thickness value is calculated to be below the maximum and the average grade is higher than the cutoff grade, it will be considered as ore; however, if the average grade is below the cutoff grade, it will be considered as waste rock and will not be included in the intercept. This maximum waste rock thickness will be referred to as the real thickness of the seam, measured perpendicular to the seam.

The intercept calculation procedure is described below, both graphically and analytically.

In order to explain the procedure of the three intersection types, as well as the interpolation over each point of the NPS, we will refer to four theoretical boreholes: S1, S2, S3, and S4, as represented in Figure 6. The algorithm calculates three different intercepts (denoted as A, B, and C), for each borehole crossing the seam, as follows:

A: Geological Intercept. This includes only those samples with grades higher than the cutoff grade. If there are no samples higher than that value, the highest-grade sample will be defined as the geological intercept. For instance, borehole S2 shows six samples 1 m in size, but only five will be used when considering a cutoff grade of 2 g Au/t, as can be seen in Figure 4 as per the coding colors.

For those samples with grades lower than the cutoff grade being inserted between samples that are above the cutoff grade, the "maximum thickness of waste rock" criteria must be taken into account, as referred to above.

B: Minimum Thickness Intercept. This will be defined as the geological intercept widened as necessary to complete a previously defined minimum real thickness. If the real thickness of the geological intercept is equal to or greater than the minimum thickness value, both intercepts will be the same.

By looking at borehole S2 in Figure 6, and considering 2 m as the minimum thickness value, it can be seen that the geological intercept already has a real thickness of over 2 m, so the minimum thickness intercept (B) will be equal to the geological intercept (A).

If the intercept is less than the minimum thickness, the algorithm will add in samples from the higher grade side. Once the limits are checked, it will re-calculate whether new samples need to be added of a higher cutoff on each side of the intercept.

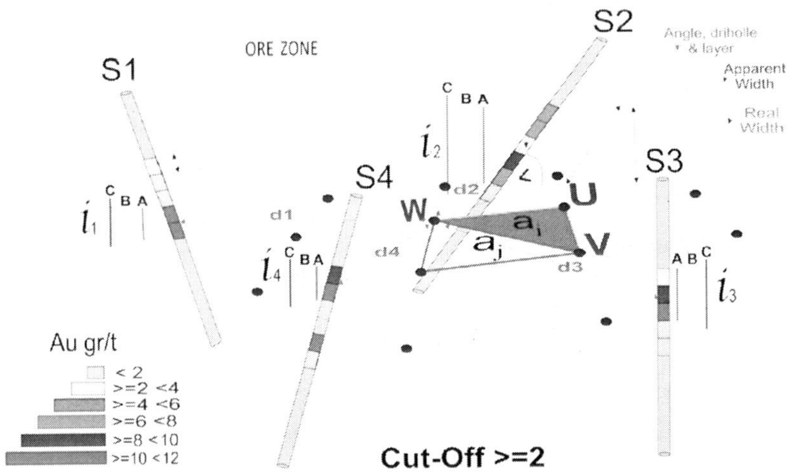

Figure 6. Intercept calculation of four boreholes (S1 to S4) through a seam-shaped mineral ore body. A (geology intercept), B (minimum thickness intercept), C (mining overbreak intercept); U, V, and W are points of the NPS; i1–i4 are the lithology intercept interval; d1–d4 distances between the intercepts center and W point.

C: Mining Overbreak Intercept. This will be defined as the minimum real thickness intercept widened at both sides by a distance defined as over excavation. In this example 50 cm will be used (25 cm on each side). This is an important parameter to consider, as it will simulate the mineral dilution that always occurs in common mining procedures that involve excavation.

All three intercepts are calculated referring to real or true thickness; therefore the angle at which the borehole and the seam intercept each other must be known. If the borehole is perpendicular to the seam, this angle is 90° and real and apparent thickness will be equal. This angle, and the intercept point coordinates, is calculated automatically by the algorithm. The intercept point will be the point where the line that represents the borehole cuts one of the triangles of the T2 surface generated from the NPS. The angle will be the one established by the plane of the intersected triangle and the segment representing the borehole.

Nevertheless, the experience of the technician is extremely important in order to define which samples to include in the body. The software allows samples to be accepted or discarded manually in order to refine the calculations. Angles defined by the intersection of boreholes and the seam shaped deposit are calculated at each intersection and can be manually adjusted according to geological knowledge if need be. In this way, possible inaccuracies in apparent thickness calculations are limited.

Interpolation parameters. In this study the Inverse Distance Weighted (IDW) interpolation method will be used; other methods could have been considered: nearest neighbor or kriging, among others. In order to use IDW, a maximum influence distance must be used, calculated through a geostatistical study, independent of the interpolation method shown in this paper and detailed below, as well as a power value. In this case the distance is fixed at 50 m and the power is 3 (Figure 7 shows the input window for these parameters). These values can be different for each intercept, if necessary.

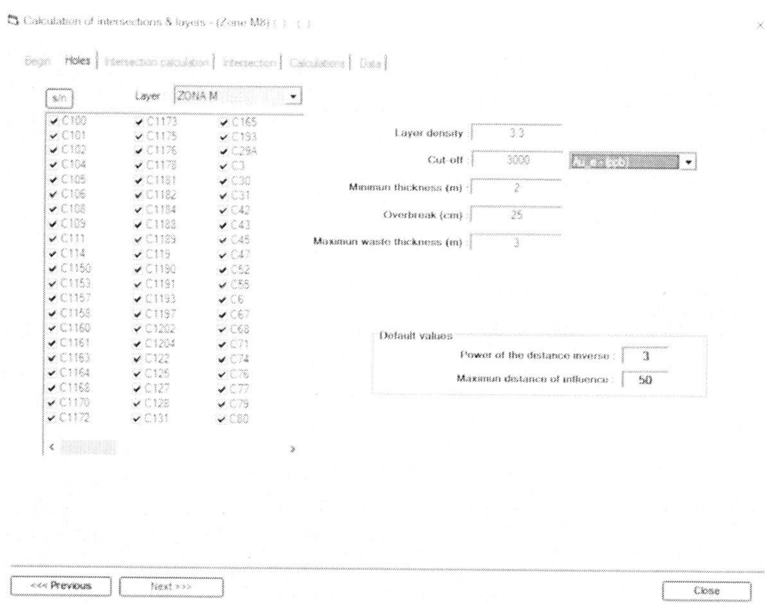

Figure 7. RecMin Iinput window for intercept parameters. Holes selection (all, in this case), cutoff, minimum thickness, overbreak (one side), and maximum waste thickness.

Once the calculation parameters are defined, the software executes the algorithm, which resolves the three intercept types in each borehole that fulfills the defined conditions. This kind of deposit is often high-grade centered, with low-grade values toward the deposit limits. When several high-grade intervals exist over the cutoff grade, the limits for each and every one of the three types of intercepts can be adjusted manually.

Figure 8 is an example of the calculation window. Intercepts are shown as graph bars: red (geological intercept), yellow (minimum thickness intercept) and white (mining overbreak intercept). The *x*-axis shows samples ("from" and "to" values) and the *y*-axis shows the different grade values in ppb (parts per billion).

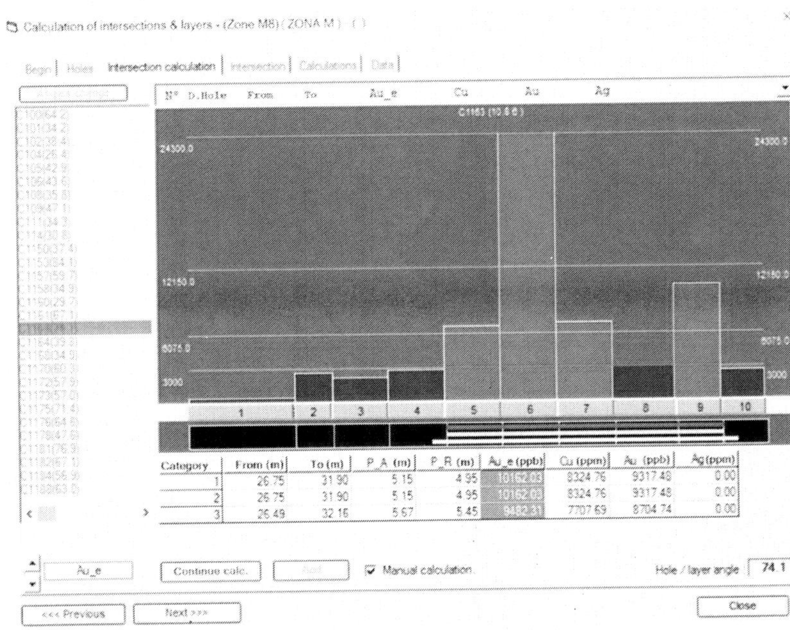

Figure 8. RecMin Output window, intercept graph bars for C1163 borehole: red (geological intercept), yellow (minimum thickness intercept), and white (mining overbreak intercept).

The calculation results (Figure 9) are stored in a database (BDS) that includes three records for each borehole, corresponding to each interpolation type.

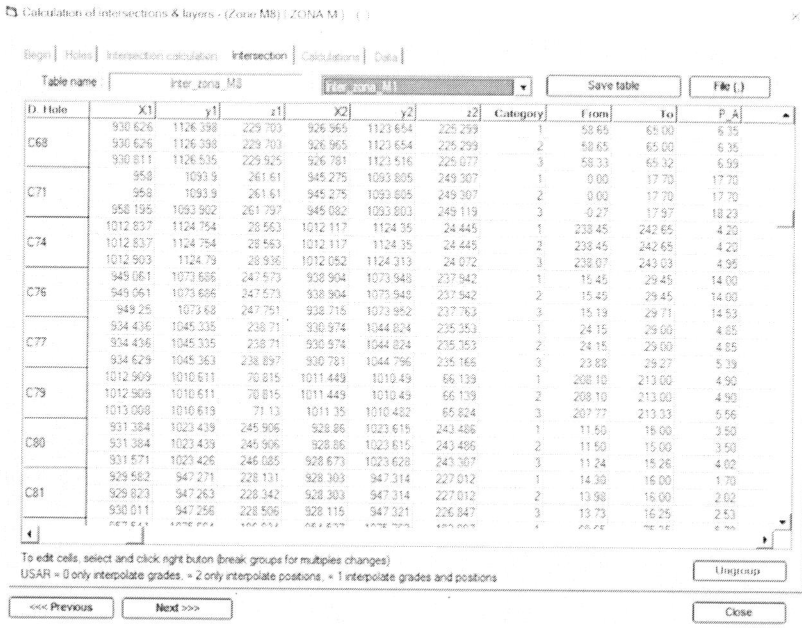

D. Hole	X1	y1	z1	X2	y2	z2	Category	From	To	P_A
C68	930.626	1126.398	229.703	926.965	1123.654	225.299	1	58.65	65.00	6.35
	930.626	1126.398	229.703	926.965	1123.654	225.299	2	58.65	65.00	6.35
	930.811	1126.535	229.925	926.781	1123.516	225.077	3	58.33	65.32	6.99
C71	958	1093.9	261.61	945.275	1093.805	249.307	1	0.00	17.70	17.70
	958	1093.9	261.61	945.275	1093.805	249.307	2	0.00	17.70	17.70
	958.195	1093.902	261.797	945.082	1093.803	249.119	3	-0.27	17.97	18.23
C74	1012.837	1124.754	28.563	1012.117	1124.35	24.445	1	238.45	242.65	4.20
	1012.837	1124.754	28.563	1012.117	1124.35	24.445	2	238.45	242.65	4.20
	1012.903	1124.79	28.936	1012.052	1124.313	24.072	3	238.07	243.03	4.95
C76	949.061	1073.686	247.573	938.904	1073.948	237.942	1	15.45	29.45	14.00
	949.061	1073.686	247.573	938.904	1073.949	237.942	2	15.45	29.45	14.00
	949.25	1073.68	247.751	938.715	1073.952	237.763	3	15.19	29.71	14.53
C77	934.436	1045.335	238.71	930.974	1044.824	235.353	1	24.15	29.00	4.85
	934.436	1045.335	238.71	930.974	1044.824	235.353	2	24.15	29.00	4.85
	934.629	1045.363	238.897	930.781	1044.796	235.166	3	23.88	29.27	5.39
C79	1012.909	1010.611	70.815	1011.449	1010.49	66.139	1	208.10	213.00	4.90
	1012.909	1010.611	70.815	1011.449	1010.49	66.139	2	208.10	213.00	4.90
	1013.008	1010.619	71.13	1011.35	1010.482	65.824	3	207.77	213.33	5.56
C80	931.384	1023.439	245.906	928.86	1023.615	243.486	1	11.50	15.00	3.50
	931.384	1023.439	245.906	928.86	1023.615	243.486	2	11.50	15.00	3.50
	931.571	1023.426	246.085	928.673	1023.628	243.307	3	11.24	15.26	4.02
C81	929.582	947.271	228.131	928.303	947.314	227.012	1	14.30	16.00	1.70
	929.823	947.263	228.342	928.303	947.314	227.012	2	13.98	16.00	2.02
	930.011	947.256	228.506	928.115	947.321	226.847	3	13.73	16.25	2.53

Figure 9. Table of calculated intercepts calculated in RecMin. Results for several boreholes. First column is the name of the borehole (down hole). Each hole has three data rows showing the intercept data.

4.5. Geostatistical Study of the Intercept Values

As with block model estimation, an influence distance or neighborhood search must be defined. This is necessary in order to consider the quantity and proximity of intercepts to be used in the interpolation and to define the parameters of the classification of the resources or reserves. This is achieved by using a variographic study, which can be performed using a variety of software to which the data must be exported. Several codes are available, both freeware and commercial.

In this example, the data were exported to Stanford Geostatistical Modeling Software (SGEMS), [20], where the variogram was modeled and the main direction and influence distance was found. RecMin can easily export both the composite and/or the intercept data to make this calculation, in both Gslib and Ascii format.

The main direction of the mineralization, as per variogram results, is indicated in graphics by a line parallel to the direction that fits between two of the vertices of the T2 surface (Figure 10). This line is saved as a file and then later selected in the software interpolation assistant in order to be considered as a continuity guide during calculation.

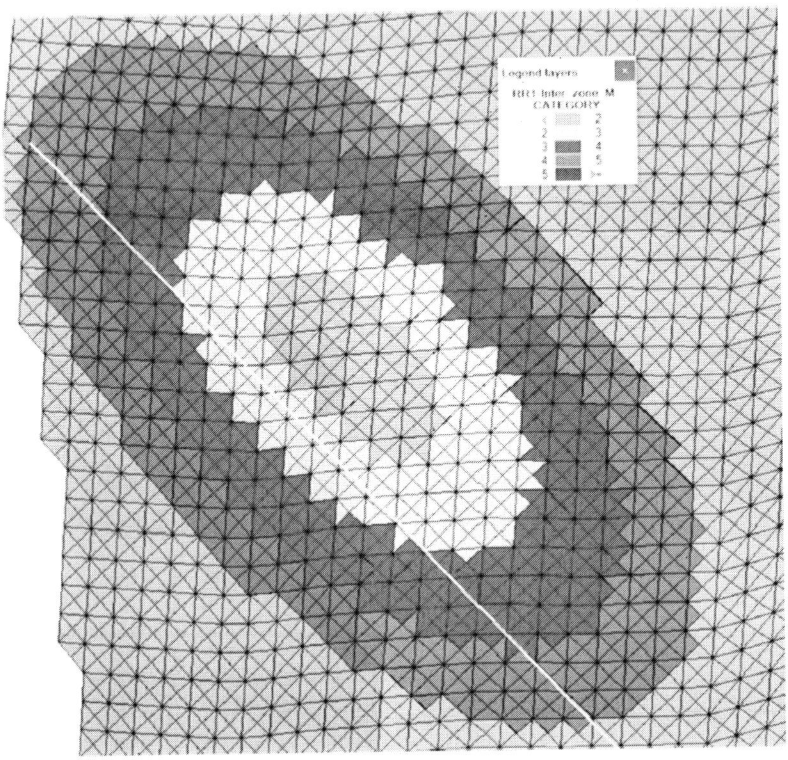

Figure 10. Definition of the guide line of the interpolation ellipse drawn over T2. A theoretical borehole crosses the seam and defines several possible categories of resources, according to the distance between the NPS points and the location of intersection center point.

The anisotropy of the variogram is reflected in the different radii of the search ellipsoid. In this case, it is a planar ellipse, as it follows a seam. Both radii of the ellipse, the long and the short, are introduced considering a ratio between them. In this example the ratio is 1:2, with 2 being the secondary direction. Figure 10 shows the shape of the ellipse for a unique theoretical borehole, crossing the seam at the center of the colored sections, as well as the main direction of the mineralization (green line), which follows the main radius of the ellipse.

During the calculation, the vector that connects the intercept and the interpolation point is decomposed in two vectors: one in the main direction and another one in the secondary direction; each vector modulus is multiplied by the corresponding factor as per input data in order to establish the calculation distance.

4.6. Interpolation of the Grade and the Thickness at Each Point of the Cloud of Points (NPS)

In order to understand the interpolation process, please refer to Figure 6 and study the triangle ai with its three vertices U, V, and W. The intercepts are calculated for each vertex with their own influence radius. Assuming that the intercepts of the four boreholes represented in the figure are inside that radius, the following calculations (Table 1) are worked out:

The interpolation in case of the vertex **W** of the triangle $\boldsymbol{a_i}$ will be:

$$(L.P)_{WA} = \frac{\frac{(L.P)_{A1}}{d_1^x} + \frac{(L.P)_{A2}}{d_2^x} + \ldots + \frac{(L.P)_{An}}{d_n^x}}{\left(\frac{1}{d_1}\right)^x + \left(\frac{1}{d_2}\right)^x + \ldots + \left(\frac{1}{d_n}\right)^x} \quad (1)$$

$$P_{WA} = \frac{\frac{P_{A1}}{d_1^x} + \frac{P_{A2}}{d_2^x} + \ldots + \frac{P_{An}}{d_n^x}}{\left(\frac{1}{d_1}\right)^x + \left(\frac{1}{d_2}\right)^x + \ldots + \left(\frac{1}{d_n}\right)^x} \quad (2)$$

$$(L.P)_{WB} = \frac{\frac{(L.P)_{B1}}{d_1^x} + \frac{(L.P)_{B2}}{d_2^x} + \ldots + \frac{(L.P)_{Bn}}{d_n^x}}{\left(\frac{1}{d_1}\right)^x + \left(\frac{1}{d_2}\right)^x + \ldots + \left(\frac{1}{d_n}\right)^x} \quad (3)$$

$$P_{WB} = \frac{\frac{P_{B1}}{d_1^x} + \frac{P_{B2}}{d_2^x} + \ldots + \frac{P_{Bn}}{d_n^x}}{\left(\frac{1}{d_1}\right)^x + \left(\frac{1}{d_2}\right)^x + \ldots + \left(\frac{1}{d_n}\right)^x} \quad (4)$$

$$(L.P)_{WC} = \frac{\frac{(L.P)_{C1}}{d_1^x} + \frac{(L.P)_{C2}}{d_2^x} + \ldots + \frac{(L.P)_{Cn}}{d_n^x}}{\left(\frac{1}{d_1}\right)^x + \left(\frac{1}{d_2}\right)^x + \ldots + \left(\frac{1}{d_n}\right)^x} \quad (5)$$

Table 1. Calculations worked out for each vertex with their own influence radius.

Borehole S_1	Borehole S_2	Borehole S_n
A_1: Geological Intercept	**A_2: Geological Intercept**	**A_n: Geological Intercept**
$(L.P)_{A1}$: Grade multiplied by thickness	$(L.P)_{A2}$: Grade multiplied by thickness	$(L.P)_{An}$: Grade multiplied by thickness
P_{A1}: thickness	P_{A2}: thickness	P_{An}: thickness
B_1: Minimum Thickness Intercept	**B_2: Minimum Thickness Intercept**	**B_n: Minimum Thickness Intercept**
$(L.P)_{B1}$: Grade multiplied by thickness	$(L.P)_{B2}$: Grade multiplied by thickness	$(L.P)_{Bn}$: Grade multiplied by thickness
P_{B1}: thickness	P_{B2}: thickness	P_{Bn}: thickness
C_1: Mining Intercept	**C_2: Mining Intercept**	**C_n: Mining Intercept**
$(L.P)_{C1}$: Grade multiplied by thickness	$(L.P)_{C2}$: Grade multiplied by thickness	$(L.P)_{Cn}$: Grade multiplied by thickness
P_{C1}: thickness	P_{C2}: thickness	P_{Cn}: thickness

$$P_{WC} = \frac{\frac{P_{C1}}{d_1^x} + \frac{P_{C2}}{d_2^x} + \ldots + \frac{P_{Cn}}{d_n^x}}{\left(\frac{1}{d_1}\right)^x + \left(\frac{1}{d_2}\right)^x + \ldots + \left(\frac{1}{d_n}\right)^x}. \quad (6)$$

where:

- **L**: Metal grade, in this case: Au_eq.
- **P**: Real thickness of the intercept.
- **d**: Distance between the NPS point to be interpolated and the center of the intercept.
- **x**: Inverse distance exponent (in the example, $x = 3$).
- **V**: Pentahedron volume.
- **n**: Borehole identification number.
- **a**: Seam triangle surface.

Notice that the software will interpolate six values per vertex or point of **NPS**: Grade multiplied by thickness and thickness for each type of intercept. In this case, the four borehole data (**S_1**, **S_2**, **S_3**, and **S_4**) will be used as they are considered to be inside its area of influence.

The same calculations will be repeated for the U and V vertices as well as for all the points inside the **NPS** point network over the seam. Then the three average grades of the $\boldsymbol{a_i}$ triangle will be:

$$a_{iA} = \frac{\frac{(L.P)_{WA}}{P_{WA}} + \frac{(L.P)_{UA}}{P_{UA}} + \frac{(L.P)_{VA}}{P_{VA}}}{3} \quad (7)$$

$$a_{iB} = \frac{\frac{(L.P)_{WB}}{P_{WB}} + \frac{(L.P)_{UB}}{P_{UB}} + \frac{(L.P)_{VB}}{P_{VB}}}{3} \quad (8)$$

$$a_{iC} = \frac{\frac{(L.P)_{WC}}{P_{WC}} + \frac{(L.P)_{UC}}{P_{UC}} + \frac{(L.P)_{VC}}{P_{VC}}}{3}. \quad (9)$$

The database will be filled with calculations as shown in Table 2. In this case the vertices have identification numbers (30038000070, 30038500060, and 30039000060). In these intercepts, five boreholes have been included.

Table 2. Calculation breakdown and interpolation of an NPS triangle

NPSID	Dist	%	D.Hole	P_R	<Au>	<Ag>	<Cu>	<As>	<Au_e>
30038000070	23.9	36.36%	C1043	1.06	1975	2.4	1200	253	1975.18
30038000070	26.9	25.50%	C1041	0.98	800	0.1	0	3900	800.00
30038000070	30.1	18.20%	C1042	3.03	9992	6.3	5433	15114	9992.76
30038000070	32.4	14.59%	C1012	0.59	2050	0.5	570	1387	2050.08
30038000070	45.3	5.34%	C1089	0.71	4850	8.4	4000	4400	4850.60
30038000070	-	100.00%	-	1.31	3299.1	2.6	1722.1	4275.1	3299.36
30038000060	19.5	59.73%	C1043	1.06	1975	2.4	1200	253	1975.18
30038000060	30.5	15.61%	C1041	0.98	800	0.1	0	3900	800.00
30038000060	34.6	10.69%	C1042	3.03	9992	6.3	5433	15114	9992.76
30038000060	36.5	9.11%	C1012	0.59	2050	0.5	570	1387	2050.08
30038000060	45	4.86%	C1089	0.71	4850	8.4	4000	4400	4850.60
30038000060	-	100.00%	-	1.2	2795.3	2.6	1544	2716.1	2795.57
30038000060	17.4	66.80%	C1043	1.06	1975	2.4	1200	253	1975.18
30038000060	29.5	13.71%	C1041	0.98	800	0.1	0	3900	800.00
30038000060	35.9	7.61%	C1042	0.59	2050	0.5	570	1387	2050.08
30038000060	36.5	7.24%	C1012	3.03	9992	6.3	2533	15114	9992.76
30038000060	42.3	4.65%	C1089	0.71	4850	8.4	4000	4400	4850.60
30038000060	-	100%	-	1.1	2562.8	2.5	1442.1	2158.1	2563.00
TOTAL	-	-	-	1.22	2885.75	2.56	1569.37	3049.37	2885.98

The following procedure is applied for attaching a volume to each triangle of T2 and obtaining a volumetric representation of the seam.

In each interpolation point a segment of equal length to the real thickness interpolated at that same point is defined; the direction of this segment will be the average of the directions perpendicular to each of the triangles sharing that point as a vertex and the segment itself will be located so that the interpolation point is at the middle of the segment.

If we connect the ends of the three segments on each corresponding side, two new triangles appear; these triangles, together with the three segments, define a volume that is called the "calculation unit" (Figure 11a).

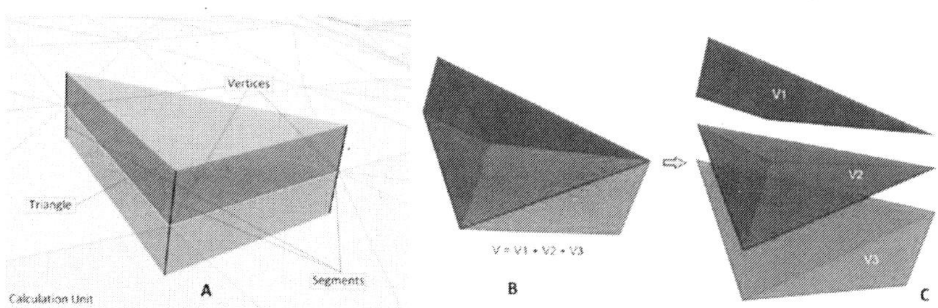

Figure 11. Calculation unit as used in the pentahedral method (**A**). Volume calculation (**B**) and calculation graphic exploded (**C**).

The volume is calculated by dividing each calculation unit (irregular octahedron) in three irregular tetrahedra, as can be seen in Figure 9b. As the coordinates of each of the four vertices of the irregular tetrahedron are known, its volume can be calculated as shown in Equation (10), where x_i, y_i, z_i are the vertex coordinates:

$$V = \frac{1}{3!}\begin{vmatrix} x_1 & y_1 & z_1 & 1 \\ x_2 & y_2 & z_2 & 1 \\ x_3 & y_3 & z_3 & 1 \\ x_4 & y_4 & z_4 & 1 \end{vmatrix} \tag{10}$$

It must be taken into account that each pair of triangles that share an edge will also share the two segments generated at each vertex edge and the corresponding faces of those two triangles will also be the same. In this way, all calculation units match and fill the entire volume that represents the seam (Figure 12).

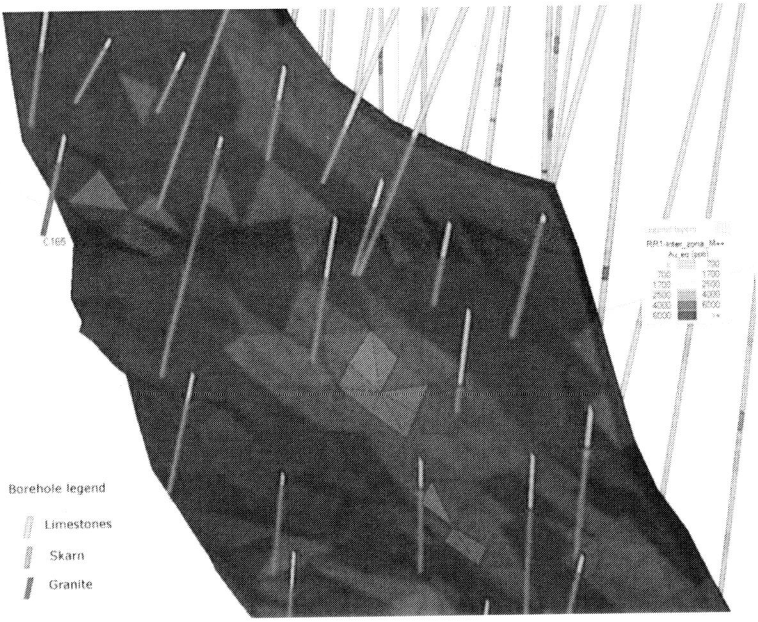

Figure 12. Mineral seam filled with calculation units, as automatically obtained by the software. Boreholes are colored by lithology. The legend refers to grade over seam (ppb).

Finally, the weight will be obtained by multiplying the calculated volume by the seam density for each one of the three calculation types.

Before initiating the calculation, all the parameters are introduced into the software (Figure 13).

4.7. Interpolation of the Positions of the Seam Surface Center

As we have seen in the calculation of intersections, once the calculations are complete and depending on the cutoff grade, minimum thickness or overbreak dilution, the intercept center does not have to match the theoretical center where the borehole cuts the seam center (T1). Differences at these locations will not affect a volume and grade calculation. Nevertheless, it will be of significance if the calculations are going to be used in the planning or exploitation mining phases; the way to correct these positions is by using position interpolation. In this way, the software relocates the center of the seam, moving it depending on how the nearest intercept centers have been moved, and the information is then stored in database fields named MOVX, MOVY, and MOVZ. This procedure relocates the NPS interpolation points by applying the inverse distance weighting method with a certain user-selectable power.

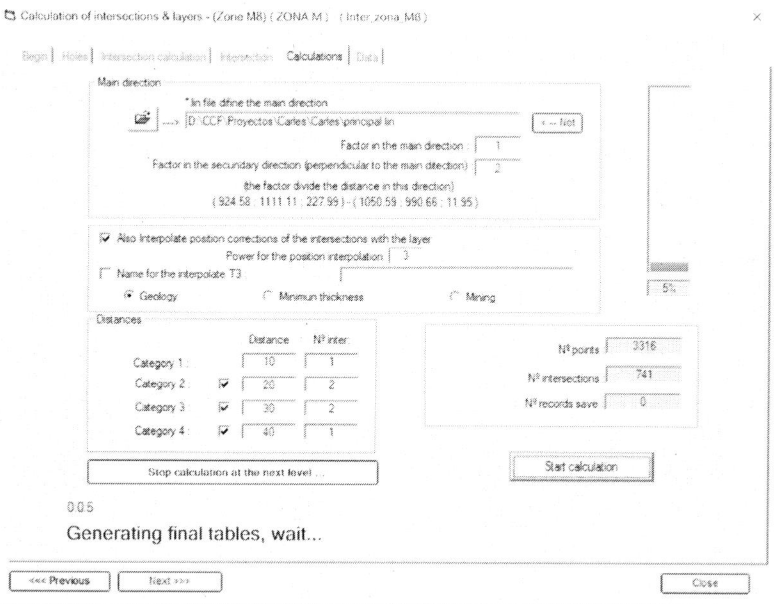

Figure 13. Calculation parameters in the interpolation of the pentahedral method.

The software can be instructed to store the new point mesh positions in a new table so they can be reused in a subsequent calculation and to obtain graphical representations (Figure 14).

The change in position of the seam center will be the same in the case of "Minimum thickness" and "Mining" interpolations, as they have the same center; in contrast, it will be different in the case of the "Geological" type, so in this case it does not need to be the same, as explained previously.

4.8. Definition of the Estimation Categories and Generation of the Calculation Units

As in any other resource estimation method, the interpolation level of confidence must be defined. This is done through the use of the so-called categories. The algorithm uses up to four categories (measured, indicated, inferred or proven, and probable). The position of samples in the octant can be used if necessary, avoiding and/or adjusting calculations according to the number of samples present in opposite octants surrounding the calculation point. In the example shown here, four numerical categories are used:

- Category 1: calculation units with an intercept at less than 10 m
- Category 2: calculation units with two intercepts at less than 20 m
- Category 3: calculation units with two intercepts at less than 30 m
- Category 4: calculation units with one intercept at less than 40 m

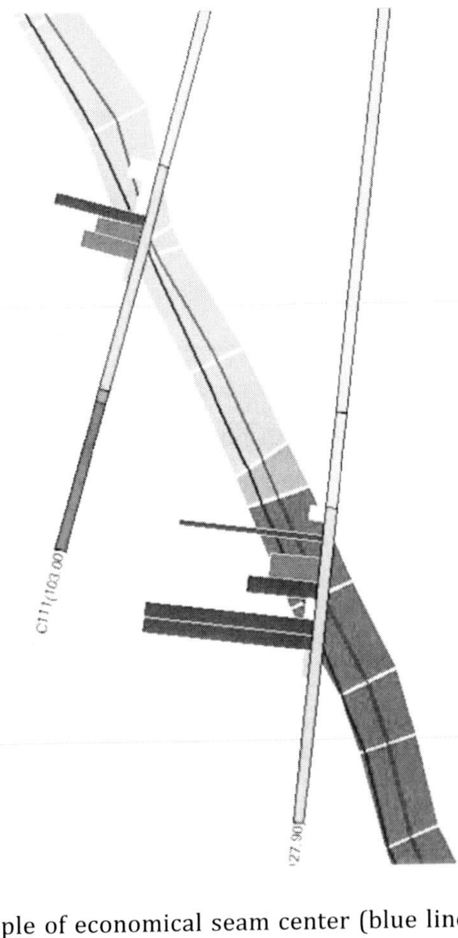

Figure 14. Example of economical seam center (blue line) vs. theoretical seam center (green line).

In these category calculations, the distance reduction defined by the main and secondary axis of the ellipsoid is also taken into account.

4.9. Calculation Units Generation and Categories Definition

Some parts of the calculation method are based on the generation of provisional data sheets created through the addition of data queries. This method allows fast operations across the entire data set.

The SQL language was also used to make the selections and differentiate, quickly and with low coding effort, between the different calculation types.

The generated information was stored in a relational database (BDT2), from where it is retrieved by the drawing module. Figure 15 shows the last tab of the calculation window, where the total amounts corresponding to a certain cutoff grade can be seen, as well as the total for each category.

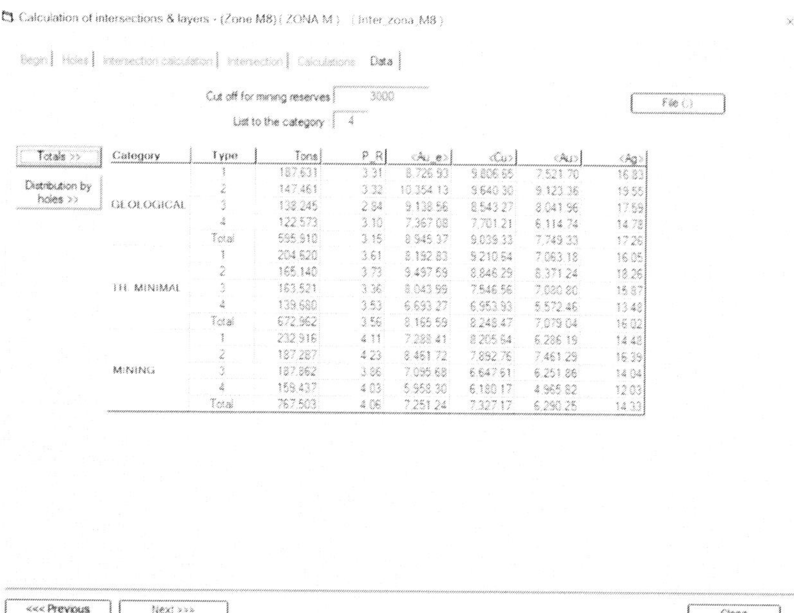

Category	Type	Tons	P_R	<Au_e>	<Cu>	<Au>	<Ag>
GEOLOGICAL	1	187,631	3.31	8,726.93	9,806.65	7,521.70	16.83
	2	147,461	3.32	10,354.13	9,640.30	9,123.36	19.55
	3	138,245	2.94	9,138.56	8,543.27	8,041.96	17.59
	4	122,573	3.10	7,367.08	7,701.21	6,114.74	14.78
	Total	595,910	3.15	8,945.37	9,039.33	7,749.33	17.26
TH. MINIMAL	1	204,620	3.61	8,192.83	9,210.64	7,063.18	16.05
	2	165,140	3.73	9,497.59	8,846.29	8,371.24	18.26
	3	163,521	3.36	8,043.99	7,546.56	7,030.80	15.87
	4	139,680	3.53	6,693.27	6,953.93	5,572.46	13.48
	Total	672,962	3.56	8,165.59	8,248.47	7,079.04	16.02
MINING	1	232,916	4.11	7,288.41	8,205.64	6,286.19	14.48
	2	187,287	4.23	8,461.72	7,892.76	7,461.29	16.39
	3	187,862	3.86	7,095.68	6,647.61	6,251.86	14.04
	4	159,437	4.03	5,958.30	6,180.17	4,965.82	12.03
	Total	767,503	4.06	7,251.24	7,327.17	6,290.25	14.33

Figure 15. Numerical results of the "Pentahedral Method," classified by intercept type and category.

The influence by percentage of each borehole of the total amount of weight and interpolated elements can also be obtained (Table 3). This allows for the checking of the results, specifically searching for balanced results and avoiding overweighed boreholes. The main results can be guided by a very low number of boreholes, which can result in a significant variation if sample mistakes are present. Figure 16 shows SQL guided results analysis, which allows for powerful, quick, and easy selection of data. Figure 17A shows the 3D results of the SQL selection and the associated legend colors. Please notice that the seam width variation is adequately simulated, as shown in Figure 17B.

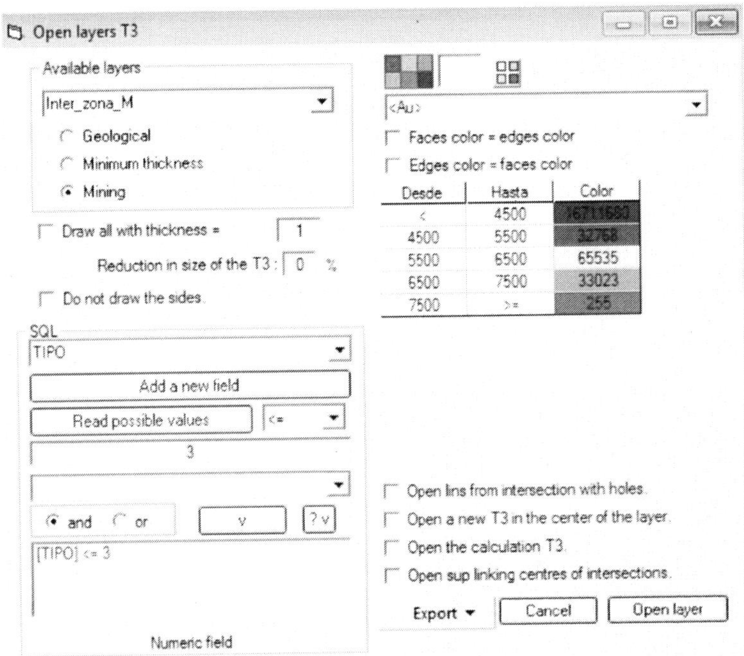

Figure 16. Assistant module used to open the pentahedral files and obtain a 3D visualization, as per SQL database conditions restrictions.

5. METHOD SUITABILITY

This method has been called the "Pentahedral Method" due to the fact that the calculation unit has a shape close to a pentahedron: two triangular faces connected by three four-sided faces. Nevertheless, it is clear that it is not a pentahedron, but rather an octahedron, as the faces with four sides could have their vertices on different planes; that is, each of them is made up of two triangular faces (Figure 11A).

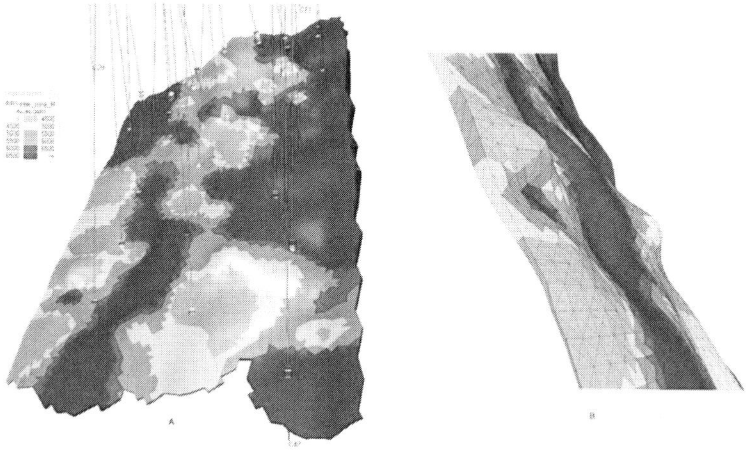

Figure 17. 3D view, pentahedral filled seam representing orebody "Zona M" of the Carlés deposit (**A**). Detail of the variable width of the calculated seam (**B**). The legend shows the Au grade, expressed in ppb.

Table 3. Influence percentage by borehole.

D.Hole	Tons	Au_e	Cu	Au	Ag
CN1014	1.5%	1.90%	0%	2%	1%
CN1015	2.9%	14.90%	9.7%	15.4%	16.6%
CN1016	1.9%	2.00%	1.2%	2.1%	1.5%
CN1017	0.4%	1.40%	0.8%	1.5%	1.2%
CN1018	0.2%	1.30%	0.9%	1.3%	0.8%
CN1019	0.6%	0.40%	0.5%	0.4%	1.0%
CN1020	1.1%	0.60%	0.7%	0.6%	0.9%
CN1021	0.7%	0.90%	1.0%	0.9%	1.4%
CN1022	0.6%	0.10%	0.0%	0.1%	0.2%
CN1023	1.5%	1.50%	1.6%	1.4%	2.1%
CN1024	1.2%	0.70%	0.0%	0.8%	0.0%
CN1025	0.6%	0.50%	0.6%	0.5%	0.9%
CN1026	0.6%	0.50%	0.7%	0.4%	0.9%
CN1027	0.9%	0.60%	1.1%	0.6%	0.8%
CN1028	0.9%	0.50%	0.6%	0.5%	0.4%
CN1029	0.6%	0.70%	0.0%	0.8%	0.0%
CN1030	0.7%	0.10%	0.0%	0.1%	0.1%
CN1031	0.7%	0.20%	0.0%	0.2%	0.0%
CN1032	0.7%	0.60%	0.8%	0.6%	0.5%

This method is a calculation alternative in the case of these seam-shaped mineral structures, as it can work in three dimensions and be completely computer-programmable.

It is suited to seams with extremely pronounced folds, as seen in Figure 18.

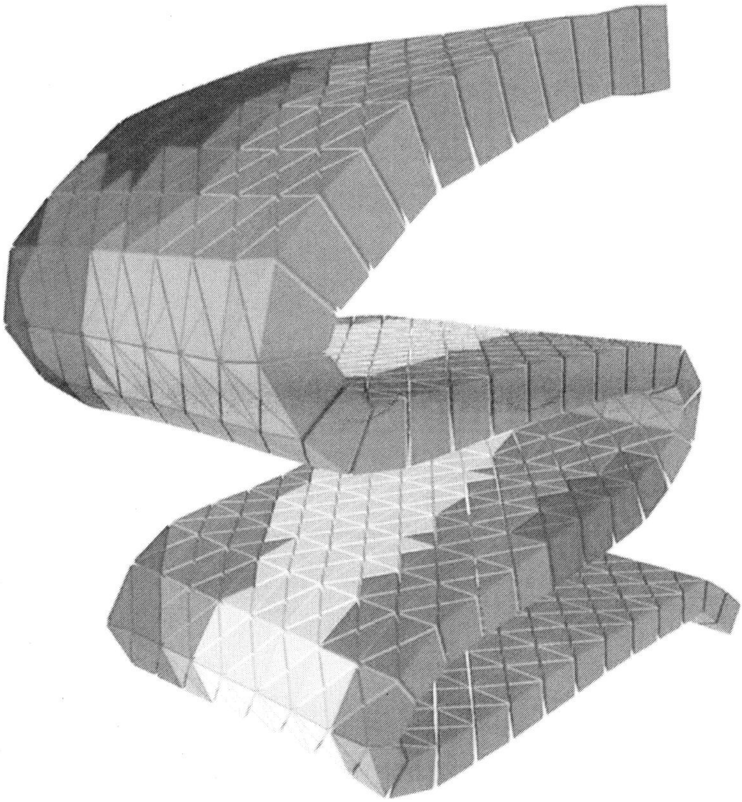

Figure 18. Extremely pronounced fold in mineral seam.

Compared to classic block models the pentahedral method offers several advantages, such as:

- Any change in the calculation parameters does not need the calculation units to be redefined.
- When defining the calculation geometry units, these can be separately prepared for subsequent exploitation and scheduling, even exporting them to other software.

- It provides a fairly accurate representation of the thickness of the seam-shaped structure even in the case of low thickness seams, which is extremely difficult to achieve with traditional block models (Figure 19).
- It interpolates intersections, not samples taken from boreholes.
- Geometrical results closely follow the deposit limits, as opposed to block models that create a stair-shaped profile (see Figure 19).
- It includes the possible internal dilution, as well as the lateral dilution due to mining overbreak.
- A minimum thickness can be defined according to mining technical conditions.

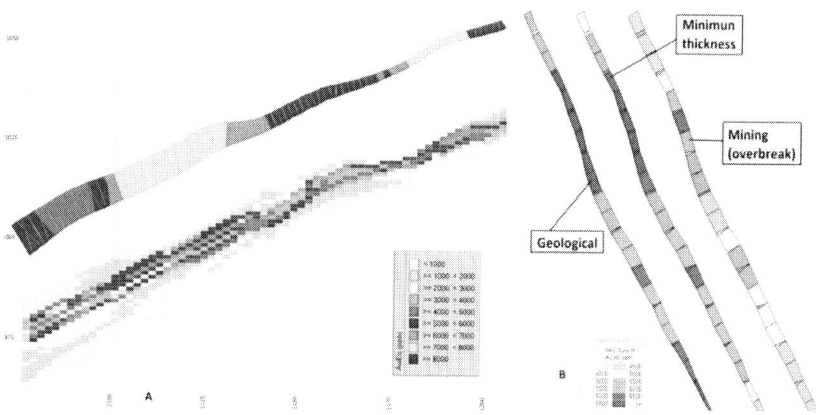

Figure 19. Graphic comparison between the block model (conventional method) and the "Pentahedral Method". Left (**A**): plane section of the blocks that define the seam colored according to grade and the same calculated seam; Right (**B**): sections of the three different types of calculation done in the same seam.

This method could be also used to model larger ore deposits or even regional scale geological seams, such as the ones shown in [21].

6. DISCUSSION AND CONCLUSIONS

This paper shows that, in the case of mineral ore bodies that are seam-shaped, using the new "Pentahedral Method" it is possible to obtain an evaluation of the reserves and resources with a higher degree of accuracy than with the classic block method.

Compared to other methods such as the polygon method, the triangle method, or the section one, this methodology is more advanced, allowing interpolation with the intercepts. This method can be applied to all types of seams, even those with extremely pronounced folds, and it allows work to be conducted in 3D and to be a good representation of the ore body to be obtained. Figure 20 is a U3D model that can be embedded in the pdf of the paper, which shows interactively what has been explained throughout this paper.

Figure 20. 3D model of the calculated seam colored by grade, including in the user selectable layers the topography and boreholes colored according to geology and sample grading.

This method also offers several additional advantages compared to the classic block model concept, mainly in its three-dimensional representation capabilities. Regarding interpolation, it also allows any method to be used, even kriging, as well as the use of search ellipsoids.

The use of SQL databases to store the results allows reports and results analysis to be issued very quickly.

The introduction of calculation variables such as "overbreak" "dilution" or "minimum thickness" gives more information from the resources data to the technician in charge of the reserves definition. These variables are related to the mining process and will simplify the reserve calculation as it will eliminate some of the resources that are not economically exploitable.

These results can be used for mine planning and design where final reserves will be defined, and have already been included in some versions of the Recmin software. At this stage the method has not been prepared to be used with other software, although it could be adapted to generate data and share it through simple ASCII files where block data are coded.

The method has passed the preliminary exam of the international patent PCT/ES03/00117, achieving positive results; the research is recognized as innovative, it implies research activity and it can be applied in the industry.

ACKNOWLEDGMENTS

We thank all those who collaborated in the development and improvement of the RecMin software. RECMIN (from: "REC ursos MIN eros" which translates as "mineral resources" in English) has been used throughout this research in order to manage data and visualization (Castañón, 2006b). RecMin is free software used in mineral deposit modeling, developed entirely by the authors over the past 25 years and thoroughly tested in real working conditions in dozens of mining projects. We also thank John Hardwick for his help in the careful English reviewing of the paper.

AUTHOR CONTRIBUTIONS

César Castañón is the corresponding author, main researcher and method and software developer. Daniel Arias and Agustin Martin-Izard have developed the geological model. Isidro Diego has checked method suitability to industry standards and has written the text. Yhonny Ruiz has done the example calculations.

REFERENCES

1. Wang, G.; Zhang, S.; Yan, C.; Song, Y.; Li, D. Mineral potential targeting and resource assessment based on 3D geological modelling in Luanchuan region, China. Comput. Geosci. **2011**, 37, 1976–1988. Minerals **2017**, 7, 72 21 of 21
2. Canadian Institute ofMining. CIMDefinition Standards onMineral Resources andMineral Reserves; CIMDefinition Standards (2014); Canadian Institute of Mining, Metallurgy and Petroleum: Quebec, QC, Canada, 2014.
3. Annels, A.E. Mineral Deposit Evaluation: A Practical Approach; Chapman & Hall: London, UK, 1991.
4. Clark, I. Practical Geostatistics; Applied Science Publishers Ltd.: London, UK, 1979.
5. Clark, I.; Harper, W.V. Practical Geostatistics 2000; Ecosse North America Llc.: Columbus, OH, USA, 2001; 342p.
6. Rendu, J.M. An Introduction to Geostatistical Methods of Mineral Evaluation; South African Institute of Mining and Metallurgy: Johannesburg, South Africa, 1981.
7. Castañón, C. Nuevo Método de Cálculo de Recursos Geológicos y Reservas Mineras para Yacimientos Tabulares y Aplicación Informática para la Gestión Integral de un Proyecto Minero. Ph.D. Thesis, Universidad de Oviedo, Asturias, Spain, 2006.
8. Silva, D.S.F.; Boisvert, J.B. Mineral resource classification: A comparison of new and existing techniques. J. S. Afr. Inst. Min. Metall. **2014**, 114, 265–273.
9. 9.Will Gemcom Produce a Leaped Frog out of Leapfrog? Available online: http://www.orefind.com/blog/orefind_blog/2012/11/14/will-gemcom-produce-a-leaped-frog-out-of-leapfrog-
10. Rossi, M.E.; Deutsch, C.V. Mineral Resource Estimation; Springer: London, UK, 2014; p. 332.
11. Boisvert, J.B.; Deutsch, C.V. Programs for kriging and sequential Gaussian simulation with locally varying anisotropy using non-Euclidean distances. Comput. Geosci. **2011**, 37, 495–510.
12. Martín-Izard, A.; Fuertes, M.; Cepedal, A.; Maldonado, C.; Pevida, L.R.; Spiering, E.; González, S.; Varela, A. Geochemical characteristisc of the Río Narcea gold belt intrusives and timing of development of the different magmatic-hydrothermal processes based on K/Ar dating. In Gold Exploration and Mining in NW Spain; Arias, D., Martín-Izard, A., Paniagua, A., Eds.; University of Oviedo: Asturias, Spain, 1998; pp. 35–42.
13. Martín-Izard, A.; Fuertes-Fuente, M.; Cepedal, A.; Moreiras, D.; García-Nieto, J.; Maldonado, C.; Pevida, L.R. The Río Narcea Gold Belt intrusions: Geology, petrology, geochemistry and timing. J. Geochem. Explor. **2000**, 71, 103–117.
14. Spiering, E.D.; Pevida, L.R.; Maldonado, C.; González, S.; García, J.; Varela, A.; Arias, D.; Martín-Izard, A. The gold belts of western Asturias and Galicia (NW Spain). J. Geochem. Explor. **2000**, 71, 89–101.
15. Pevida, L.R.; Maldonado, C.; Spiering, E.; González, S.; García, J.; Varela, A.; Martín-Izard, A.; Cepedal, A.; Fuertes, M. Geology and exploration guides along the Río Narcea gold belt. In Gold Exploration and Mining in NW Spain D; Arias, D., Martín-Izard, A., Paniagua, A., Eds.; University of Oviedo: Asturias, Spain, 1998; pp. 27–36.
16. Martın-Izard, A.; Paniagua, A.; García-Iglesias, J.; Fuertes-Fuente, M.; Boixet, L.L.; Maldonado, C.; Varela, A. The Carlés copper-gold-molybdenum skarn (Asturias, Spain): Geometry, mineral associations and metasomatic evolution. J. Geochem. Explor. **2000**, 71, 153–175.
17. Castañón, C. Manual de Ayuda del Software Minero RecMin. 2006. Available online: http://recmin.com/WP/?page_id=255 (accessed on 8 May 2017).
18. Meinert, L.D. Skarns and skarn deposits. In Ore Deposit Models; Sheahan, P.A., Cherry, M.E., Eds.; Geological Association of Canada: St. John's, NL, Canada, 1993; Volume II, pp. 117–134.
19. Bustillo, M.; López-Jimeno, C. Manual de Evaluación y Diseño de Explotaciones Mineras; Entorno Gráfico S.L.: Madrid, Spain, 1997.

20. The Stanford Geostatistical Modeling Software (SGeMS). Available online: http://sgems.sourceforge.net/
21. Galera, C.; Bennis, C.; Moretti, I.; Mallet, J.L. Construction of coherent 3D geological blocks. Comput. Geosci. **2003**, 29, 971–984.

CHAPTER 14

Pore Structure Characterization of Shale Using Gas Physisorption: Effect of Chemical Compositions

Yosep Han [1], Daewoong Kwak [1], Siyoung Q. Choi [2], Changhoon Shin [3], Youngsoo Lee [1,*] and Hyunjung Kim [1,*]

[1] *Department of Mineral Resources and Energy Engineering, Chonbuk National University, 567, Baekje-daero, Deokjin-gu, Jeonju, Jeonbuk 54896, Korea*

[2] *Department of Chemical and Biomolecular Engineering, KAIST, Daejeon 305-701, Korea*

[3] *Gas Resource Technology Center, Korea Gas Corporation, 1248 Suin-ro, Sangnok-Gu, Ansan-si, Kyounggi-do 15328, Korea*

ABSTRACT

In this study, the pore structure characteristics of Canadian Horn River basin shales with various chemical compositions were evaluated using gas physisorption analyses. The samples used in this research were obtained from two different regions (shallow and deep regions) of rock cuttings during the drilling of the shale gas field located in Horn River basin. The pore size, specific surface area, total pore volume, micropore surface area, and micropore volume of the shale samples were measured using both nitrogen and CO_2. The results indicated that the pore size was not a function of chemical composition, while distinct trends were observed for other macroscopic and microscopic pore-related properties. In particular, the greatest specific surface area and total pore volume were observed for silica-rich carbonate shales, while clay-rich siliceous shales exhibited the greatest micropore volume and micropore surface area. The trends clearly suggested that macroscopic and microscopic pore-related properties of the Canadian Horn River basin shales were closely related to their chemical composition. Furthermore, a stronger correlation was observed between the quartz content and the micropore-related physical properties of shales (i.e., the micropore surface area and micropore volume) in comparison to other properties.

Keywords: shale; pore structure; gas physisorption; micropore volume; chemical composition

1. INTRODUCTION

With an increasing natural gas consumption worldwide, and in the face of challenges involved in finding massive conventional gas resources, novel and unconventional gas resources have been actively explored. As a representative unconventional gas resource, the shale gas resources of the world are known to be abundant. More recently, the amount of gas production from shale gas has increased in Canada, the USA, and other countries, owing to the advancements in cutting-edge technologies including horizontal drilling and multi-stage hydraulic fracturing [1]. Nevertheless, the porous shale medium that stores hydrocarbon has complicated and anisotropic geological structures.

Accordingly, the condition of this gas significantly varies depending on the region [2,3,4]. Thus, the same evaluation criteria for estimating the gas in a given place can hardly be applied to a variety of regions in different countries, indicating that it is difficult to provide information about the gas found in each territory. The international considerations for effectively using shale gas include its successful supply and economic feasibility. Moreover, understanding of the storage and flow mechanism of shale gas is essential in optimizing the development plan and subsequent economic production of gas.

Shale usually has a wide range of pore size distribution [5]; therefore, various methods are required for evaluation [6,7]. The pore structure in the matrix of shale gas reservoirs is considerably complicated. Furthermore, the total matrix porosity mainly comprises nanopore distributions [6,7,8,9,10,11]. In particular, considerable amount of shale gas is adsorbed onto the nanosize pore surface, where organic matter and clay are present [10,11,12,13,14]. To estimate the shale gas in place, the total organic carbon (TOC) and clay contents [15] as well as the nanosize pore size distribution must be accurately evaluated. Thus, appropriate methods for analyzing and interpreting the shale's pore size distribution are required.

Considerable effort has been made to analyze the pore structure of shales and interpret the correlation of the results with methane (CH_4) adsorption to shales. For instance, Yuan et al. [16] used pore size distribution data determined by N_2isotherm for CH_4 adsorption and diffusion in shale. The parameters for the evaluation of CH_4 adsorption and diffusion were the size and moisture content of shale samples. Yang et al. [13] investigated the pore structures of shale samples from a fractal basis. They found that micropores had a larger fractal dimension than mesopores and macropores. Additionally, the shale samples with better fractal dimensions exhibited a larger amount of CH_4 adsorption.

Ji et al. [12] conducted methane adsorption tests using clay-rich shales and reported that montmorillonite showed greater CH_4 adsorption than other clay-rich shales due to an abundant presence of micro-mesopores, which provided extensive Brunauer–Emmett–Teller (BET) surface area. Ross and Bustin [10] reported the correlation between the shale's composition and pore structure and the corresponding CH_4 adsorption amount. Greater CH_4 adsorption was observed in shale with a higher TOC content, while the BET surface area or micropore volume had no correlation with the amount of CH_4adsorption into the shale. Moreover, the total porosity of quartz-rich shale was lower than that of clay-rich shale due to the tight-rock characteristics.

Considering these outcomes, the amount of CH_4 gas in shale seems to be affected by the pore structure of the shale. Conversely, the pore structure values (e.g., BET surface area and micropore-mesopore volume) would function as critical values for estimating the amount of CH_4 gas in place. Since the pore structure of shale changes with the mineral composition of the shale, systematic studies regarding the correlation between the chemical composition and the pore structure of shale are required. In this study, therefore, the chemical composition and the micropore and mesopore size distribution of Canadian shales with different sampling depths were analyzed. The pore size distribution was evaluated using a gas (nitrogen and carbon dioxide) physisorption method. Furthermore, the correlations of pore structure and chemical composition are provided.

Table 1. Mineralogical composition and total organic carbon (TOC; wt %) of the shale with different sampling depths.

Sampling Depth (m)	Depth Index (m)	Quartz	Feldspar/ Calcite	Dolomite/ Plagioclase	Pyrite/ Barite	Apatite/ Illite	Layer	TOC
2325–2330	2325	33	7/6	33/0	3/0	0/18	Muskwa	1.4
2335–2340	2335	40	6/2	4/6	4/0	0/38	Muskwa	0.8
2325–2360	2355	20	11/18	14/0	2/0	0/23	Upper Otterpark	1.2
2375–2380	2375	18	0/54	6/6	1/0	0/10	Middle Otterpark	0.5
2405–2410	2405	26	4/5	17/2	1/1	1/31	Middle Otterpark	1.6
2435–2440	2435	35	6/3	22/3	10/0	0/19	Lower Otterpark	1.3
2445–2450	2445	58	4/2	2/4	4/1	0/28	Evie	0.3
2465–2470	2465	37	2/50	2/0	1/0	0/8	Evie	1.7

2. MATERIALS AND METHODS

2.1. Shale Sample and Preparation

The samples used in this study were obtained from rock cuttings during drilling of the shale gas field located in Horn River basin, Canada. Twenty-four samples were obtained from depths of 2300 m to 2500 m, which was also divided into two depth regions (shallow region: 2325 m to 2380 m underground, and deep region: 2405 m to 2470 m underground). The entire depth region was separated into eight sub-regions using a depth interval of 5 m for each sub-region. Table 1 and Table 2 provide the depth index that represents the lowest depth value of each section. More detailed information of Horn River basin and the stratigraphic column on each shale sample can be found elsewhere [14,17]. Three representative samples were chosen from each sub-region to increase the reliability of the results. Total organic carbon values were determined by subtracting total inorganic carbon (IC) from total carbon values (TC); TOC = TC − IC. A TOC analyzer manufactured by Shimadzu TOC-VCPH (Shimadzu, Tokyo, Japan) was used to analyze 50 mg samples.

2.2. X-ray Diffraction (XRD) Analysis

For XRD analysis, the shale samples were dried at 50 °C for 5 h after being milled to a fine particle size, and random grain mounts were prepared and analyzed. XRD patterns were recorded with a Bruker D8 HRXRD X-ray diffractometer (Bruker, Karlsruhe, Germany) using Ni-filtered Cu Kα radiation at 40 kV and 40 mV, as well as a scan rate of 1°/min with a step size of 0.01° in 10°–80° 2-theta. The mineralogy was quantified by Rietveld refinement analyses using the TOPAS 3 program [18,19].

2.3. Low-Pressure N_2 and CO_2 Isotherm Analysis

Low pressure gas physisorption analyses using both nitrogen (N_2) and carbon dioxide (CO_2) gases were used to measure the pore size distribution between the pore sizes of 0.35 and 150 nm and describe the pore space of organic, microporous shales, as average pore size distributions are in the nanometer scale, into which gas can effectively penetrate. Between 1 and 2 g of cutting samples (size < 1 mm) prepared for gas physisorption analysis were first outgassed for a minimum of 4 days in a vacuum oven [7].

Table 2. Lithofacies and physical properties of the shales with different sampling depths.

Depth Index (m)	Shale Lithofacies/ Sample Name	Surface Area (m^2/g) [a]	Total Pore Volume (cm^3/g) [b]	BJH Pore Size (nm) [c]	Micropore Volume (mm^3/g) [d]	Micropore Surface Area (m^2/g) [d]
			Shallow Region			
2375	Silica-rich carbonate shale/sample-1	1.61 ± 0.12	0.0123 ± 0.0012	20.3 ± 1.1	0.23 ± 0.08	1.51 ± 0.21
2325	Carbonate-siliceous shale/sample-3	0.82 ± 0.09	0.0045 ± 0.0008	24.5 ± 1.2	0.22 ± 0.04	1.98 ± 0.35
2335	Clay-rich siliceous shale/sample-5	1.00 ± 0.20	0.0084 ± 0.0010	21.8 ± 0.9	0.30 ± 0.09	2.17 ± 0.32
2355	Mixed shale/sample-7	1.25 ± 0.17	0.0074 ± 0.0009	18.3 ± 2.1	0.28 ± 0.11	1.79 ± 0.32
			Deep Region			
2465	Silica-rich carbonate shale/sample-2	1.79 ± 0.13	0.0132 ± 0.0009	20.9 ± 1.7	0.41 ± 0.02	2.57 ± 0.39
2435	Carbonate-siliceous shale/sample-4	1.28 ± 0.11	0.0094 ± 0.0006	19.8 ± 2.2	0.37 ± 0.05	2.56 ± 0.53
2445	Clay-rich siliceous shale/sample-6	1.13 ± 0.22	0.0091 ± 0.0011	20.7 ± 1.5	0.66 ± 0.11	4.38 ± 0.28
2405	Mixed shale/sample-8	1.65 ± 0.09	0.0101 ± 0.0005	18.6 ± 0.9	0.83 ± 0.09	3.88 ± 0.33

[a] Calculated from the N_2 adsorption isotherms in Figure 2 using by BET model. [b] Determined at the relative pressure of 0.99 from the N_2 adsorption isotherms in Figure 2. [c] Determined by averaging the BJH pore size distribution curve in Figure 3. The data in Figure 3 were determined by the BJH model from the adsorption branch of the N_2 isotherm in Figure 2. [d] Determined at the micropore size of 2 nm from the cumulative pore volume-micropore size curve in Figure 5. The data in Figure 5 were calculated by the D–R method from the adsorption branch of CO_2 isotherm in Figure 4.

N_2 adsorption-desorption isotherms were collected at 77 K using a Micromeritics 3Flex. The total pore volume was determined from the amount of N_2 adsorbed at a relative pressure, P/P_0, of 0.99; where P is the gas vapor pressure in the system and P_0 is the vapor pressure at the temperature of interest. The meso-macropore size distributions (2.0–150 nm) were analyzed from adsorption data in the partial relative pressure range of 0.3–0.99 using the Barrett–Joyner–Halenda (BJH) method [20,21]. The specific surface area was assessed from N_2 adsorption data in the partial relative pressure range of 0.05–0.30 using the BET method [22,23] based on the following equation:

$$\frac{1}{w\left(\frac{P_0}{P}\right)} = \frac{1}{W_m C} + C - \frac{1}{W_m C\left(\frac{P_0}{P}\right)} \quad (1)$$

where W is the weight of the nitrogen gas at relative pressure P_0/P, W_m is the weight of the monolayer adsorbent (nitrogen), and C is the BET constant which relates to the sorption energy between the adsorbent and adsorbate.

Carbon dioxide (CO_2) adsorption data (273 K) were also collected on the same apparatus. For mircropore analysis, CO_2 adsorption isotherms were measured over a relative pressure range of 4.5×10^{-5}–3.2×10^{-2} at 273 K and monolayer capacities were determined using the Dubinin–Radushkevich (D–R) equation [24]:

$$\log V = \log V_0 - S\log^2\left(\frac{P}{P_0}\right) \quad (2)$$

where V is the volume of the sorbed gas at equilibrium pressure (cm^3/g, Standard Temperature and Pressure, STP), V_0 is the total micropore volume (cm^3/g, STP), S is a constant, P is pressure and P_0 is saturation vapor pressure. The lower limit of micropore depends on the kinetic diameter of the probing gas molecule, such as 0.35 nm for CO_2. Micropore volumes (pore size < 2.0 nm) were measured by the volume of adsorbed CO_2 (which completely fills the micropores), expressed in terms of bulk liquid at atmospheric pressure and 273 K [24,25]. As discussed by Dubinin and Stoeckli [26], the theory of volume filling of micropores is a rational basis for the description of microporous structures in carbonaceous materials. The measurements for N_2 and CO_2 isotherms were repeated the times for each shale sample at the same sampling depth.

3. RESULTS AND DISCUSSION

3.1. Mineralogy and Shale Lithofacies

The chemical composition of eight representative samples corresponding to each sub-region is summarized in Table 1. All the samples include comparatively high levels of quartz, calcite, dolomite, and illite contents. To increase the reliability of the results regarding the changes in the pore structure dependent on chemical composition, two shales having a similar chemical composition to each other were grouped from two different depth regions (shallow versus deep). Note that four major chemical compositions (i.e., quartz dominated shale, carbonaceous shale, clay dominated shale, and mixed shale) were eventually selected based on the information of the sCore Lithofacies Classification [27]. Specifically, the Horn River basin can be divided into three groups: Muskwa, Otter Park, and Evie formation [28]. Evie and Otter Park were known to be mainly composed of deep to marginal tropical shales and shallow tropical carbonates, while Muskwa consisted of deep to marginal tropical shales. Their ternary diagram based on the weight percent of minerals (Table 1) is also shown in Figure 1.

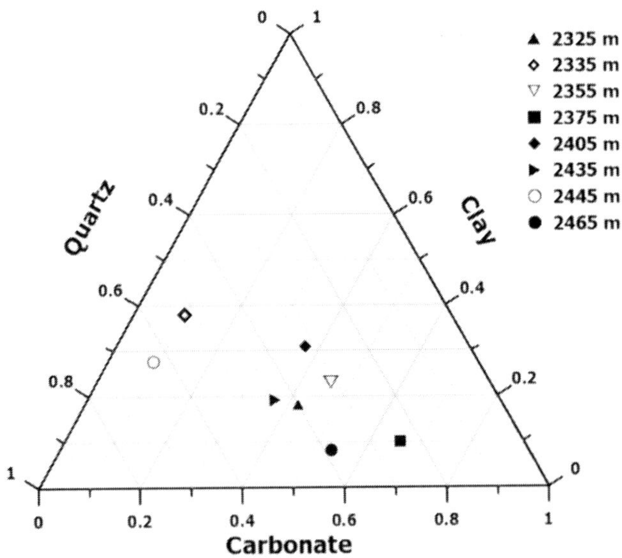

Figure 1. Ternary diagram of the shales. The plotting is based on the quartz–clay–carbonate system, and the plotted data were obtained by normalizing the weight percent information summarized in Table 1.

As shown in Figure 1, two shale samples classified as the same category have a similar chemical composition but are from different depths. In the shale samples obtained from 2375 m and 2465 m underground, the quartz level was less than 40%, and carbonate was the primary component (50–80%). The sample classification and lithofacies of this study are shown in Table 2. In this study, all samples were categorized into four groups (i.e., four different chemical compositions) based on a 50 m or more difference in the sampling depth (i.e., shallow versus deep) (see Figure 1 and Table 2).

3.2. Pore Characteristics of the Shale with Chemical Composition

Figure 2 shows the adsorption-desorption isotherm curves for every sample, which were observed as isotherm type II [6,24]. In all the shale samples, a complete monolayer adsorption was observed up to the relative pressure level of approximately 0.1 before a multi-layer adsorption was initiated [24]. Therefore, a multi-layer adsorption occurred on the shale surface, and the meso-macropore-level pores were found to be slit-shaped [24]. The hysteresis loop of all the shale samples was confirmed as H3 (hysteresis loop type 3). Thus, the meso-macropore size distribution was determined using the desorption branch based on the BJH model [24].

Figure 3 shows the distribution of the meso-macropore pore size for all the samples. The shale samples that had four different chemical compositions exhibited slightly different curve shapes. Nevertheless, their mean pore size ranged from 18 nm to 25 nm without significant difference (see Table 2). This indicated that the chemical composition did not influence the meso-macropore size distribution, regardless of the sampling depth. However, a distinct trend in the specific surface area and the total pore volume was observed among the samples with different chemical compositions, which is valid for two different sampling regions.

Regardless of the sampling region, silica-rich carbonate shale samples consistently exhibited greater values of specific surface area than other samples (i.e., carbonate-siliceous shale, clay-rich siliceous shale and mixed shale) (Table 2). Quantitatively, for the shallow region, the BET specific surface area of sample-1 was determined to be 1.61 m^2/g, while sample-3, sample-5, and sample-7 were 0.82, 1.00 and 1.25 m^2/g, respectively. The same trend was also observed for the deep region. For the total pore volume in the shallow region, sample-1 exhibited 0.0123 cm^3/g. This value is much greater than those of other samples. Similar to the specific surface area, this observation is also true for the deep region. This trend is likely due to the difference in the amount of carbonate in the samples. Previous studies reported that carbonate minerals are easily dissolved relative to other minerals, resulting in the greater development of meso- or macropores [29].

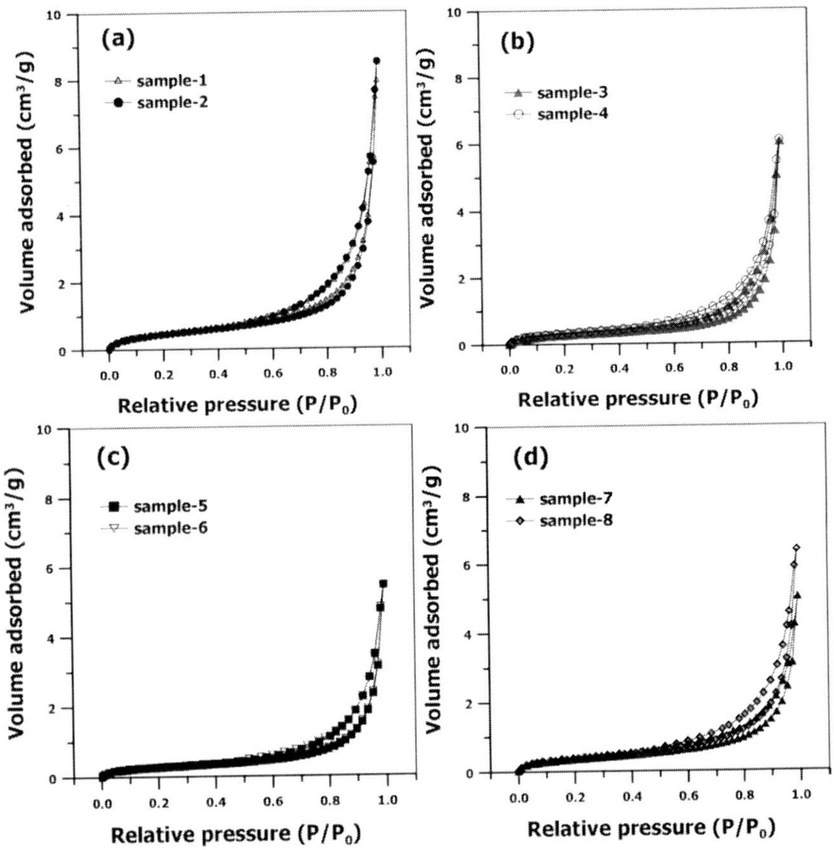

Figure 2. An average plot of nitrogen adsorption–desorption isotherms of the shale samples with different chemical compositions at two different depth regions: (**a**) silica-rich carbonate shale; (**b**) carbonate-siliceous shale; (**c**) clay-rich siliceous shale; and (**d**) mixed shale.

Figure 4 shows the carbon dioxide adsorption isotherms of all the samples with different chemical compositions. The corresponding micropore properties (i.e., micropore volume and micropore surface area) are also presented in Figure 5 and Table 2. Unlike the observations for the specific surface area and total pore volume, the clay-rich siliceous shale samples (sample-5 and 6) exhibited the greatest values compared with the silica-rich carbonate shale (sample-1 and 2) and carbonate-siliceous shale (sample-3 and 4) samples. Note that the mixed shale samples (sample-7 and 8) were excluded from this comparison, since no clear trend was observed at either the shallow or deep regions. Specifically, for the shallow region, the micropore volume of sample-5 was found to be 0.30 mm^3/g, which is relatively greater than those for sample-1 and sample-3 (0.23 and 0.22 mm^3/g, respectively).

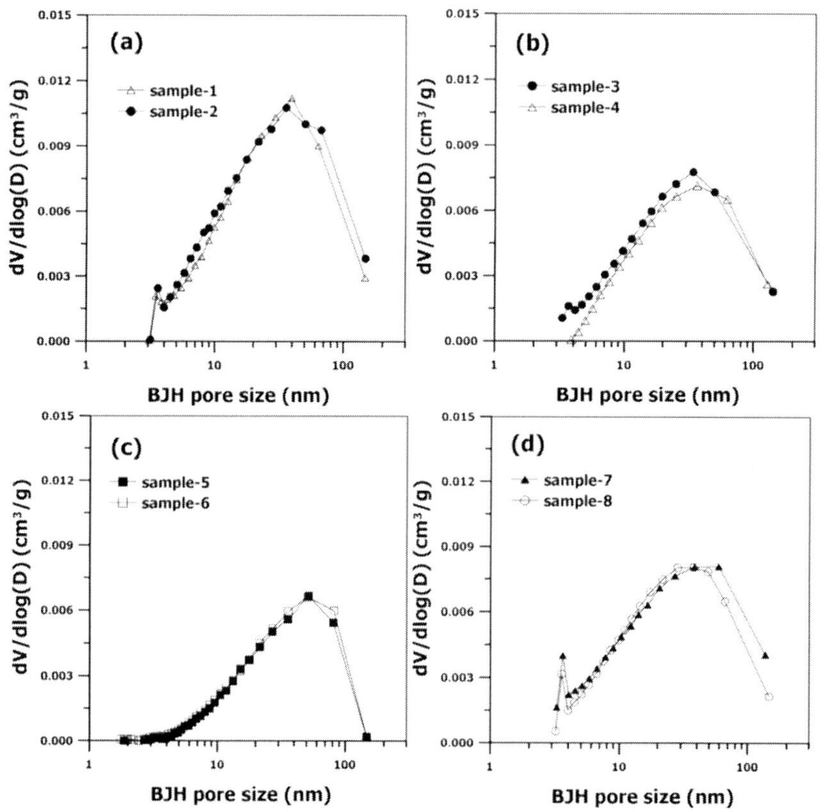

Figure 3. An average plot of Barrett–Joyner–Halenda (BJH) pore size distribution curves of the shale samples with different chemical compositions at two different depth regions: (**a**) silica-rich carbonate shale; (**b**) carbonate-siliceous shale; (**c**) clay-rich siliceous shale; and (**d**) mixed shale. The dV/dlog (D) represents the BJH differential pore volume of the shale samples.

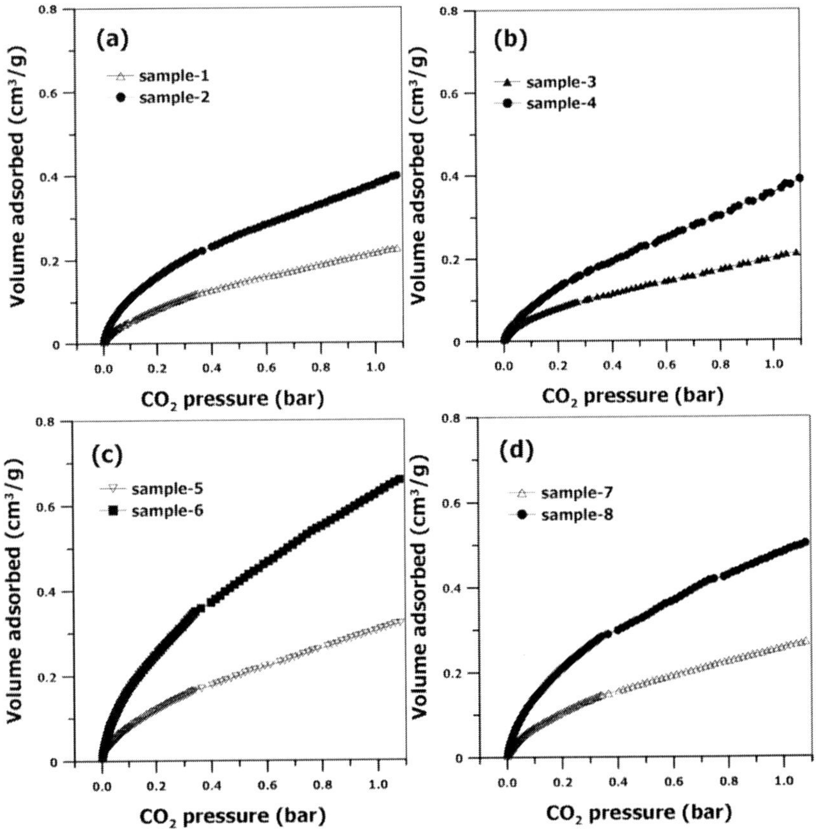

Figure 4. An average plot of carbon dioxide adsorption isotherms of the shale samples with different chemical compositions at two different depth regions: (**a**) silica-rich carbonate shale; (**b**) carbonate-siliceous shale; (**c**) clay-rich siliceous shale; and (**d**) mixed shale.

A similar trend was observed for the micropore surface area, where a greater value for sample-5 (2.17 m^2/g) was observed than that for sample-1 (1.51 m^2/g) and sample-3 (1.98 m^2/g). In addition, the differences in the two micropore properties became more obvious for the deep region. Sample-6 showed a 1.6–1.8 increase in the micropore volume and micropore surface area than that for sample-2 and sample-4. This suggests that the shales containing high amount of clay minerals (i.e., sample-5 and -6) show better development of micropore structures, an observation that could be explained by the unique crystalline structure of the clays (i.e., a layer structure with a stack of sheet) [30]. Overall, the results from the specific surface area, the total pore volume, the micropore volume, and the micropore surface area suggest that the nanopore formation of the Canadian shale samples used in this study is likely to be affected by the chemical composition.

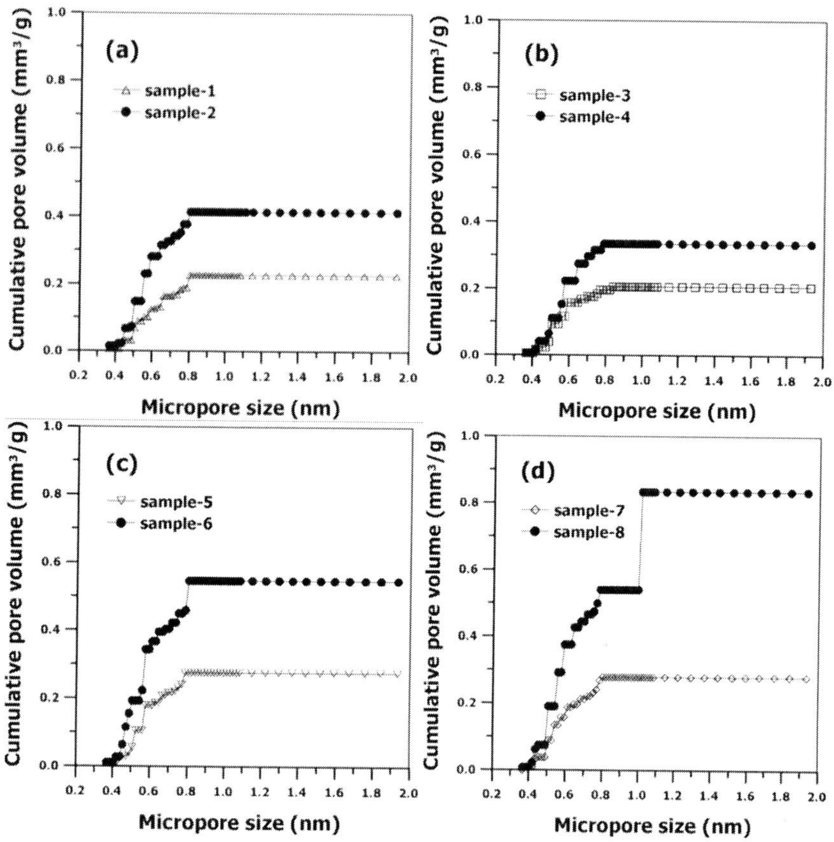

Figure 5. An average plot of cumulative micropore volume of the shale samples with different chemical compositions at two different depth regions: (**a**) silica-rich carbonate shale; (**b**) carbonate-siliceous shale; (**c**) clay-rich siliceous shale; and (**d**) mixed shale.

3.3. Correlation of the Pore Structure of the Shale with Chemical Composition

Since the purpose of the present study is to examine the correlation of both the micropore surface area and the micropore volume of shales with its chemical composition, we plotted each physical property as a function of the contents of quartz, carbonate, and clay. The results are presented in Figure 6. Figure 6a shows the correlation between the micropore surface area and the chemical composition. The coefficients of determination (R-squared value) for the fitting results as a function of quartz, carbonate, and clay were determined to be 0.6359, 0.6461 and 0.1809, respectively. This indicates that the quartz content has the strongest correlation with the micropore surface area of shales.

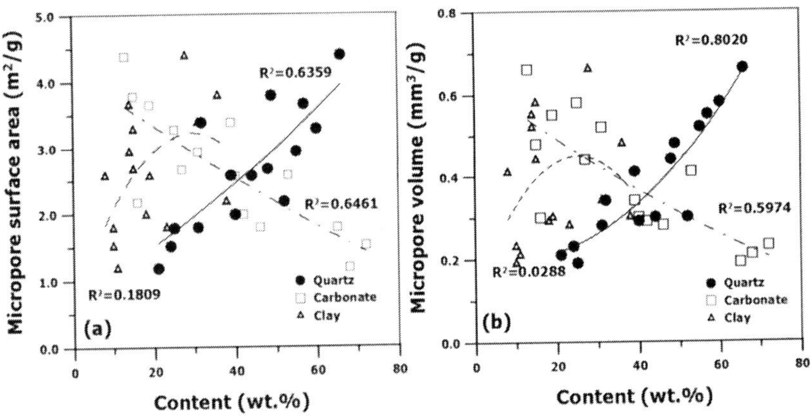

Figure 6. Correlation plots between two physical properties of shales ((**a**) micropore surface area and (**b**) micropore volume) and main, individual chemical components (quartz (black circle), carbonate (void square), and clay (void triangle)). Coefficient of determination (R^2) values for each correlation plot are also presented for comparison.

Also note that the carbonate content exhibits a similar coefficient of determination (~0.6461) with the quartz, but, unlike quartz, it shows a negative correlation. The correlation results between the micropore volume and the chemical composition of shales are also presented in Figure 6b. The coefficient of determination of the fitting results as a function of the quartz content was determined to be 0.8020, which was greater than those of the carbonate and clay contents. This result is consistent with that for the micropore surface area.

While the quartz content has a strong relationship with the micropore volume, the clay content shows even worse correlation than the micropore surface area (the coefficient of determination = 0.0288). The coefficient of determination of carbonate was 0.5974, indicating a negative tendency similar to the micropore surface area. Accordingly, overall results suggest that the quartz content might be a good indicator for estimating the amount of shale gas adsorbed compared to the carbonate and clay contents, and that the presence of clay clearly has no correlation with the two physical properties investigated herein.

Unlike our results, a previous study reported that no micropores or mesopores existed in siliceous/quartz-rich shale (high Si/Al), and that the porosity was as low as 1% [10]. Conversely, clay-rich shale (low Si/Al) had a high porosity. However, it should be noted that the reported results were not obtained from the CO_2 adsorption but from Hg intrusion analysis. Additionally, the shale samples used in their study had a wide range of specific surface areas unlike those used in our study, which could lead to the difference in the correlation trend.

Since micropore volume has a positive correlation with specific surface area [8,10,31], our results have an important significance regarding the feasibility of assessing the amount of shale gas from ore-related properties of shales. Nevertheless, it may not be appropriate to generalize that the higher quartz content of shales results in the greater formation of micropores. To enhance the reliability of our outcomes, repetitive tests using a larger sample size are necessary. Additionally, diverse shale samples from various regions in Canada and other countries must be analyzed.

4. CONCLUSIONS

In the present study, pore-related physical properties such as the BET surface area, pore volume, and pore size of shales were thoroughly tested to examine the correlation with chemical composition of shales obtained from rock cuttings during drilling of the shale gas field located in Horn River basin. The key findings are presented below:

1) The N_2 and CO_2 gas physisorption technique was found to be suitable for analyzing the pore-related properties of shales (e.g., BET surface area, pore volume, and pore size).

2) Chemical composition did not play a critical role in the BJH pore size. However, the chemical composition was critical for the specific/micropore surface area and the total/micropore volume. The silica-rich carbonate shale showed the greatest specific surface area and total pore volume, while the micropore volume and micropore surface area was found to be the greatest for the clay-rich siliceous shale samples.

3) Both the micropore surface area and micropore volume were found to be a strong function of the quartz content as compared to the carbonate and clay contents.

The outcomes from the present study suggest that pore-related properties (especially micropore properties) might be a good candidate for estimating shale gas adsorption characteristics. However, acquiring more datasets for shales samples with diverse chemical compositions and from different regions, as well as correlating their microscopic properties (e.g., total/micropore volume, specific/micropore surface area) with their macroscopic properties (e.g., core porosity, core permeability) is guaranteed for universal application.

ACKNOWLEDGMENTS

This work was supported by the Energy Efficiency & Resources Core Technology Program of the Korea Institute of Energy Technology Evaluation and Planning (KETEP), granted financial resource from the Ministry of Trade, Industry & Energy, Republic of Korea (No. 20132510100060), the Ministry of Education, the Korea Energy and Mineral Resources Engineering Program (KEMREP), and the research funds of Chonbuk National University in 2013.

AUTHOR CONTRIBUTIONS

Yosep Han conceived the idea and wrote the paper; Yosep Han and Daewoong Kwak performed experiments; Siyoung Q. Choi analyzed the numerical results. Changhoon Shin contributed to XRD analysis and provide samples; Youngsoo Lee reviewed the paper and gave feedback for improvement; and, Hyunjung Kim drafted the paper and supervised the entire research. All authors commented on the paper.

REFERENCES

1. Kuuskraa, V. United States energy information administration. In *World Shale Gas Resources: An Initial Assessment of 14 Regions Outside the United States*; US Department of Energy: Washington, DC, USA, 2011.
2. Bustin, A.M.; Bustin, R.M. Importance of rock properties on the producibility of gas shales. *Int. Coal Geol.* **2012**, *103*, 132–147.
3. Chen, S.; Zhu, Y.; Wang, H.; Liu, H.; Wei, W.; Fang, J. Shale gas reservoir characterisation: A typical case in the southern sichuan basin of China. *Energy* **2011**, *36*, 6609–6616.

4. Hu, Q.-H.; Liu, X.-G.; Gao, Z.-Y.; Liu, S.-G.; Zhou, W.; Hu, W.-X. Pore structure and tracer migration behavior of typical American and Chinese shales. *Pet. Sci.* **2015**, *12*, 651–663.
5. Zhang, P.; Hu, L.; Meegoda, J.N. Pore-scale simulation and sensitivity analysis of apparent gas permeability in shale matrix. *Materials* **2017**, *10*, 104.
6. Clarkson, C.R.; Solano, N.; Bustin, R.; Bustin, A.; Chalmers, G.; He, L.; Melnichenko, Y.B.; Radliński, A.; Blach, T.P. Pore structure characterization of north american shale gas reservoirs using usans/sans, gas adsorption, and mercury intrusion. *Fuel* **2013**, *103*, 606–616.
7. Chalmers, G.R.; Bustin, R.M.; Power, I.M. Characterization of gas shale pore systems by porosimetry, pycnometry, surface area, and field emission scanning electron microscopy/transmission electron microscopy image analyses: Examples from the barnett, woodford, haynesville, marcellus, and doig units. *AAPG Bull.* **2012**, *96*, 1099–1119.
8. Loucks, R.G.; Reed, R.M.; Ruppel, S.C.; Jarvie, D.M. Morphology, genesis, and distribution of nanometer-scale pores in siliceous mudstones of the mississippian barnett shale. *J. Sediment. Res.* **2009**, *79*, 848–861.
9. Kuila, U.; Prasad, M. Specific surface area and pore-size distribution in clays and shales. *Geophys. Prospect.* **2013**, *61*, 341–362.
10. Ross, D.J.; Bustin, R.M. The importance of shale composition and pore structure upon gas storage potential of shale gas reservoirs. *Mar. Pet. Geol.* **2009**, *26*, 916–927.
11. Zhang, T.; Ellis, G.S.; Ruppel, S.C.; Milliken, K.; Yang, R. Effect of organic-matter type and thermal maturity on methane adsorption in shale-gas systems. *Org. Geochem.* **2012**, *47*, 120–131.
12. Ji, L.; Zhang, T.; Milliken, K.L.; Qu, J.; Zhang, X. Experimental investigation of main controls to methane adsorption in clay-rich rocks. *Appl. Geochem.* **2012**, *27*, 2533–2545.
13. Yang, F.; Ning, Z.; Liu, H. Fractal characteristics of shales from a shale gas reservoir in the sichuan basin, China. *Fuel***2014**, *115*, 378–384.
14. Kim, J.; Kim, D.; Lee, W.; Lee, Y.; Kim, H. Impact of total organic carbon and specific surface area on the adsorption capacity in Horn River shale. *J. Pet. Sci. Eng.* **2017**, *149*, 331–339.
15. Gasparik, M.; Ghanizadeh, A.; Bertier, P.; Gensterblum, Y.; Bouw, S.; Krooss, B.M. High-pressure methane sorption isotherms of black shales from the Netherlands. *Energy Fuels* **2012**, *26*, 4995–5004.
16. Yuan, W.; Pan, Z.; Li, X.; Yang, Y.; Zhao, C.; Connell, L.D.; Li, S.; He, J. Experimental study and modelling of methane adsorption and diffusion in shale. *Fuel* **2014**, *117*, 509–519.
17. Kam, P.; Nadeem, M.; Novlesky, A.; Kumar, A.; Omatsone, E.N. Reservoir characterization and history matching of the Horn River shale: An integrated geoscience and reservoir-simulation approach. *J. Can. Pet. Technol.* **2015**.
18. Rietveld, H. Line profiles of neutron powder-diffraction peaks for structure refinement. *Acta Crystallogr.* **1967**, *22*, 151–152.
19. Lee, E.; Han, Y.; Park, J.; Hong, J.; Silva, R.A.; Kim, S.; Kim, H. Bioleaching of arsenic from highly contaminated mine tailings using Acidithiobacillus thiooxidans. *J. Environ. Manag.* **2015**, *147*, 124–131.
20. Barrett, E.P.; Joyner, L.G.; Halenda, P.P. The determination of pore volume and area distributions in porous substances. I. Computations from nitrogen isotherms. *J. Am. Chem. Soc.* **1951**, *73*, 373–380.
21. Han, Y.; Kim, H.; Park, J.; Lee, S.-H.; Kim, J.-Y. Influence of Ti doping level on hydrogen adsorption of mesoporous Ti-SBA-15 materials prepared by direct synthesis. *Int. J. Hydrogen Energy* **2012**, *37*, 14240–14247.
22. Brunauer, S.; Emmett, P.H.; Teller, E. Adsorption of gases in multimolecular layers. *J. Am. Chem. Soc* **1938**, *60*, 309–319.
23. Han, Y.; Choi, J.; Tong, M.; Kim, H. Synthesis and characterization of high-surface-area millimeter-sized silica beads with hierarchical multi-modal pore structure by the addition of agar. *Mater. Charact.* **2014**, *90*, 31–39.
24. Lowell, S.; Shields, J.E.; Thomas, M.A.; Thommes, M. *Characterization of Porous Solids and Powders: Surface Area, Pore Size and Density*; Springer Science & Business Media: Berlin, Germany, 2012; Volume 16.

25. Rouquerol, J.; Avnir, D.; Fairbridge, C.; Everett, D.; Haynes, J.; Pernicone, N.; Ramsay, J.; Sing, K.; Unger, K. Recommendations for the characterization of porous solids (technical report). *Pure Appl. Chem.* **1994**, *66*, 1739–1758.
26. Dubinin, M.; Stoeckli, H. Homogeneous and heterogeneous micropore structures in carbonaceous adsorbents. *J. Colloid Int. Sci.* **1980**, *75*, 34–42.
27. Wang, G.; Carr, T.R. Marcellus shale lithofacies prediction by multiclass neural network classification in the appalachian basin. *Math. Geosci.* **2012**, *44*, 975–1004.
28. Ferri, F.; Hickin, A.S.; Reyes, J. Horn river basin–equivalent strata in besa river formation shale, northeastern british columbia (NTS 094K/15). *Geosci. Rep.* **2012**, 1–15.
29. Abbaszadeh, M.; Nasiri, M.; Riazi, M. Experimental investigation of the impact of rock dissolution on carbonate rock properties in the presence of carbonated water. *Environ. Earth Sci.* **2016**, *75*, 1–6.
30. Chiou, W.; Faas, R.; Kasprowicz, J.; Li, H.; Lomenick, T.; OBrien, N.; Pamukcu, S.; Smart, P.; Weaver, C.; Yamamoto, T. *Microstructure of Fine-Grained Sediments: From Mud to Shale*; Springer Science & Business Media: Berlin, Germany, 2012.
31. Labani, M.M.; Rezaee, R.; Saeedi, A.; Al Hinai, A. Evaluation of pore size spectrum of gas shale reservoirs using low pressure nitrogen adsorption, gas expansion and mercury porosimetry: A case study from the perth and canning basins, western australia. *J. Pet. Sci. Eng.* **2013**, *112*, 7–16.

CHAPTER 15

Mineralogy and origin of coarse-grained segregations in the pyrometallurgical Zn-Pb slags from Katowice-Wełnowiec (Poland)

R. Warchulski[1] A. GawędaJ[1]. Janeczek[1] M. Kądziołka-Gaweł[2]

1 Faculty of Earth Sciences, University of Silesia, Będzińska 60, 41-200 Sosnowiec, Poland

2 Faculty of Mathematics, Physics and Chemistry, University of Silesia, Bankowa 14, 40-007 Katowice, Poland

ABSTRACT

The unique among pyrometallurgical slags, coarse-grained (up to 2.5 cm) segregations (up to 40 cm long) rimmed by "aplitic" border zones occur within holocrystalline historical Zn-smelting slag in Katowice, S Poland. Slag surrounding the segregations consists of olivine, spinel series, melilite, clinopyroxene, leucite, nepheline and sulphides. Ca-olivines, kalsilite and mica compositionally similar to oxykinoshitalite occur in border zones in addition to olivine, spinel series and melilite. Miarolitic and massive pegmatite-like segregations are built of subhedral crystals of melilite, leucite, spinel series, clinopyroxene and hematite. Melilite, clinopyroxenes and spinels in the segregations are enriched in Zn relatively to original slag and to fine-grained border zones. The segregations originated as a result of crystallization from residual melt rich in volatiles (presumably CO_2). The volatile-rich melt was separated during fractional crystallization of molten slag under the cover of the overlying hot (ca. 1250 °C) vesicular slag, preventing the escape of volatiles. That unique slag system is analogous to natural magmatic systems.

1. INTRODUCTION

Pyrometallurgical slags from base-metal smelting have recently been studied extensively mainly with the purpose of assessing their environmental impact for many of them contain elevated concentrations of potentially toxic metals, including As, Cd, Cu, Pb and Zn (e.g. Ettler et al. 2001; Puziewicz et al. 2007; Álvarez-Valero et al. 2009, Piatak and Seal 2010; Vítková et al. 2010; Kierczak et al. 2010; Ettler and Johan 2014). Those metals may be partitioned among phases with different leaching potential during weathering. Therefore, the detailed knowledge of phase composition of slags is prerequisite for understanding their leaching behaviour. Pyrometallurgical slags are also interesting from the petrological perspective, as their structures, textures and phase composition are analogous to those in mafic volcanic rocks (e.g. Vernon 2004; Puziewicz et al. 2007).

The chemical and mineral composition of slags depend on the composition of primary ore, additives used in the smelting processes (e.g. fluxing material) and on the smelting technology. Slags originated from Zn and Pb smelting are composed mainly of silicates (melilite series, clinopyroxenes, olivine, Ca-olivines, and Ca-plagioclases) and spinel series. Numerous other phases may also be present either as dominant or minor constituents depending on local conditions of their formation.

The grain size of crystalline phases in slags depends on the thermal history of a slag bed but most crystalline slags are fine to medium-grained with skeletal/dendritic crystals. In this paper we report the occurrence of coarse-grained segregations in holocrystalline slags deposited at a site of a historic, zinc smelter in Katowice, the capital of the Upper Silesia region in southern Poland. The main goal of this paper is to decipher the origin of the segregations and to determine the crystallization sequence in different zones of the segregations, the partition of elements to different crystalline phases in different slag zones and finally to construct the model of formation for the unusual pegmatite-like segregations. It is a common practise to use mineral names for designating analogous crystalline phases that compose slags. We will follow that convention throughout the article.

1.1 Site description

Samples were collected from remnants of a slag bed at a site of historic zinc and lead smelting plant located in Wełnowiec, a district of Katowice (Fig. 1a). The zinc smelter was built in 1873 in place of a former iron smelter. The smelting plant in Katowice-Wełnowiec was among the largest single zinc producers in Europe. The smelting plant used the method invented at the end of the eighteenth century by Johann Christians Ruberg and is called the Silesian process (Greiner 2004). The ore concentrate was loaded to 288 muffles (retorts) horizontally inserted into the continuously heated furnaces. Zinc vaporized at 1200–1300 °C and condensed upon cooling in a clay tube (condenser) attached to each muffle. Finally, metal blocks weighing 20 kg each were formed from liquid zinc. The horizontal muffles allowed a constant loading of batch and fuel (coke) without cooling the system. The last furnace of that type was decommissioned in the Katowice-Wełnowiec plant in 1980. Initially, the smithsonite ore (calamine) was smelted; however, it was soon substituted by the Zn and Pb-sulphide ore mined in the Upper Silesian Mississippi Valley-type deposits hosted by the mid-Triassic dolomites. The ore consists of collomorphic sphalerite (zinc blende), galena, pyrite, marcasite, and locally wurtzite and brunckite (e.g. Heijlen et al. 2003). Dolomite, barite, and calcite are gangue minerals. Of environmental concern is the fact, that the light-coloured collomorphic aggregates of sphalerite are enriched in Cd up to 1.5 wt.%; whereas, As and Tl tend to concentrate in galena (up to 4.9 wt.% As and 0.08 wt.% Tl) and in marcasite/pyrite (up to 1.6 wt.% As and 0.09 wt.% Tl) (Mayer and Sass-Gustkiewicz 1998). The question how these elements migrate to slag phases is still open and would be investigated in the subsequent paper.

Slags from Katowice-Wełnowiec have been used for road constructions and only remnants of the bottom portion of the slag-pile that covered an area of 24 ha are preserved as an industrial heritage (Fig. 1). The exact age of the investigated slag is unknown. Most probably it was deposited at the beginning of the twentieth century, when the Silesian method of zinc smelting was a common practise.

1.2 Experimental

Thirty samples were collected at the Katowice-Wełnowiec site. Samples were selected on the basis of macroscopic observations to represent all distinguishable slag types. Before further preparations samples were cleared from the soil and dried in 22 °C for seven days. Thin sections of slags were examined using an Olympus BX-51 polarizing microscope and an analytical scanning electron microscope (SEM; FET Philips XL30) equipped with an energy-dispersive spectrometer. Electron probe micro-analyses

(EPMA) and backscattered electron (BSE) imaging of phases were performed using electron probe micro-analyser CAMECA SX 100 (Inter-Institutional Laboratory of Microanalysis of Minerals and Synthetic Materials, the Warsaw University).

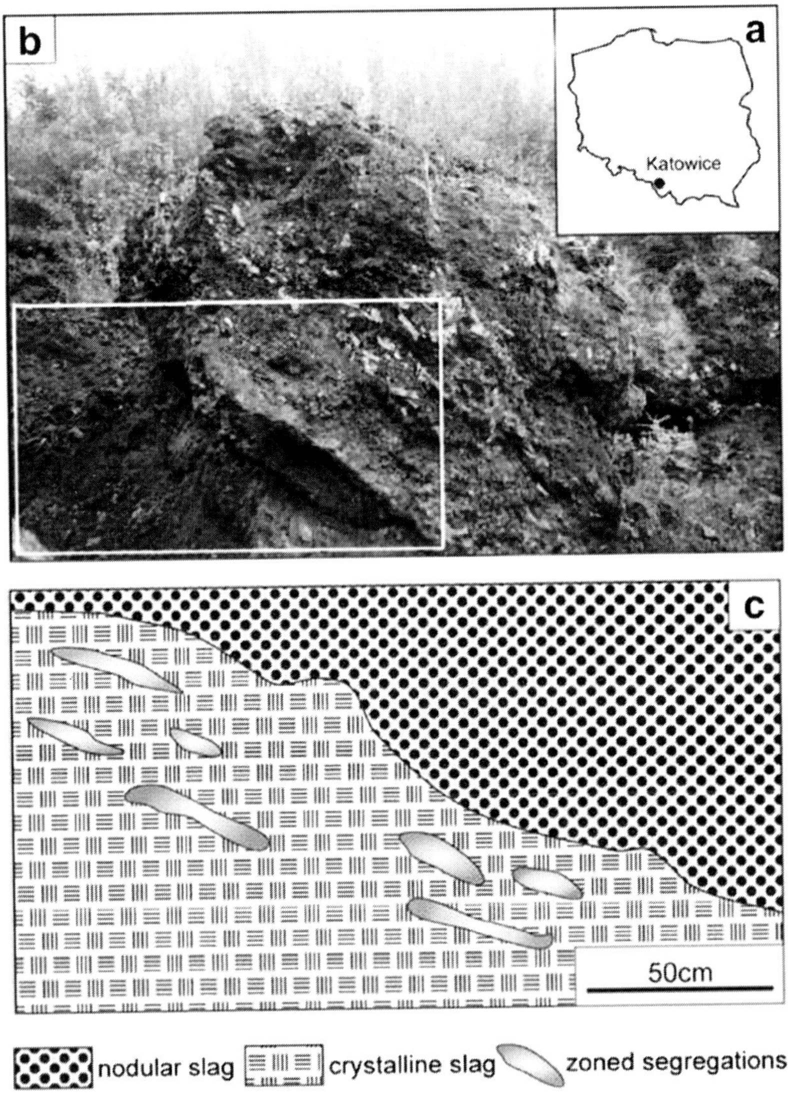

Figure. 1 Remnants of the slag pile in the Katowice-Wełnowiec slag-dump. Inset shows the location of Katowice in Poland (a). The framed area in (b) is explained in the generalized schematic illustration (c)

Analyses were performed at 15 keV accelerating voltage, 10–20.1 nA beam current, and beam diameter of up to 5 μm. Standards included: Ag – Ag_2Te; Al – $KAlSi_3O_8$; As – GaAs; Ba – $BaSO_4$; Bi – Bi_2Te_3; Ca – $CaSiO_3$; Cd – CdS; Ce – CeP_5O_{14}; Cl – $Na_4AlBeSi_4O_{12}Cl$; Co – CoO; Cr – Cr_2O_3; Cu – $CuFeS_2$; F – $Ca_5(PO_4)_3(F,Cl,OH)$; Fe – Fe_2O_3; K – $KAlSi_3O_8$; La – LaB_6; Mg – $MgCaSi_2O_6$; Mn – $MnCO_3$; Na – $NaAlSi_3O_8$; Nb – Nb; Nd – $NdGaO_3$; Ni – NiO; P – $Ca_5(PO_4)_3(F, Cl, OH)$; Pb – PbS, $PbCrO_4$; Pr – Pr synthetic glass; S – $BaSO_4$, $CuFeS_2$, ZnS; Sb – Sb_2Te_3, InSb; Se – Bi_2Se_3; Si – $CaSiO_3$; Sr – $SrTiO_3$; Ti – TiO_2; V – V_2O_5; Y – YP_5O_{14}; Zn - ZnS.

X-ray powder diffraction patterns were obtained using a diffractometer PANalytical X'PERT PRO – PW 3040/60 (X'PERT HighScore Plus software) and Co Kα1 radiation (Fe filter). Operating conditions were 45 kV and 30 mA. The X'PERT HighScore Plus software was applied for data processing. The elimination algorithm of Ladell et al. (1975) was used for Kα2 stripping.

Mössbauer spectra of bulk samples were recorded at room temperature using a constant acceleration spectrometer with a ^{57}Co:Pd source. A metallic α-Fe powder absorber was used for velocity and isomer shift calibrations. The Sterns et al. (1998) methodology was used for the interpretation of the obtained Mössbauer spectra.

2. RESULTS

2.1 Generalities

The investigated slag is holocrystalline, fine- to medium-grained (0.2–1.0 mm) and is composed of olivine, pyroxene, spinel series, melilite series and subordinate leucite, nepheline and sulphides (Warchulski and Szopa 2014). The holocrystalline slag is overlain by fine-grained vesicular slag mixed with debris of building materials (Fig. 1b). Numerous elongate, coarse-grained segregations occur in the uppermost layer of the holocrystalline slag aligned parallel to the contact with the overlying vesicular slag (Fig. 1c). The segregations are ellipsoidal or flattened irregular with large aspect ratio (3:1 to 5:1) and they range in size from 10 to 40 cm along their major axis (Fig. 2). There are two types of segregations: (a) with crystal lined cavities, referred hereafter as miarolitic segregations (Fig. 2a, b), and (b) massive segregations with subhedral crystals (Fig. 2c). Melilite, spinel, and pyroxene are major phases in both types of segregations. Subordinate hematite is confined to the miarolitic segregations; whereas, leucite has been observed in the massive segregations.

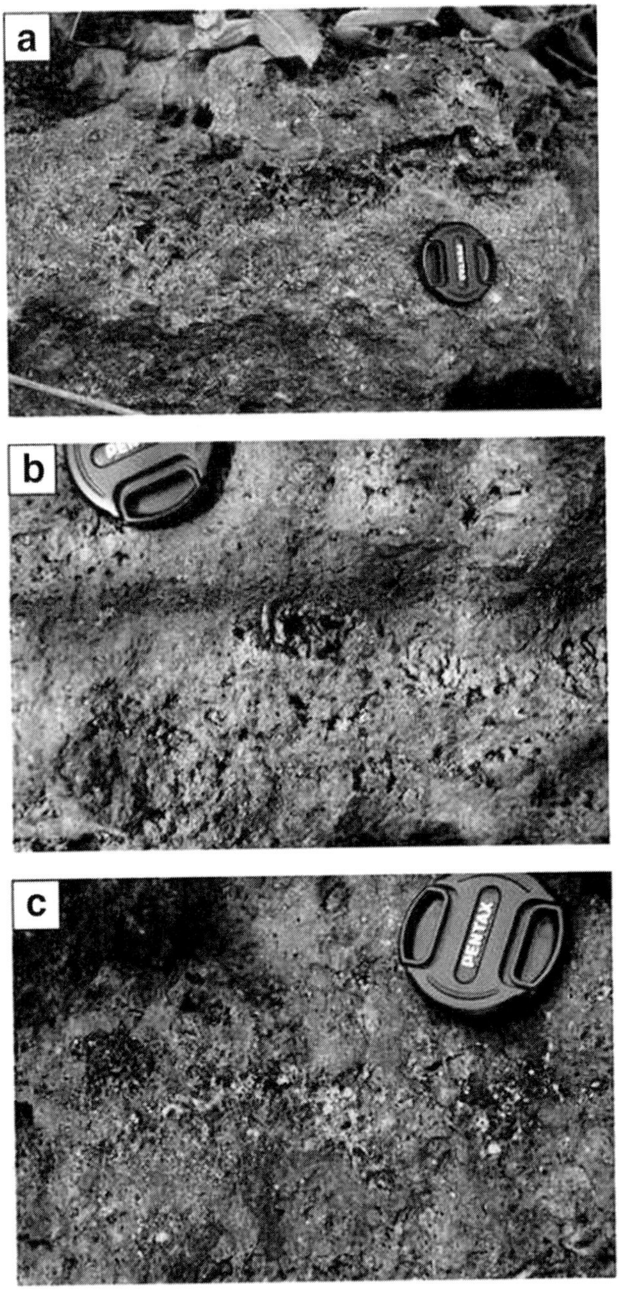

Figure. 2 Miarolitic segregations (a, b) and massive segregation with leucite-rich core (c) in slag

There is a fine-grained (0.4–0.6 mm) equigranular border zone between slag and the segregations composed of olivine, spinel, melilite, pyroxenes, monticellite-kirschsteinite, and sulphides. The transition from the border zone to the segregations is marked by a sharp increase in crystal size of up to 25 mm across. Walls of cavities in the miarolitic segregations are lined with euhedral crystals of melilite overgrown by pyroxene, leucite, spinel and hematite (Fig. 3a-c). Locally, apatite has been observed. While phase assemblages of both types of segregations and their corresponding border zones are almost identical they differ in mineral chemistry.

2.2 Mineral chemistry

2.2.1 Oxides

Spinel-group phases occur in the holocrystalline slag and in the segregations as euhedral, zoned crystals up to 400 μm across (Fig. 4a, b). They are composed of a multicomponent solid solution of franklinite ($ZnFe_2O_4$) - gahnite ($ZnAl_2O_4$) - magnetite ($FeFe_2O_4$) - hercynite ($FeAl_2O_4$) (Table 1; Fig.5a) enriched in Mg, Mn and Ti. There are significant differences in chemical compositions between spinel phases in the segregations, their border zones and the surrounding slag (Table 1). Spinels from the regular slag are represented by Mg-spinel with magnetite compound (Fig. 5), and are enriched in Ti relative to spinels from the segregations (Table 1). Spinels in the border zone of segregation span a wide range of compositions from Mg- and Al-rich to enriched in magnetite component (Fig. 5). Spinels in the border zones of miarolitic segregations and in the massive segregations show different compositional trends (Fig. 5). While there are increasing concentrations of magnetite components in both groups of spinels, from around 10 mol% to over 80 mol% with a gap between 40 and 60 mol%, spinels from the border zone of the miarolitic segregations are relatively enriched in Mg and poorer in Zn. Trends in the border zones corresponds to spinel zoning distinguished in this slag parts: the Mg-Al core and the rim enriched in Fe. Spinels in both types of segregations are Zn-rich and can be grouped into the franklinite-dominated and gahnite-dominated compositions (Fig. 5a).

Mössbauer spectroscopic measurements were applied to distinguish different oxidation states of iron. Mössbauer spectra of spinel-pyroxene aggregates from the segregations show the overlapping magnetic and paramagnetic components (Fig. 6). The relatively small amount of Fe^{3+} in the magnetic component (Table 2) probably results from the extensive substitution of Fe^{3+} by Al^{3+}. This is confirmed by the Al distribution in zoned crystals (Fig. 4a). The chemical composition of the core (M1) is approximated by the formula (Mg, Zn)Al_2O_4, whereas rims (M2) are Al-depleted and their chemical compositions can be expressed as

$(Fe^{2+},Mn^{2+})(Fe^{3+},Ti, Mn^{4+})_2O_4$. The occurrence of hematite as the latest phase in the segregations suggests oxidative conditions. Some of Fe+3 in hematite is substituted by Al, Ti, Mg, Mn and Zn (Table 1).

Figure. 3 Mineral phases in miarolitic cavity: (a) platy crystals of melilite overgrown by clinopyroxene with different morphology (b, c)

2.2.2 Silicates and aluminosilicates

Melilite is common phase in the holocrystalline slag and it is a predominant phase in the segregations. Its yellowish, platy or locally dendritic crystals up to 25 mm across, show perfect cleavage (Figs. 3a). They are overgrown by pyroxene and oxide minerals (Figs. 3a, c and 4d). The mean composition of the melilite series corresponds to the alumo-akermanite - hardystonite solid solution: $(Ca_{1.93}Na_{0.07}K_{0.02})(Mg_{0.51}Fe_{0.19} Zn_{0.19}Al_{0.10}Mn_{0.01}) (Si_{1.96}Al_{0.04})O_7$ (Table 1; Fig. 7). Melilite crystals from the massive segregations have the highest concentrations of Zn and represent the $Ca_2MgSi_2O_7$ - $Ca_2ZnSi_2O_7$ solid solution with less than 10 % of the $Ca_2FeSi_2O_7$ endmember (Fig. 7). Melilite from the miarolitic segregations is relatively depleted in Zn and shows variable concentrations of Fe and Mg (Fig. 7). Melilite from the border zone of the massive segregations has intermediate concentrations of Zn between Zn-rich melilite from the massive segregations and Zn-poor melilite from miarolitic segregations. Melilite from the border zone of the miarolitic segregations is Mg-rich (>70 % $Ca_2MgSi_2O_7$) and Zn-poor.

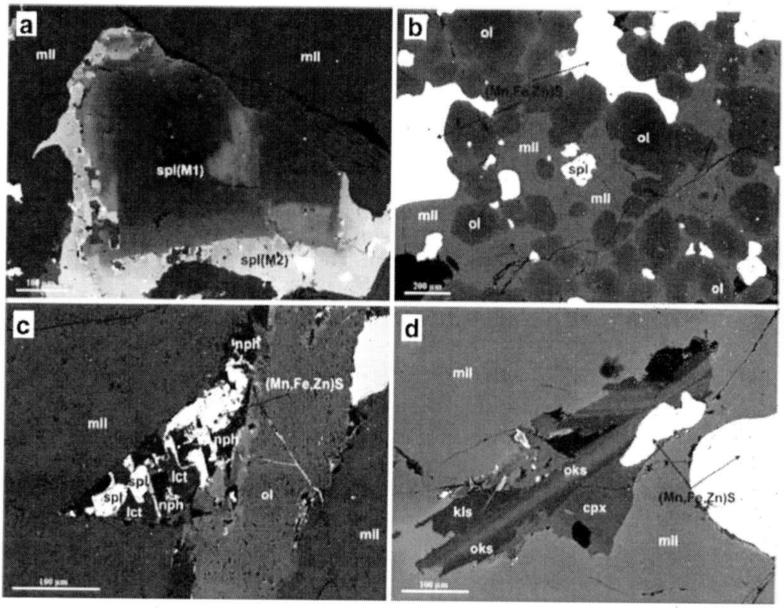

Figure. 4 BSE images of: (a) zoned magnetite within melilite; (b) rounded crystals of zoned olivine surrounded by mellite and blebs of (Mn,Fe,Zn)S; (c) nepheline, spinel, leucite and (Mn,Fe,Zn)S in the interstices between melilite and olivine in slag; (d) oxykinoshitalite-like mica, clinopyroxene and kalsilite within melilite. Abbreviations: cpx – clinopyroxene; kls – kalsilite; lct – leucite; mll – melilite; nph – nepheline; oks – oxykinoshitalite; ol – olivine; spl – spinel; spl(M1–2) – refer to spinel compositions (see text)

Table 1 EPMA data (wt.%) for spinel, hematite, melilite and clinopyroxene from slag, border zones and segregations

	spl I	spl II	spl III	hem	mll I	mll II	mll III	cpx I	cpx II	cpx III
SiO_2	—	—	—	—	41.95	40.72	42.01	49.22	46.11	45.54
TiO_2	2.18	0.33	0.13	1.19	0.06	0.05	0.12	0.46	0.21	0.26
Al_2O_3	14.93	49.36	3.68	2.49	3.92	1.68	2.16	0.90	7.22	7.95
Fe_2O_{3calc}	49.49	18.00	67.01	95.16	–	–	–	6.10	9.90	6.05
FeO_{calc}	25.06	11.25	2.70	0.63	5.21	6.24	0.75	11.04	0.10	5.53
ZnO	2.12	5.77	18.43	0.32	1.84	5.37	8.49	0.14	0.58	0.61
MnO	1.99	0.46	2.40	0.21	0.46	0.47	0.24	3.09	0.60	0.47
MgO	4.18	14.6	4.78	0.22	7.44	6.29	7.93	6.12	10.14	9.81
CaO	–	–	–	–	36.66	38.28	37.66	20.10	24.57	23.84
Na_2O	–	0.21	0.46	–	1.82	0.13	0.70	1.80	1.05	0.04
K_2O	–	–	–	–	0.09	0.43	0.54	–	–	–
Total	99.95	99.98	99.59	100.22	99.43	99.66	100.58	98.96	100.46	100.09

a.p.f.u. (atoms per formula units)

Si	–	–	–	–	1.97	1.96	1.98	1.95	1.73	1.73
Ti^{4+}	0.06	0.01	0.00	0.02	0.00	0.00	0.01	0.01	0.01	0.01
Al	0.61	1.61	0.16	0.08	0.21	0.10	0.12	0.04	0.32	0.36
Fe^{3+}_{calc}	1.29	0.38	1.88	1.87	–	–	–	0.18	0.28	0.17
Fe^{2+}_{calc}	0.72	0.26	0.08	0.01	0.20	0.25	0.03	0.37	0.00	0.18
Zn	0.05	0.12	0.51	0.01	0.06	0.20	0.29	0.00	0.02	0.02
Mn	0.05	0.01	0.07	0.00	0.02	0.02	0.01	0.10	0.02	0.01
Mg	0.22	0.60	0.27	0.01	0.52	0.45	0.56	0.36	0.57	0.56
Ca	–	–	–	–	1.84	1.98	1.9	0.85	0.99	0.97
Na	–	0.01	0.03	–	0.17	0.01	0.07	0.14	0.08	0.00
K	–	–	–	–	0.01	0.03	0.03	–	–	–
O^{2-}	4.00	4.00	4.00	3.00	7.00	7.00	7.00	6.00	6.00	6.00

spl–spinel, hem–hematite, mll–melilite, cpx–clinopyroxene, I–slag, II–border zones, III-segregations

All melilite crystals are compositionally homogeneous (Fig. 4). Unit cell parameters of melilite from the miarolitic segregations calculated for the space group P–¯42₁m are a_0 = 7.84 Å, c_0 = 5.01 Å. They are similar to unit cell parameters of akermanite and hardystonite in accordance with EPMA data.

Pyroxene chemical compositions (Table 1) are dominated by a diopside – hedenbergite solid solution with elevated concentrations of the wollastonite compound (Fig. 8a). Ca-rich pyroxene locally occurs in the massive segregations (Fig. 8a). The chemical composition of pyroxenes is further complicated by the substitution of Mg and Fe by Al, Zn, Na, Mn, and Ti (Table 1). Zinc concentration in pyroxenes increases from Fe- and Mg-dominated compositions in the slag, through border zone of the segregations to Zn-rich internal part of the segregations (Fig. 8b). Approximately 30 % of Fe in pyroxenes from the segregations is trivalent as suggested by the Mössbauer spectra of pyroxene-spinel aggregates (Fig. 6). Unit cell parameters of idiomorphic, prismatic pyroxene crystals from the miarolitic segregation calculated for the space group C2/c are: a_0 = 9.76 Å; b_0 = 8.92 Å, c_0 = 5.28 Å; β = 106°. Pyroxene crystals are zoned with the core enriched Is–isomer shift, Qs–quadrupole splitting, H–hyperfine splitting, A–contribution in Na, Mn, Fe^{2+}, Mg, and Si; whereas, rim has elevated concentrations of Fe^{3+}, Al, and Ti.

Olivine occurs in the slag and in the border zones. Olivine in the slag is fayalite-rich (Fa/Fo = 0.76) and is enriched in Mn relative to olivine in the border zone, in which forsteritic olivine prevails (Fa/Fo = 0.34) (Table 3). Olivine in the border zone is slightly enriched in Zn (2.76 wt.%) compared to olivine from the slag (1.0 wt.% ZnO). Calcium content in all examined olivines is similar, ranging from 3.3 mol% in the slag to 3.8 mol% in the border zone (Table 3). Some olivine crystals are zoned (Fig. 4b) with the core enriched in Fe, Mn, and Ca relative to the Mg-rich rim.

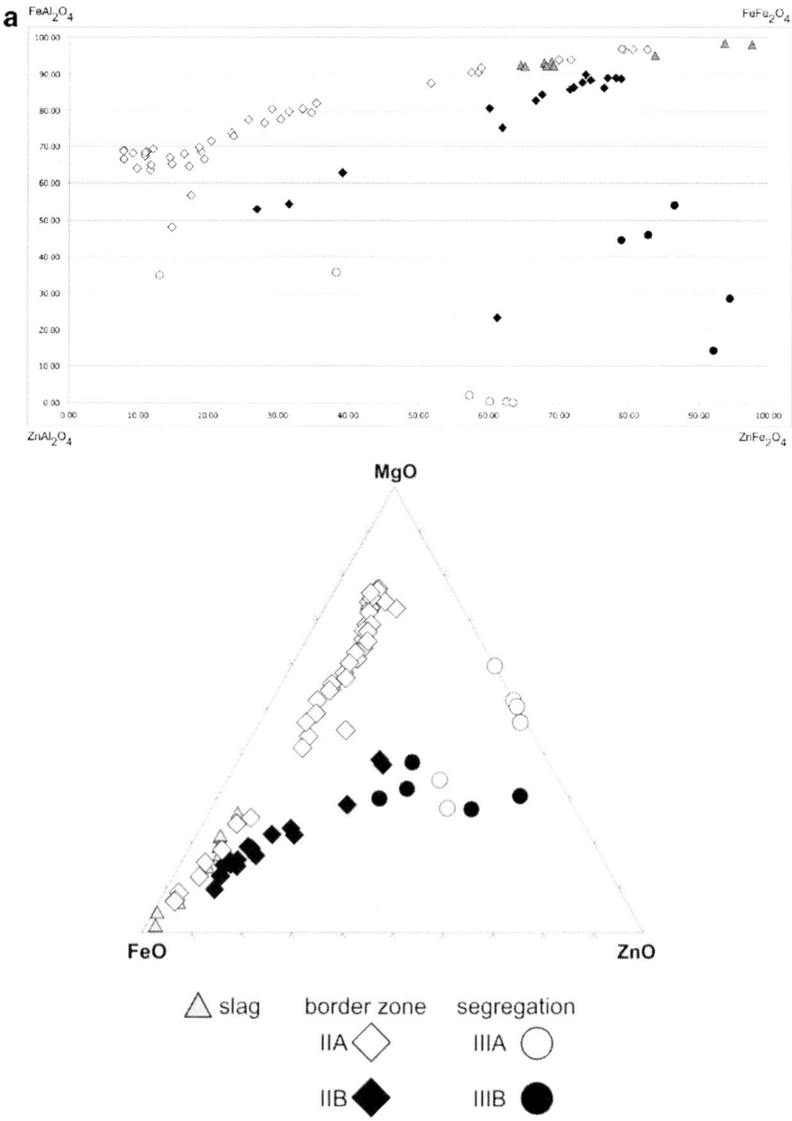

Figure. 5 (a) The gahnite $ZnAl_2O_4$ - hercynite $FeAl_2O_4$ - franklinite $ZnFe_2O_4$ - magnetite $FeFe_2O_4$ quadrilateral diagram showing compositions of spinel series in crystalline slag, border zones and segregations. (b) Compositional variations of spinel series in the ternary diagram MgO-FeO-ZnO. Abbreviations: IIa – border zone of miarolitic segregations; IIb – border zone of massive segregations; IIIa – miarolitic segregation; IIIb – massive segregations

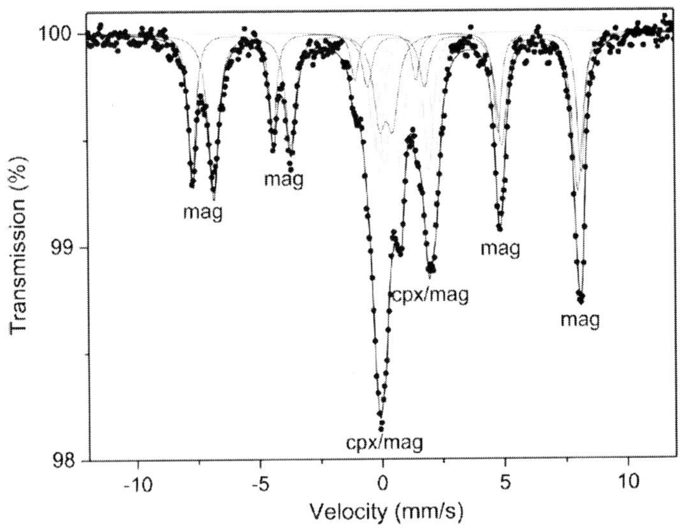

Figure. 6 Mössbauer spectrum of the magnetite-pyroxene aggregate from miarolitic segregation. Solid dots – experimental data; solid lines – fitted curves; cpx- clinopyroxene, mag – magnetite

Table 2 Hyperfine interaction parameters, obtained from Mössbauer spectra

	Is (mm/s)	Qs (mm/s)	H (T)	A (%)
Pyroxene	0.558 ± 0.018	0.646 ± 0.019	-	10.7
	1.056 ± 0.013	2.504 ± 0.048	-	11.4
	1.000 ± 0.024	2.052 ± 0.050	-	12.9
Magnetite	0.290 ± 0.0061	0.536 ± 0.096	-	9.8
	0.672 ± 0.006	0.025 ± 0.006	46.27 ± 0.01	33.3
	0.277 ± 0.005	−0.011 ± 0.005	49.28 ± 0.04	21.9

Kirschsteinite – monticellite solid solution was distinguished in the border zone only. Crystals of this phase has simple composition with minor substitutions of Mn, Na and Zn for Mg and Fe (Table 3).

Kalsilite and nepheline are rare in the slag (Fig. 4c and d). Both have a simple chemical composition that does not deviate significantly from the ideal composition (Table 3).

White leucite crystals occur in interstices between other phases in the slag and in the border zone of the segregations (Figs. 2c). Leucite chemical composition is complex with the noticeable presence of Ce (Table 3).

Mica crystals up to 400 μm across occur in the border zone of the segregations (Fig. 4d). The average chemical composition of the mica corresponds to oxykinoshitalite and can be expressed as: $(K_{0.86}Ba_{0.15})(Fe^{2+}{}_{0.39}Mn_{0.02}ZnO_{0.04}Mg_{2.49}Na_{0.06})$ $(Si_{2.82}Al_{1.20})O_{10}(O,OH)_2$. The elements Ca and Sb were detected in some EPM analyses of mica (Table 4). The examined mica is zoned with Ba and Al concentrated in the core, and K, Si and Fe in the rim (Fig. 4d; Table 4).

2.2.3 Phosphates

Acicular crystals of a phase compositionally close to fluorapatite $(Ca_{5.15}[PO_4]_{3.00}F_{0.76})$ with some LREE and Y (Table 4) occur as inclusions in late pyroxene crystals from the miarolitic segregations (Fig. 9).

2.2.4 Sulphides

Sulphides representing MnS-ZnS-FeS solid solution are dispersed within a groundmass of slag and the border zone. The Mn content is up to 4.53 wt.% (Table 5). They significantly differ from sulphides in the host ore-bearing rocks, which are almost pure ZnS, with subordinate Fe substitutions (Rajchel 2008).

Pyrrhotite has been observed in slag surrounding the segregations and rarely in the border zones. Pyrrhotite has rather uniform chemical composition with minor substitutions of Mn, Cu, Co and Ni for Fe (Table 5). Chalcopyrite occurs as a subordinate phase only in slag surrounding the segregations. Small amounts of Ag, Zn, Mn, Co, Ni have been detected in chalcopyrite (Table 5). Only one crystal of galena has been found in the analysed slags. Its chemical composition is complex with numerous substitutions for Pb (Table 5).

3. DISCUSSION

3.1 The sequence of crystallization

Calcium-rich silicates: melilite, clinopyroxenes, and Ca-olivines are predominant phases in the slags and in the segregations. The enrichment in Ca most probably resulted from the Ca-bearing fluxing material added to the charge and, consequently partitioning of Ca into the silicate melt during metallurgical process. Iron sulphides (pyrite and marcasite) associated with the Zn-Pb ore were the source of Fe in spinels, olivines and clinopyroxenes (Warchulski et al. 2015).

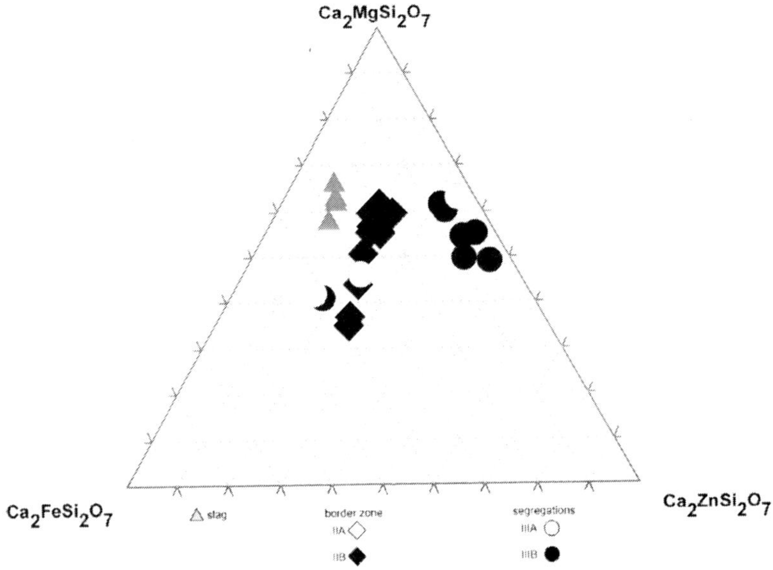

Figure. 7 The melilite ternary diagram $Ca_2MgSi_2O_7$ – $Ca_2FeSi_2O_7$ – $Ca_2ZnSi_2O_7$ showing compositions of melilite in slag, border zone and segregations. Abbreviations: IIa – border zone of miarolitic segregations; IIb – border zone of massive segregations; IIIa – miarolitic segregation; IIIb – massive segregations

Based on spatial relations between mineral phases the following paragenetic sequence has been established in the slag: olivine and spinel → melilite → clinopyroxene (Fig. 10). Leucite, nepheline and sulphides crystallized from residual melt as the latest phases. Similar sequence of crystallization commonly occurs in pyrometallurgical slags regardless of Zn smelting technology (e.g. Ettler et al. 2001; Puziewicz et al. 2007; Warchulski et al. 2015).

The sequence of crystallization in the border zone of segregations is similar to the slag. Crystallization began with olivine followed by spinel and melilite. The interstices between those phases were filled by monticellite – kirschteinite solid solution, second generation of spinel and sulphides. Locally, mica (similar to oxykinoshitalite), pyroxene and kalsillite crystallized between melilite crystals. Oxykinoshitalite was described as a new species from olivine nephelinite composed of olivine, clinopyroxene, ol-olivine, mtc-kir-monticellite–kirschsteinite solid solution, kls–kalsilite, nph–nepheline, lct–leucite, I–slag, II–border zones, III-segregations

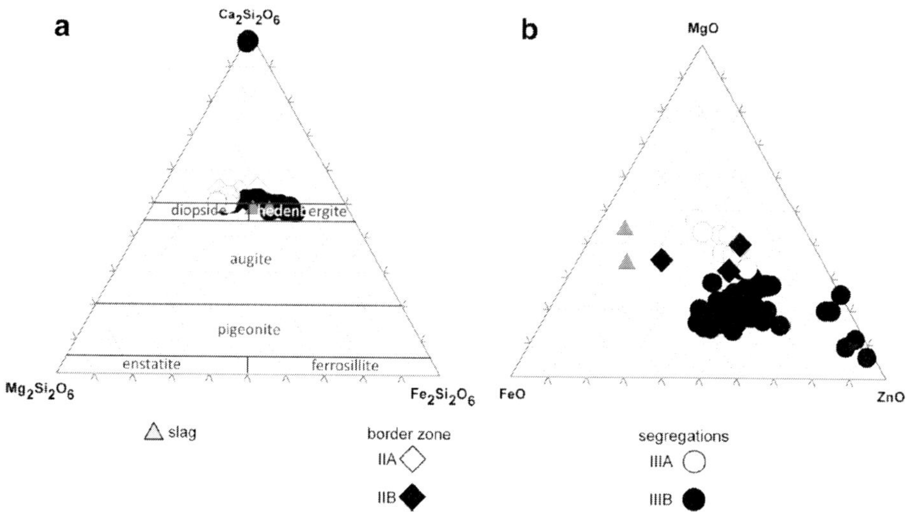

Figure. 8 (a) The pyroxene ternary diagram showing compositions of pyroxenes from slag, border zones and segregations. (b) Pyroxene compositions plotted in the ternary diagram MgO-FeO-ZnO. Abbreviations: IIa – border zone of miarolitic segregations; IIb – border zone of massive segregations; IIIa – miarolitic segregation; IIIb – massive segregations

Fe–Ti oxide, nepheline, calcite, apatite and K-rich feldspar (Kogarko et al. 2005), an assemblage analogous to the examined slag. The occurrence of oxykinoshitalite in the border zone suggests local enrichment in water, which is a rather uncommon feature in pyrometallurgical slags. This may be caused by interaction with rainwater after and/or during slag deposition.

The crystallization conditions of the monticellite – kirschteinite solid solution in the border zone can be inferred from the univariant pressure-temperature curves determined by Walter (1963) and experimentally confirmed by Gustafson (1974) and defined by the following reactions:

$$2Ca_2MgSi_2O_7+Mg_2SiO_4 \Leftrightarrow 3CaMgSiO_4+CaMgSi_2O_6 \quad (1)$$

$$6CaSiO_3+2Fe_3O_4 \Leftrightarrow 6CaFeSiO_4+O_2 \quad (2)$$

Table 3 EPMA data (wt.%) for olivine, monticellite – kirschsteinite solid solution, kalsilite, nepheline and leucite from slag, border zones and segregations

	ol I	ol II	mtc-kir	kls	nph	lct I	lct II	lct III
Sb_2O_5	–	–	–	0.70	–	–	–	–
SiO_2	34.7	37.61	36.63	39.95	42.51	53.76	54.78	54.68
Al_2O_3	–	–	–	30.81	27.15	23.76	23.23	23.77
Ce_2O_3	–	–	–	–	–	0.46	0.01	0.21
Fe_2O_{3calc}	–	–	–	–	–	0.50	1.45	0.89
FeO_{calc}	32.94	21.19	10.19	0.50	9.66	–	–	–
ZnO	1.00	2.76	0.79	–	0.12	–	–	–
MnO	4.92	1.46	0.65	–	0.33	–	–	–
MgO	23.88	34.61	20.72	0.12	1.03	–	–	–
CaO	2.37	2.86	30.88	–	0.75	–	–	–
BaO	–	–	–	–	0.21	1.30	0.29	0.92
Na_2O	–	–	–	0.12	13.48	–	–	–
K_2O	–	–	–	28.62	5.69	20.74	20.91	20.81
Total	99.81	100.49	99.86	100.82	100.93	100.52	100.67	101.28
a.p.f.u. (atoms per formula units)								
Sb^{5+}	–	–	–	0.01	–	–	–	–
Si	0.99	1.00	1.00	1.04	1.06	1.97	1.99	1.99
Al	–	–	–	0.94	0.80	1.02	0.99	1.02
Ce^{3+}	–	–	–	–	–	0.01	0.00	0.00
Fe^{3+}_{calc}	–	–	–	–	–	0.01	0.04	0.02
Fe^{2+}_{calc}	0.79	0.47	0.23	0.01	0.20	–	–	–
Zn	0.02	0.05	0.02	–	0.00	–	–	–
Mn	0.11	0.03	0.01	–	0.01	–	–	–
Mg	1.02	1.37	0.84	0.00	0.04	–	–	–
Ca	0.07	0.08	0.9	–	0.02	–	–	–
Ba	–	–	–	–	0.00	0.02	0.00	0.01
Na	–	–	–	0.01	0.65	–	–	–
K	–	–	–	0.95	0.18	0.97	0.98	0.96
O^{2-}	4.00	4.00	4.00	4.00	4.00	4.00	4.00	4.00

Those reactions occur between ~900 and 1400 °C at atmospheric pressure (Kushiro and Yoder 1964) and were already noted in other slags from Upper Silesia (Warchulski et al. 2015; Puziewicz et al. 2007).

The sequence of crystallization in the massive segregations differs from the miarolitic segregations by the lack of hematite. Spinel compositionally close to franklinite was the first phase to crystallize in the massive segregations possibly due to the increased oxygen fugacity during the reaction [2] in the border zone. Melilite crystallized next and was followed by pyroxene and leucite. Crystallization in the miarolitic segregations began with centimetre-sized euhedral melilite followed by often intergrown Zn-rich phases: pyroxene and spinel (Fig. 3).

3.2 The origin of coarse-grained segregations in pyrometallurgical slags

The coarse-grained segregations in the smelting slag from Katowice-Wełnowiec resemble miarolitic segregations found in basalts and other volcanic rocks (e.g. Sparks 1978; Shea et al. 2010). By analogy to the origin of natural miarolitic cavities in igneous rocks, we suggest that residual liquid enriched in volatiles and Ca was originated during the fractional crystallization of molten slag. The residual volatile-rich liquid moved upwards until it encountered a solidified uppermost portion of the slag that cooled at the contact with air. That explains the ellipsoidal or irregular flattened shape of the segregations and their orientation parallel to the contact with the overlying vesicular slag (Fig. 1).

Table 4 EPMA data (wt.%) for oxykinoshitalite-like mica and fluorapatite

	oks core	oks rim	fap	fap
Sb_2O_5	0.23	0.20	–	–
P_2O_5	–	–	39.66	40.57
SiO_2	35.52	39.15	1.23	1.09
	14.33	12.94	–	–
Y_2O_3	–	–	0.23	0.24
La_2O_3	–	–	0.32	0.36
Ce_2O_3	–	–	0.56	0.62
Pr_2O_3	–	–	0.19	0.00
Nd_2O_3	–	–	0.17	0.19
FeO	5.87	7.13	0.48	0.15
ZnO	0.42	0.72	–	–
MnO	0.16	0.37	–	–
MgO	22.3	22.19	–	–
CaO	–	–	54.5	54.16
BaO	7.64	2.60	–	–
Na_2O	0.73	1.16	–	–
K_2O	8.13	9.87	–	–
F	–	–	2.92	2.52
Total	95.31	96.31	100.26	99.9

a.p.f.u. (atoms per formula units)

Sb^{5+}	0.01	0.01	—	—
P^{5+}	—	—	2.98	3.03
Si	2.96	3.14	0.11	0.1
Al	1.41	1.22	—	—
Y^{3+}	—	—	0.01	0.01
La^{3+}	—	—	0.01	0.01
Ce^{3+}	—	—	0.02	0.02
Pr^{3+}	—	—	0.01	0.00
Nd^{3+}	—	—	0.01	0.01
Fe^{2+}	0.41	0.48	0.04	0.01
Zn	0.03	0.04	—	—
Mn	0.01	0.02	—	—
Mg	2.77	2.65	—	—
Ca	—	—	5.18	5.12
Ba	0.25	0.08	—	—
Na	0.12	0.18	—	—
K	0.86	1.01	—	—
F	—	—	0.82	0.70
O^{2-}	12.00	12.00	13.00	13.00

oks–oxykinoshitalite, fap–fluorapatite

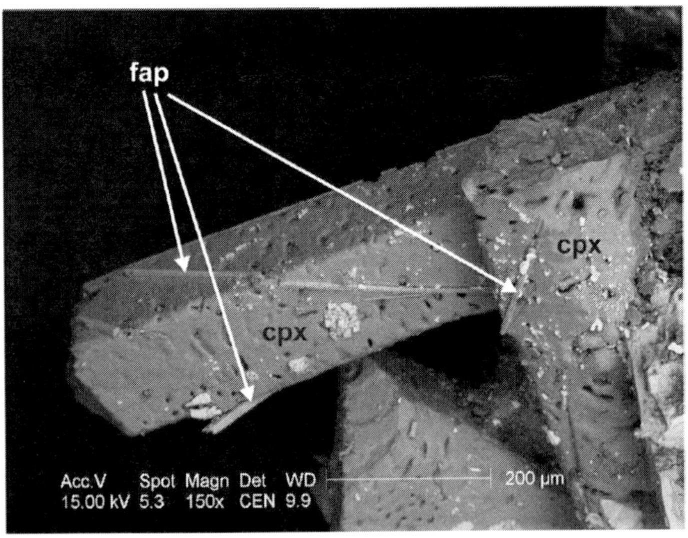

Figure. 9 Fluorapatite as inclusions pyroxene crystal. Abbreviations: cpx – clinopyroxene; fap – fluorapatite

Perhaps CO_2 was the major volatile component of the molten slag due to the thermal dissociation of carbonates added as a fluxing material. The water activity was insignificant except for the border zone, in which hydrous mica crystallized. Fluorine and phosphorus activity was negligible and probably restricted to the latest stages of crystallization in miarolitic cavities leading to the crystallization of fluorapatite.

The high concentration of Zn in silicates and oxides in slags resulted from the non-efficient metal extraction in smelters (Warchulski et al. 2015; Warchulski 2015). All of major phases in the segregations (melilite, pyroxene, spinel) are enriched in Zn, relative to the bulk slag. Zinc concentrations in spinel series and pyroxenes in the segregations are an order of magnitude higher than in their counterparts in the surrounding slag (Fig. 11, Table 1). Intermediate concentrations of Zn occur in the border zone. The relative enrichment in Zn of segregations reflects the Zn preference for volatile phase as seen in magmatic systems (e.g. Nadeau et al. 2013 and references therein).

Considering both the reaction (1, 2) and the enrichment in Zn, the estimated temperature of crystallization of the segregations is above 1000 °C. This estimate is consistent with the temperature of the molten slag (1200–1300 °C) that was transported to the slag bed. The temperature of the overlain vesicular slag calculated from the olivine-bulk chemical composition is in the range of 1200–1350 °C, while calculations based on the clinopyroxene-bulk chemical composition give 1140–1260 °C (Warchulski 2015.).

Puziewicz et al. (2007) by analogy to rapidly cooled, but holocrystalline silica-undersaturated lavas assumed that low viscosity is the key factor causing crystallization of slag-forming phases. In the border zone of the segregations the viscosity of the liquid increased due to the escape of volatiles into the cavities. That caused the increase in the nucleation rate in the border zone leading to the development of the fine-grained "aplitic" rim around the segregations (Figs. 2a-c). The accumulation of volatiles in the segregations caused the breakdown of Si-O and Al-O bonds in the residual liquid. Some Al was also bound to Ca, Na, K and acted as melt modifier preventing the polymerization of Si-O-Si. That resulted in the decrease in viscosity of the melt and subsequently the decrease in the nucleation density resulting in the growth of the centimetre-sized euhedral or subhedral crystals (Fig. 3). Similarly to naturally occurring fractionation processes, the fractionated liquid became oxidized towards the centre of the segregation as suggested by the presence of trivalent Fe in leucite, spinels and hematite. The abundant sulphides in slag and border zones of the segregations reflect the enrichment in S caused by the thermal decomposition of ore sulphides under reducing conditions.

Table 5 EPMA data (wt.%) for sulphides

	(Mn.Zn.Fe)S	(Mn.Zn.Fe)S	(Mn.Zn.Fe)S	(Mn.Zn.Fe)S	po	po	ccp	ccp	gn
Fe	14.98	14.6	27.97	28.21	58.76	58.54	38.00	38.41	0.29
Cu	0.17	0.48	0.12	-	-	0.25	23.56	23.3	0.04
Zn	46.21	46.5	21.26	21.08	-	0.15	1.22	1.49	-
S	33.59	33.46	49.34	49.27	38.83	39.48	35.07	34.42	13.06
Mn	4.12	4.53	0.99	0.91	0.16	0.05	0.67	0.76	0.13
As	0.14	0.09	-	-	0.01	-	0.04	-	0.11
Se	0.01	0.03	0.01	0.02	-	0.05	0.00	0.02	0.33
Pb	-	-	-	-	-	-	-	-	85.25
Ag	0.13	-	0.01	-	-	-	0.38	0.38	0.89
Sb	0.00	0.03	-	-	0.02	-	-	-	1.25
Co	0.02	-	-	0.06	0.13	0.01	0.12	0.11	0.14
Ni	-	-	-	-	0.14	0.07	0.13	-	-
Total	99.37	99.72	99.7	99.55	98.05	98.6	99.19	98.89	101.49

a.p.f.u. (atoms per formula units)

Fe	0.26	0.25	0.56	0.57	0.87	0.85	0.62	0.64	0.01
Cu	0.00	0.01	0.00	-	-	0.00	0.34	0.34	0.00
Zn	0.67	0.68	0.43	0.42	-	0.00	0.02	0.02	-
S	1.00	1.00	1.00	1.00	1.00	1.00	1.00	1.00	0.99
Mn	0.07	0.08	0.02	0.02	0.00	0.00	0.01	0.01	0.01
As	0.00	0.00	-	-	0.00	-	0.00	-	0.00
Se	0.00	0.00	0.00	0.00	-	0.00	0.00	0.00	0.01
Pb	-	-	-	-	-	-	-	-	1.00
Ag	0.00	-	0.00	-	-	-	0.00	0.00	0.02
Sb	0.00	0.00	-	-	0.00	-	-	-	0.02
Co	0.00	-	-	0.00	0.00	0.00	0.00	-	-
Ni	-	-	-	-	0.00	0.00	0.00	0.00	0.00

(Mn,Zn,Fe)S-(Mn,Zn,Fe)S solid solution, po–pyrrhotite; ccp–chalcopyrite; gn–galena

They probably originated as a result of recrystallization of original sulphides, coupled with Fe-Mn incorporation into the new sulphide structure. The occurrence of sulphides in the slag suggests strong variability of the oxygen fugacity oscillating between MH and QMF buffers during melt crystallization.

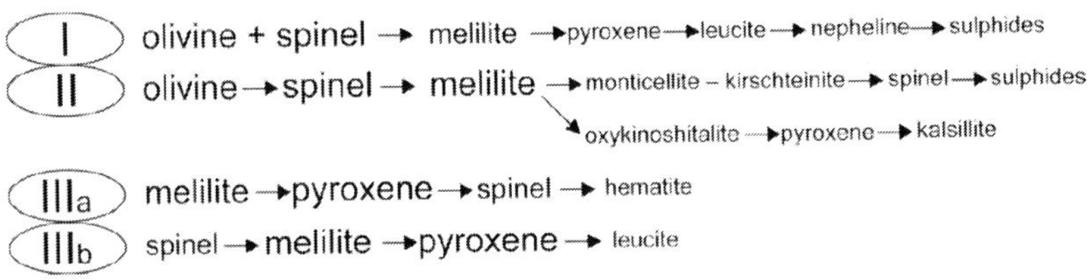

Figure. 10 The sequence of the crystallization in the crystalline slag (I), border zones (II), miarolitic segregations (IIIa), and in massive segregations (IIIb)

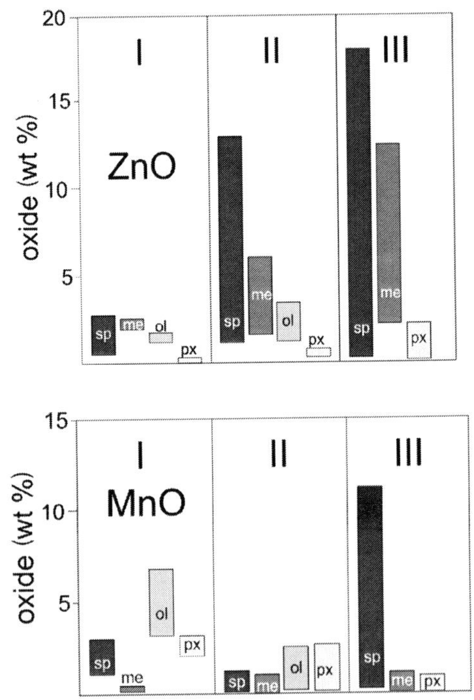

Figure. 11 Partitioning of Zn and Mn (as oxides) between mineral phases in slag (I), border zone (II) and segregations (III)

Phase compositions of the slag and the segregations are generally the same. However, the chemical compositions of the phases are significantly different (Fig. 11, Tables 1,and 3). Chemical zoning observed in major phases of pyrometallurgical slags (spinels, olivine, pyroxene) is indicative of the fractional crystallization (e.g. Ettler et al. 2001; Puziewicz et al. 2007; Kucha et al. 1996; Warchulski et al. 2015). The sector zoning in pyroxene crystals common in the miarolitic segregations is indicative of the disequilibrium growth, with different crystal faces attracting different cations as a result of different proportion of bonds at the crystal – melt interface (Dowty 1976) or can be an effect of different diffusion rates for the specific components in relation to crystal growth rate (Watson and Liang 1995). Both processes may have operated in slag and in segregations.

The deposition of the new portion of the hot slag (ca. 1250 °C) onto the surface of the investigated slag prevented the latter from rapid cooling and from volatiles to escape, enabling the formation of the holocrystalline slag with numerous coarse-grained segregations. The vesicular texture and the absence of the coarse-grained segregations in the overlying slag provide an evidence for its rapid degassing at the contact with air. The observed cross-section of two slags in Katowice-Wełnowiec resembles the layered intrusions formed by repeated magma batches maintaining the super-solidus conditions (e.g. Diaz-Alvarado et al. 2013) and, as in the case of the tabular intrusions – prevent volatiles from escaping forming the internal pegmatoidal structures (e.g. Gawęda et al. 2013).

CONCLUSIONS

The occurrence of coarse-grained segregations with euhedral crystals is a unique feature of the investigated slag from Katowice – Wełnowiec compared to other pyrometallurgical slags worldwide. The segregations originated from volatile-rich residual melt separated from molten slag during fractional crystallization. The formation of coarse-grained segregations was enabled by the fast deposition of vesicular slag on the top of the holocrystalline slag, acting as a thermal and physical barrier preventing the latter from rapid cooling and escape of volatiles.

Metallurgical silicate slags can be considered analogous to igneous rocks and as such they can provide additional insight into the behaviour of natural melts regardless their origin.

ACKNOWLEDGMENTS

We thank the reviewers for their constructive comments. This study was supported by National Science Center (NCN) through grant no. 2014/13/B/ST10/02403 to AG, and by the Polish Ministry of Science and Higher Education scholarship for young researchers: "Restoration of the historic smelting process at zinc smelter Katowice - Szopienice on the base of petrographic and geochemical data" to RW.

REFERENCES

1. Álvarez-Valero AM, Sáez R, Pérez-López R, Delgado J, Nieto JM (2009) Evaluation of heavy metal bio-availability from almagrera pyrite-rich tailings dam (Iberian Pyrite Belt, SW Spain) based on a sequential extraction procedure. J Geochem Explor 102:87–94
2. Diaz-Alvarado J, Fernandez C, Castro A, Moreno-Ventas I (2013) SHRIMP U-Pb geochronology and thermal modelling of multilayer granitoid intrusions: implications for the building and thermal evolution of the Central System batholiths, Iberian massif, Spain. Lithos 175-176:104–123
3. Dowty E (1976) Crystal structure and crystal growth: II. sector zoning in minerals. Am Mineral 61:460–469
4. Ettler V, Johan Z (2014) 12 years of leaching of contaminants from Pb smelter slags: geochemical/mineralogical controls and slag recycling potential. Appl Geochem 40:97–103
5. Ettler V, Legendre O, Bodénan F, Touray J-C (2001) Primary phases and natural weathering of old lead-zinc pyrometalurgical slag from Příbram, Czech Republic. Can Mineral 39:873–888
6. Gawęda A, Müller A, Stein H, Kądziołko-Gaweł M, Mikulski S (2013) Age and origin of the tourmaline-rich hydraulic breccias in the tatra granite, Western Carpathians. J Geosci 58:133–158
7. Greiner P (2004) My land Katowice (moja ziemia Katowice). Publisher, Bractwo Gospodarcze Związku Górnośląskiego, Katowice, pp. 42–48
8. Gustafson WI (1974) The stability of andradite, hedenbergite and related minerals in the system Ca – Fe – Si – O – H. J Petrol 15:455–496
9. Heijlen W, Muchez P, Banks DA, Schneider J, Kucha H, Keppens E (2003) Carbonate-hosted Zn-Pb deposits in upper Silesia, Poland: origin and evolution of mineralizing fluids and constraints on genetic models. Econ Geol 98:911–932
10. Kierczak J, Bril H, Neel C, Puziewicz J (2010) Pyrometallurgical slags in upper and lower Silesia (Poland): from environmental risk to use of slag-based product – a review. Arch Environ Prot 36(3):111–126
11. Kogarko LN, Uvarova YA, Sokolova E, Hawthorne FC, Ottolini L, Grice JD (2005) Oxykinoshitalite, a new species of mica from Fernando de Noronha island, Pernambuco, Brazil: occurrence and crystal structure. Can Mineral 43:1501–1510
12. Kucha H, Martens A, Ottenburgs R, De Vos W, Viaene W (1996) Primary minerals of Zn-Pb mining and metallurgical dumps and their environmental behavior at Plombières, Belgium. Environ Geol 27:1–15
13. Kushiro I, Yoder HS (1964) Experimental studies on the basalt- eclogite transformation. Carnegie inst. Washington. Ann Rept Dir Geophys Lab 1964-65:89–94
14. Ladell J, Zagofsky A, Pearlman S (1975) Cu Kα2 elimination algorithm. J Appl Crystallogr 8:499–506

15. Mayer W, Sass-Gustkiewicz M (1998) Geochemical characterization of sulphide minerals from the olkusz lead-zinc ore cluster, upper Silesia, (Poland), based on laser ablation data. Mineral Pol 29:87–105
16. Nadeau O, Williams-Jones AE, Stix J (2013) Magmatic-hydrothermal evolution and devolatilization beneath merapi volcano, Indonesia. J Volcanol Geotherm Res 261:50–68
17. Piatak NM, Seal RR II (2010) Mineralogy and the release of trace elements from slag from the hegeler zinc smelter, Illinois (USA). Appl Geochem 25:302–320
18. Puziewicz J, Zainoun K, Bril H (2007) Primary phases in pyrometallurgical slags from a zinc-smelting waste dump, Świętochłowice, upper Silesia, Poland. Can Mineral 45:1189–1200
19. Rajchel B (2008) Geochemia Tl, As, Cd, Pb w rudach oraz odpadach hutniczych Zn-Pb Górnego Śląska. AGH University, Cracov, Doctoral thesis
20. Shea T, Houghton BF, Gurioli L, Cashman KV, Hammer JE, Hobden BJ (2010) Textural studies of vesicles in volcanic rocks: an integrated methodology. J Volcanol Geotherm Res 190:271–289
21. Sparks RSJ (1978) The dynamics of bubble formation and growth in magmas: a review and analysis. J Volcanol Geotherm Res 3:1–37
22. Sterns JG, Khasanov AM, Miller JW, Pollak H, Zhe L (1998) Mössbauer mineral handbook. Mössbauer Effect Data Center, Asheville, USA
23. Vernon RH (2004) A practical guide to Rock microstructure. Cambridge University Press
24. Vítková M, Ettler V, Johan Z, Kříbek B, Šebek O, Mihaljevič M (2010) Primary and secondary phases in copper-cobalt smelting slags from the Copperbelt Province, Zambia. Mineral Mag 74:581–600
25. Walter LS (1963) Experimental studies on Bowen's decarbonation series. I: P-T univariant equilibria of the 'monticellite' and 'akermanite' reactions. Am J Sci 261:488–500
26. Warchulski R (2015) Zn-Pb slag crystallization: evaluating temperature conditions on the basis of geothermometry. Eur J Mineral. doi:10.1127/ejm/2015/0027-2496
27. Warchulski R, Szopa K (2014) Phase composition of Katowice - Wełnowiec pyrometallurgical slags: preliminary SEM study. CTGeo 3:76–81
28. Warchulski R, Gawęda A, Kądziołka-Gaweł M, Szopa K (2015) Composition and element mobilization in pyrometallurgical slags from the Orzeł Biały smelting plant in the Bytom – piekary Śląskie area, Poland. Mineral Mag 79:459–483
29. Watson EB, Liang Y (1995) A simple model for sector zoning in slowly grown crystals: implications for growth rate and lattice diffusion, with emphasis on accessory minerals in crustal rocks. Am Mineral 80:1179–1187

CHAPTER 16

Mineralogy and Geochemistry of Geophagic Clays from Share Area, Northern Bida Sedimentary Basin,Nigeria

Sc Odewumi*

Department of Science Laboratory Technology, University of Jos, P.M.B. 2084, Jos, Nigeria

ABSTRACT

The Share area occurs within the northern Bida Sedimentary Basin in Nigeria. The focus is on the geological settings, mineralogical and chemical compositions of Share geophagic clays with the aim of inferring their depositional environments, provenance and possible human health implications. The field method include collection of representative samples at appropriate clay stratum and the samples of the geophagic clays were subjected to petrographic, X-Ray fluorescence and X-Ray diffraction analyses. XRD revealed that Geophagic clays are polymineralic and the main reflections of the clay minerals identified were kaolinite, nacrite, gibbsite and dickite. The peaks of other traces are hematite, quartz and pyrite. Geochemical data indicated that geophagic clays composed of SiO2 content ranging from 64.40% to 46.30%, Al_2O_3 values varies from 36.54% to 20.54%, Fe_2O_3 values varies from 4.71% to 1.42%, MgO content varies from 0.08% to 0.06%, CaO ranges between 1.30% and 0.03%, TiO2 values varies from 9.17% to 0.20%, and MnO content ranges from 0.631%. to 0.09%.

These values indicate that the geophagic clays are essentially siliceous aluminosilicates.

The presence of sedimentary structures typical of tidal sedimentation, low abundance of MgO and K_2O indicate lack of expandable clays, and the presence of ironstones further attests to abundant oxygen due to sub aerial exposure resulting to the oxidation of Fe^{2+} to Fe^{3+}. The presence of basal conglomerate, rootlets structures, iron capping, fining upward sequence and abundances of kaolinite indicate a fluvial (continental) sedimentation. Geophagic clays may introduce metals like Fe, Zn, Cu, Ni, Ti, and Zr to the gastro-intestinal system of consumers, causing adverse effects like increase in the gastro-intestinal pH, and the binding of plant toxins and pathogens which create a surfacial coating on the stomach with inferred pharmaceutical implications. Negative impact of ingesting geophagic clays may lead to electrolyte disturbance, intestinal obstruction, constipation, hypertension, peritonitis, eclampsia, iron deficiency, anemia, microbiological infections, helminthiasis and heavy metal poisoning while coarse sandy quartz particles in the clays could affect dental enamel and also lead to the possible rupturing of the Sigmoid colon due to the abrasive nature of the particles.

Keywords: Aluminosilicate; Kaolinite; Hematite; Quartz; Continental and Granite

1. INTRODUCTION

Geophagia, the deliberate ingestion of earthy material by both animals and humans [1-3]. It is practiced by women of child bearing age in developing countries of Africa [4]. Geophagia has been reported in Nigeria, Uganda, South Africa and Swaziland on the African continent [1,2]. An underlying reason for geophagic practices being sustained in developing countries is that the geophagic materials are readily available at little or no cost. Geophagic materials have varied mineralogical and chemical compositions [5]. The materials are usually soils sediments that are clayey in particle size and contain at least one clay mineral as mineralogical constituent [2,6]. Women indulge in the practice because they believe in the relieving effects resulting from soil consumption, which include supplementation of minerals and nutrients and antacid, anti-emetic, and anti-diarrheal properties [4,7-8].

Clays are composed of mixtures of finer grained clay minerals and clay-sized crystals of other minerals such as quartz, carbonate and metal oxides. Clays and clay minerals are found mainly on or near the surface of the Earth

[9]. Clay minerals in the earth crust can be used as indicators of the environment during weathering, allothiand authi-genesis in the sediments and in the study of source area of the detrital supply [10]. In Nigeria, geophagy is almost synonymous with clay consumption and geophagia was reported in Share area, Odewumi [11] and Ibadan, Asaba, Benin City and Aramoko-Ekiti [12]. There is a very limited documentation of the geology, mineralogy and geochemistry of the geophagic clays in Nigeria. This paper's primary objective is to examine the geological settings, mineralogical and chemical compositions of geophagic clays from Share area, with the aim of inferring depositional environments, provenance and possible human health implications.

2. GEOLOGY AND PETROGRAPHY

The basal conglomerate was directly on the basement complex which comprises of gneisses and metasedimentary rocks that have been intruded by granites of probable Pan African (ca. 600Ma) age. Its composition and texture are clearly indicative of little or no textural/ compositional maturation as viewed around NYSC Orientation Camp Yipata, Kwara State. These beds of conglomerate are interlayered with pebbles, clast size varies from granules to pebbles. Roundness is equally variable while perfectly rounded pebbles are mixed with angular ones. The fabric is isotropic, since no imbrications or any other mode of systematic clast arrangement is apparent. The conglomerate is mixed with sand grade which can be regarded as matrix supported. The quartz occurs as colourless, transparent and shows no cleavage.

Sandstone unit was observed along the Agbonna hill, size sorting is very poor and overlying the basal conglomerates. There is an extreme variation in the structural composition of the crystals, although there appear to be bimodal distribution. The depositional environment of the sandstones appears to fall in a mid-fan facies setting [13]. The monocrystalline grains are invariably sharp in extinction. The polycrystalline grains show undulose extinction, with the elongate crystals usually oriented parallel to each other. The grains are generally sub-rounded, inclusions in the quartz grains are mainly microlites, muscovite and euhedral zircon. The inner layer consists of euhedral quartz crystals which have grown into a core of loose, white and fine quartz sand. The microtexture of the sandstone grains range in size from 0.2 to 1.6mm i.e. fine to very coarse with an average of 0.55mm which is coarse. Quartz grains are present with very few feldspar.

Claystone occurs on the slopes of Agbonna hill within Share town and constitute the geophagic material. It overlies coarse sandstone and consists entirely of clay with hardly any silt, as shown by the absence of any gritty feel to the teeth. Rootlet structures were also observed in the claystone unit, the secondary structure is jointing and there is a profuse permeation of both beds by short tight joints in all directions. These have given rise to intense iron staining, presents as yellow and brown patches and fridges around the joints. In petrographic studies, the claystones have sheet structures and a single perfect cleavage, they are optically negative and the slow vibration direction which invariably lies in the plane of the single perfect cleavage. Ironstone is the fourth lithological unit that directly overlying the clay stone that is kaolin rich. The ironstone serves as the capping in Share area of northern Bida sedimentary basin.

3. MATERIALS AND METHOD

The samples of Geophagic materials were collected at appropriate clay stratum in Share area of northern Bida sedimentary Basin. Eight samples of Geophagic clays from Share area were analyzed using XRD. The procedure was by preparing a strongly oriented specimen, air drying, grinding the natural sample and sucking it onto the flat surface of a porous tile disc. Mineralogical analysis was performed at the Centre for Energy Research and Development, Obafemi Awolowo University, Ile Ife (Nigeria). X-ray diffraction patterns were obtained with a diffractometer equipped with Ni filtered Cu-kα radiation, with automatic slit and on-line computer control. The samples were scanned from 15° to 72° (2θ). Mineral identification on the diffractograms and a semi quantitative mineralogical composition were processed using EVA software. Measurements were performed on strongly oriented powder preparation from bulk samples.

X-Ray diffraction patterns registered for all the reflections correspond to each mineral. When the peaks are sufficiently separated from each other, their heights may be used to determine the orientation of a particular mineral in the mixture. Recognition of different peaks and comparison of their relative heights form the basis of clay mineral analysis by X-Ray diffraction [14]. In the study of clays, XRD method incorporates a means of identifying the different clay minerals forming the rock materials as well as measuring their individual orientation. The wavelength of 1.5406 was used to calculate the diffraction angles. The procedure used to prepare disc specimen produces a very highly oriented structure, greatly emphasizing the basal plane reflections 001 and 002, and reducing the edge reflection 020.

Eight samples of geophagic materials collected from the area were selected and cut into thin sections in the workshop of the Department of Geology, University of Ilorin, Ilorin (Nigeria). Petrographic studies of these thin sections were carried out using the transmitted light petrological microscope. The X-Ray fluorescence analyses comprising major oxides and trace elements of eight Geophagic clay samples were carried out at the Centre for Energy Research and Development, Obafemi Awolowo University, Ile Ife (Nigeria).

To better characterize the geophagic materials, the chemical index of alteration (CIA, $[Al_2O_3/(Al_2O_3+CaO+Na_2O+K_2O)]\times100$) and chemical index of weathering (CIW, $[Al_2O_3/(Al_2O_3+CaO+Na_2O)]\times100$) Harnois [15], values were calculated. In the CIA index the CaO is incorporated in the silicate structure [16]. The primary provenance of the geophagic materials was inferred from TiO_2 vs Al_2O_3 segregation diagram for granite, rhyolite and basalt [17].

4. RESULTS

X-Ray diffraction patterns of Share Geophagic clays are shown in figures 1-4. The mineralogical compositions of the Share geophagic clays is rather homogeneous and is mainly kaolinite associated with gibbsite, nacrite, dickite and quartz with small amounts of pyrite, sillimanite, wollastonite and hematite. The results in wt% of semi quantitative analysis of minerals identified in Share geophagic clays are shown in table 3. Iron in the geophagic clays is in the form of pyrite and/or hematite.

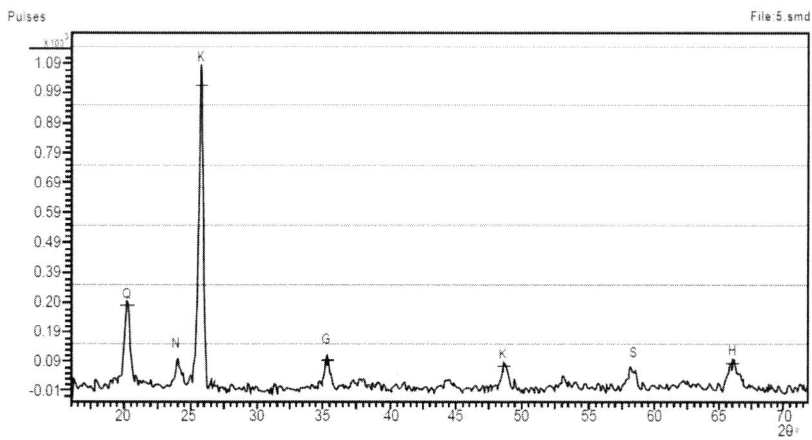

Figure 1: X-Ray diffraction traces for disc specimen of geophagic Clays (SG 1) (G=Gibbsite, K=Kaolinite, N=Nacrite, Q=Quartz, S=Sillimanite and H-Hematite).

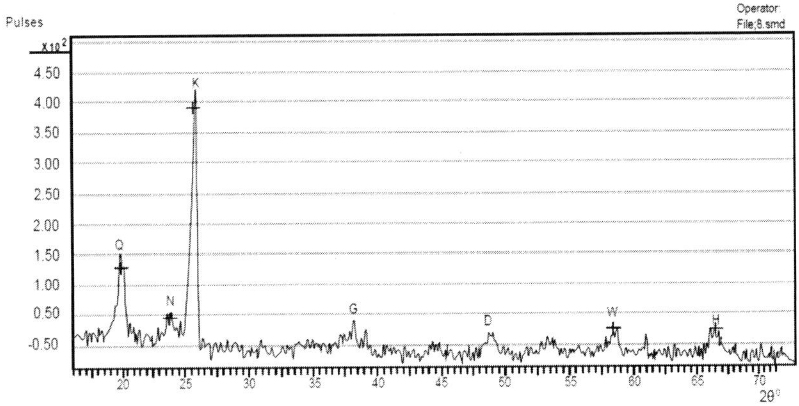

Figure 2: X-Ray diffraction traces for disc specimen of geophagic clays (SG 2) (K=Kaolinite, D=Dickite, N=Nacrite, Q=Quartz, G=Gibbsite, D=Dickite, H=Hematite and W=Wollastonite).

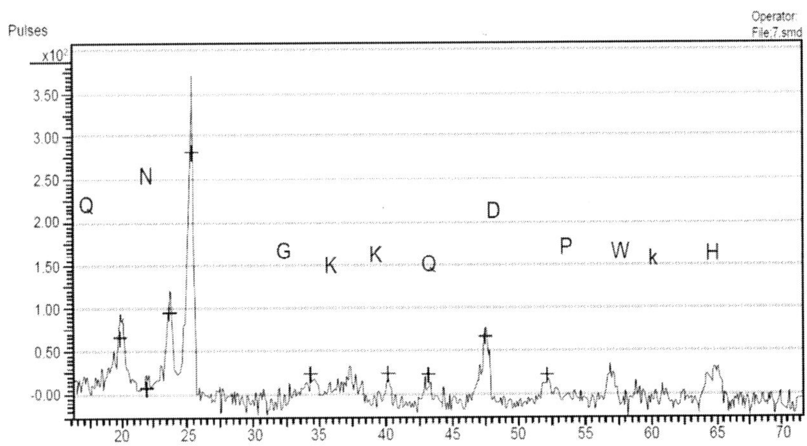

Figure 3: X-Ray diffraction traces for disc specimen of geophagic clay (SG 3) (K=Kaolinite, N=Nacrite, W=Wollastonite, Q=Quartz, P=Pyrite, D=Dickite, H=Hematite, G=Gibbsite and P=Pyrite).

The result of the geochemical analysis as shown in table 1 indicate that the geophagic clays composed of SiO_2 content of 46.30% to 64.40%; Al_2O_3 values of 20.54% to 36.57%; $Fe_2O_{3(T)}$ values of 1.42 to 4.71%; MgO content of 0.06% to 0.08%; CaO values of 0.03 to 1.30%; TiO_2 values of 0.20 to 1.30%; K_2O values of 0.01 to 0.03%; MnO values of 0.09 to 0.631%. These concentrations suggest that the clays are hydrated siliceous aluminosilicate

[18]. The relatively high Fe_2O_3 is probably due to superficial oxidation and contamination by the Fe-rich percolating water from the highly ferruginous facies capping the ridge. Low abundance of MgO and K_2O indicate lack of expandable clays.

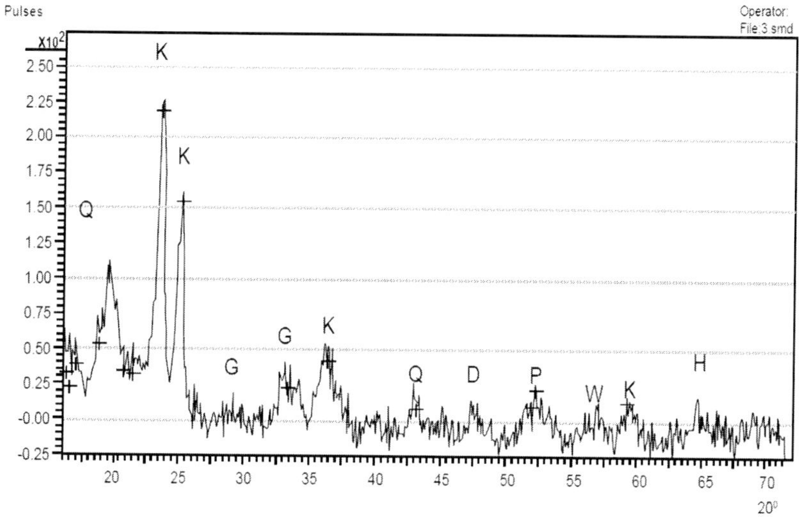

Figure 4: X-Ray diffraction traces for disc specimen of geophagic clay (SG 4) (K=Kaolinite, N=Nacrite, W=Wollastonite, Q=Quartz, P=Pyrite, D=Dickite, H=Hematite, G=Gibbsite and P=Pyrite).

The chemical compositions of geophagic clays in Share area are compared with chemical compositions of Nigerian clays in Asaba, Aramoko-Ekiti, Ibadan and Benin [12]; Itakpe [19]; Isan Ekiti and Ara- Ijero [20]; Itu area [21]; Ubulu-Uku and Awo-Omama [18] in Nigeria as shown in table 2. The SiO_2 content of the Geophagic clays in Share area is higher than the SiO_2 content of clays in Asaba, Aramoko-Ekiti, Ibadan, Benin, Ubulu-Uku, Awo-Omama and Itakpe areas but lower than the SiO_2 content of clays in Ara Ijero and Itu areas. The values of Al_2O_3 is higher in Share Geophagic clays than clays from Isan-Ekiti, Ara-Ijero and Itu area but lower than the values of Al_2O_3 in other areas in Nigeria. The values of $Fe_2O_{3(T)}$ in Share geophagic clays is higher than the values of $Fe_2O_{3(T)}$ in Aramoko, Ibadan, Benin and Awo-Omama clays but lower than the values of $Fe_2O_{3(T)}$of clays in other areas (Table 2). The values of other major oxides of Geophagic clays in Share area compared well with other clays in Nigeria with little variations.

Table 1: Chemical compositions of geophagic clays from Share area

Major Oxides (%)	SG 1	SG 2	SG 3	SG 4	HBC	PAAS	UC
SiO_2	60.30	64.40	62.88	46.30	-	62.80	66.00
Al_2O_3	21.62	20.54	22.60	36.57	-	18.90	15.20
$Fe_2O_{3(T)}$	4.71	1.42	2.57	1.56	3.00	6.5	4.5
MgO	0.06	0.06	0.07	0.08	-	2.2	2.2
CaO	1.30	0.03	0.04	0.07	0.10	1.3	4.2
TiO_2	0.20	1.31	0.30	0.92	0.20	1.0	0.5
K_2O	0.03	0.01	0.03	0.02	-	3.7	3.4
Na_2O	0.02	0.04	0.05	0.03	-	1.20	3.90
MnO	0.47	0.16	0.63	0.09	0.05	0.11	0.08
CIA	94.12	99.61	99.47	99.67	-	75.30	56.93
CIW	94.25	99.66	99.60	99.73	-	88.32	65.24
K_2O/ Na_2O	1.5	0.25	0.6	0.67	-	3.08	0.87
SiO_2/Al_2O_3	2.79	3.14	2.78	1.27	-	3.32	4.34
Trace Elements (ppm)							
Cu	3.0	7.0	6.0	8.0	1.0	-	-
Zn	31	89	104	157	33	85.0	71.0
Ni	39	98	26	124	0.1	55.0	20.0
Zr	210	411	185	610	0.3	21.0	190.0

SG1-4= Share Geophagic clays; HBC=Human Body content (FAO/WHO, 2001); PAAS=Post Archean Australian Shale and (UC) Upper Continental Crust [22].

Table 2: Chemical compositions of Share Geophagic clays compared to some other clays in Nigeria.

Oxides	SG	A	B	C	D	E	F	G	H	I	J	K
SiO_2	58.47	47.20	56.20	53.20	52.80	52.65	55.49	60.47	60.70	70.82	50.74	51.25
Al_2O_3	25.33	24.94	26.67	25.44	28.31	27.24	18.63	17.77	17.75	13.11	32.70	32.32
$Fe_2O_{3(T)}$	2.57	4.10	2.11	2.09	2.53	3.01	9.67	8.18	6.04	2.81	2.59	2.21
Na_2O	0.04	0.02	<0.01	0.28	0.03	0.37	0.46	0.44	0.23	0.07	0.06	0.06
K_2O	0.02	0.37	0.36	0.40	0.34	1.44	1.84	1.17	1.40	0.34	0.37	0.27
CaO	0.11	0.03	0.05	0.02	0.04	0.19	0.77	0.47	0.83	0.64	Tr	Tr
MgO	0.07	1.0	0.19	0.13	0.15	0.38	1.25	1.26	1.22	0.16	0.15	0.12
MnO	0.34	0.03	0.01	<0.01	<0.01	-	0.04	0.03	0.03	-	0.006	0.005
TiO_2	3.42	1.55	1.50	1.84	1.58	-	-	-	-	-	1.99	1.20

SG=Share area (average of 4 samples), A=Asaba, B=Aramoko, C=Ibadan, D=Benin Okunlola and Owoyemi [12], E=Itakpe Okunlola [19]; F=Isan (Brown), G=Isan (Red); H=Ara Ijero Elueze and Bolannwa [20]; I=Itu Elueze et al [21]; J=Ubulu-Uku, K=Awo Omama Emofunieta et al [18]

Table 3: Mineralogical compositions (wt.%) of semi quantitative analysis of Share geophagic clays.

Sample	hematite	nacrite	dickite	Quartz	Kaolinite	Gibbsite	sillimanite	wollastonite	Pyrite
SG1	2	Trace	-	25	70	3	Trace	-	-
SG2	5	4	9	20	60	2	-	Trace	-
SG3	5	13	12	18	45	2	-	3	2
SG4	2	2	25	20	45	6	-	Trace	trace

Major oxide and trace element concentrations of the geophagic clays from Share area are compared to the average upper continental crust (UC) Taylor and McLennan [22] and the average Post-Archean Australian Shale (PAAS) Taylor and McLennan [22] as shown in table 1. Chemical composition of the Share geophagic clays, PAAS and UC showed predominance of SiO_2 and Al_2O_3; the value of SiO_2/Al_2O_3 ratio in Share geophagic clays ranges between 1.27 to 3.14 which is lower than SiO_2/Al_2O_3 ratio in PAAS (3.32) and UC (4.34). Geophagic clays showed high SiO_2/Al_2O_3 ratio which is also related to high quartz content. The concentrations of major oxides apart from Al and Si were generally very low reflecting very intense weathering of primary minerals [23].

5. DISCUSSION

The CIA values obtained from the Share geophagic clays ranges from 94.12 to 99.67 (Table 1) and were higher than CIA values of 45 to 55 according to Depetris and Probst [24] are indicative of lack of weathering Samples with very high CIA values in the study area also have high kaolinite content (Figures 1-4). These CIA values are compared to the CIA value for the upper continental crust (56.93) and the CIA value for the Post Archean Australia Shale (75.30). This signifies occurrence of weathering in Share area. The CIW values of the geophagic clay samples ranges between 94.25 and 99.73 which were slightly higher than the CIW values of PAAS (88.32), but significantly higher than the CIW values of UC (65.24). Kaolinite is form in warm, moist regions as a residual weathering product or by hydrothermal alteration of feldspars while dickite and nacrite are normally restricted to hydrothermal occurrences [25]. Kaolinite, dickite and nacrite were identified in Share area and this indicates that the process of formation of geophagic clays could be by weathering and/or hydrothermal alteration.

Titanium concentration in geophagic clay samples ranges between 0.02 and 1.31wt%, this compared well with Titanium values of PAAS (1.00 wt.%) and UC (0.36 wt%). These concentrations of Ti in the geophagic clays were minute, would not be able to mineralize into neither anatase nor rutile (Table 1). Most of the TiO_2 in the geophagic clays could possibly be in the form of free Ti-oxides [26]. Some Ti may have substituted for Al in the octahedral sheet of kaolinite, knowing that no Al^{3+} substitution in the tetrahedral sheet would occur because the Al_2O_3 values in the geophagic clays (20.54 to 36.57 wt.%) is below the Al_2O_3 value for pure kaolinite 39.49 wt.%, [17].

The Share geophagic samples (SG 1-4) have a lower concentrations of Al_2O_3 (20.54-36.57 wt.%) than Al_2O_3 content (39.49 wt.%) of pure kaolinite while the SiO_2 content in the geophagic clays in Share area (54.55-64.44wt%) are higher than SiO_2 content (46.3 wt.%) of pure kaolinite. The Al_2O_3 and SiO_2 contents in this sample are not comparable to that of pure kaolinite. And this is indicative that during kaolinitization which involves the formation of hydrated minerals, predominantly kaolinite, quartz and other minerals were also incorporated from the weathering of underlying basement rocks.

The elemental composition of Geophagic clays in Share area are compared to the adequate daily intake of trace elements in the human body (as shown in Table 1), the trace element content of the geophagic clays in Share area is relatively high. The Cu content in the geophagic clays have a range of 3.0 to 8.0ppm while daily intake necessary for human body function is only 1.0ppm. The values of other trace elements like Zr (610-185ppm), Zn (31-157ppm) and Ni (26-124ppm) are higher than the values for daily intake necessary for human body function of Zr (0.3ppm), Zn (33ppm) and Ni (0.1ppm) respectively. The values of major oxides like $Fe_2O_{3(T)}$,TiO_2 and MnO are higher than the values needed for human body function (table 1). Share geophagic clays may introduce heavy metals like Fe, Zn, Cu, Mn, Ni, Ti, and Zr to the gastro-intestinal system of consumers, causing adverse effects like increase in the gastro-intestinal pH and the binding of plant toxins and pathogens which create a surfacial coating on the stomach with inferred pharmaceutical implications. This is similar to the report given by Ekosse et al. [27] on the Geophagia clays and soils from southern Africa.

This Geophagic clays may impact positively or negatively on human health when ingested. Kaolin (kaolinite, nacrite, gibbsite and dickite) can be used to treat diarrhea, gastritis, colitis, enhancement bio-activities and maintenance of normal intestinal flora. Negative impact may lead to electrolyte disturbance, intestinal obstruction, constipation, hypertension, peritonitis, eclampsia, iron deficiency, anemia, microbiological infections, helminthiasis and heavy metal poisoning [1,9]. The coarse sandy quartz particles in geophagic clays could affect dental enamel and also lead to the possible rupturing of the Sigmoid colon due to the abrasive nature of the particles [10].

The presence of sedimentary structures typical of tidal sedimentation, the presence of ironstones further attests to abundant oxygen due to sub-aerial exposure resulting to the oxidation of Fe^{2+} to Fe^{3+}. The presence of basal conglomerate, rootlets structures, iron capping, fining upward sequence and abundances of kaolinite indicate a fluvial (continental) sedimentation [28-30]. Similar kaolinite clay minerals were also reported by Mucke et al. [31], in Lokoja area of Middle Niger Embayment (southern Buda basin) indicating that the environment is not fully marine.

The absence of any marine fossils and clay minerals like illite, smectite, montmorrillonite and chlorite which were not identified in the mineralogical composition of geophagic clays. According to Fagel et al. [32], these clay minerals will correspond to trends register in most marine sediments. This further attest to non-marine depositional environment in Share area. Mineralogically, non-marine clays are characteristically less altered during diagenesis and they lack magnesia, lime and alkalies as observed in the area.

Thus, information on provenance of geophagic clays could provide clues on type of parent materials from which they were derived. The TiO_2 vs Al_2O_3 diagram figure 5; Ekosse [17] indicated that the provenance of the Share geophagic clays were predominantly from acidic rocks, as they plot in the granite/rhyolite field. The sediment provenance could thus be attributed to the granitic rocks from Share area which constitutes part of the Precambrian basement complex of southwestern Nigeria [11].

6. CONCLUSION

Geophagic clays in the area are polymineralic and the patterns of the clay minerals identified could be attributed to kaolinite, nacrite, gibbsite and dickite. The presence of basal conglomerate, rootlets structures, iron capping, fining upward sequence, abundances of kaolinite, low abundance of MgO and K_2O, the presence of hematite, pyrite and sedimentary structures typical of tidal sedimentation indicate non-marine depositional environment. The presence of ironstones further attests to abundant oxygen due to sub-aerial exposure resulting to the oxidation of Fe^{2+} to Fe^{3+}. This is typical of fluvial (continental) sedimentation.

The CIA values obtained from the Share geophagic clays were very high, ranges between 94.12 to 99.67 and exceeds the CIA values of 45- 55 that indicate lack of weathering. The CIW values of the geophagic clay samples ranges between 94.25.99 to 99.73 which were slightly higher than the CIW values of PAAS (88.32), but significantly higher than the CIW values of UC (65.24). This is indicative that the process of formation of geophagic clays in Share area could be by weathering and/or hydrothermal alteration. The provenance of the Share geophagic clays indicated that they were derived predominantly from acidic rocks (granite/rhyolite). The sediment provenance could thus be attributed to the granitic rocks from Share area which constitutes part of the Precambrian Basement Complex of southwestern Nigeria.

The geophagic clays could be beneficial or harmful, kaolin can be used to treat diarrhea, gastritis, colitis, enhancement of bio-activities and maintenance of normal intestinal flora. Harmful effects may lead to electrolyte disturbance, intestinal obstruction, constipation, hypertension, peritonitis, dental damage, eclampsia and iron deficiency. These geophagic clays may introduce metals like Fe, Zn, Cu, Mn, Ni, Ti, and Zr to the gastro-intestinal system of consumers causing adverse effects like increasing the gastro-intestinal pH and the binding of plant toxins and pathogens, and creating a surfacial coating on the stomach with inferred pharmaceutical implications.

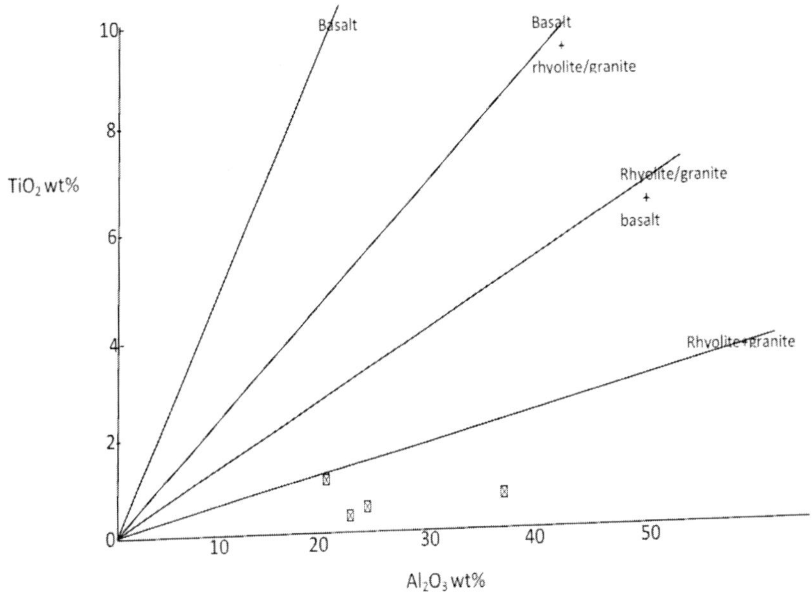

Figure 5: TiO2 vs Al2O3 binary plot of geophagic clay samples from Share area Ekosse [17].

REFERENCES

1. Abrahams PW, Parsons JA (1997) Geophagy in the tropic: an appraisal of three Geophagic materials. Environmental Geochemistry and Health 19: 19-22.
2. Ekosse GE, De Jager L, Ngole VM (2010) Traditional mining and mineralogy of Geophagic clays from Limpopo and Free State Provinces, South Africa. African Journal of Biotechnology 9: 8058-8067.
3. Geissler PW, Mwaniki D, Thiong'O F, Friis H (1998) Geophagy as a risk factor for geohelminth infections: a longitudinal study of Kenyan primary schoolchildren. Trans R Soc Trop Med Hyg 92: 7-11.
4. Brand CE, De Jager L, Ekosse GE (2010) Possible health effects associated with human geophagic practice: an overview. South African Medical Technology 23: 11-13.
5. Ferrell Jr RE (2008) Medicinal clay and spiritual healing. Clays and Clay Minerals 56: 751-760.

6. Ngole VM, Ekosse GE, De Jager L, Songca SP (2010) Physicochemical characteristics of geophagic clayey soils from South Africa andSwaziland. African. Journal of Biotechnology 9: 5929-5937.
7. Bisi-Johnson MA, Obi CL, Ekosse GE (2010) Microbiological and health related Perspectives of geophagia: an overview. African Journal of Biotechnology 9: 5784-5791.
8. Kawai K, Saathoff E, Antelman G, Msamanga G, Fawzi WW (2009) Geophagy (soil-eating) in relation to anemia and helminth infection among HIV-infected pregnant women in Tanzania. Am J Trop Med Hyg 80: 36-43.
9. Murray HH (2007) Applied clay mineralogy. Occurrences, processing and application of kaolins, bentonites, palygorskite-sepiolite, and common clays. Elsevier.
10. Konta J (1995) Clay and Man: Clay raw materials in the service of man. Applied Clay Science 10: 275-335.
11. Odewumi SC (2011) Mineralogical Characterization of Clays and Clay Mineral in Share area, Bida Sedimentary Basin Nigeria. In Ekosse G-I E, de Jager L and Ngole V.M (Editors). An innovative Perspective on the role of Clays and clay minerals in Africa. Book of Conference Proceedings. Central University of Technology, Bloemfontein, South Africa.
12. Okunlola OA, Owoyemi KA (2011) Compositional characteristics of Geophagic clays of southern Nigeria. In Ekosse G-I E, de Jager L and Ngole V.M (Editors). An innovative Perspective on the role of Clays and clay minerals in Africa. Book of Conference Proceedings. Central University of Technology, Bloemfontein, South Africa.
13. Braide SP (1992) Geological development, origin and energy mineral resources potentials of the Lokoja in the southern Bida Basin. Journal of Mining and Geology 28: 33-44.
14. Tchalenko JS, Burnett AD, Hung JJ (1998) Measurement of orientation in clay. J Sediment Petrol 20: 123-134.
15. Harnois L (1988) The CIW index: a new chemical index of weathering. Sedimentary Geology 55: 319-322.
16. Nesbitt HW, Young GM (1982) Early Proterozoic climates and plate motions inferred from major element chemistry of luttites. Nature 291: 715-717.
17. Ekosse GE (2001) Provenance of the Kgwakgwe kaolin deposit in southeastern Botswana and its possible utilization. Applied Clay Science 20: 137-152.
18. Emofurieta WO, Kayode AA, Coker SA (1992) Mineralogy, geochemistry and Economic evaluation of the kaolin deposits near Ubulu-Uku, Awo-Omama and Buan in southern Nigeria. Journal of Mining and Geology 28: 211-220.
19. Okunlola OA (2008) Compositional characteristics and functional industrial application of Itakpe clay occurrence central Nigeria. European Journal of Scientific Research 19: 453-461.
20. Elueze AA, Bolarinwa AT (1995) Assessment of functional applications of lateritic clay bodies in Ekiti environs, southwestern Nigeria. Journal of Mining Geology 31: 79-87.
21. Elueze AA, Ntekim EE, Ekwere SJ (1998) Compositional and industrial assessment of clay bodies in Itu area, southeastern Nigeria. Journal of Mining Geology 35:117-124.
22. Taylor SR, McLennan SM (1985) The Continental Crust: Its Composition and Evolution. Blackwell, USA.
23. Ekosse GE, Ngole VM (2012) Mineralogy, geochemistry and provenance of geophagic soils from Swaziland. Appl Clay Sci 57: 25-31.
24. Depetris PJ, Probst JL (1998) Variability of the chemical index of alteration (CIA) in the Paranfi River suspended load. Mineralogical Magazine 62A: 366-367.
25. Duane MM, Robert CR Jr (1997) X-Ray Diffraction and the Identification and Analysis of Clay Minerals(eds). Oxford University Press, USA.
26. Dolcater DL, Syers JK, Jackson ML (1970) Titanium as free oxide and substituted forms in kaolinites and other soil minerals. Clays Clay Miner 18: 71-79.
27. Ekosse GE, Mulaba-Bafibiandi AF (2008) Granulometric evaluation of continental Bentonites and kaolin for ceramic applications. Journal of Applied Sciences 8: 1021-1027.
28. Belal SA (1992) Sedimentology and mineral composition of the Kurnub sandstone in Wadi Osieb, S.W. Jordan. Sedimentary Geology. Elsevier Science Publishers BV Amsterdam 78: 267-283.

29. Henning D, Helge N (1992) The Diagenesis of continental (Karoo-Tertiary) siliciclastics from an East African rift valley (Rukwa-Tukuyu area) Tanzania. Sedimentary Geology 78: 251-256.
30. Kelvin AL, David RG (1992) Fluvial, coastal, nearshore and shelf deposition. In the Upper Preterozoic to Lower Cambrian Addy Quartzite, northeastern Washington. Sedimentary Geology. Elsevier Science Publishers. B.V. Amsterdam 77: 15-35.
31. Mucke A, Badejoko TA, Akande SO (1999) Petrographic-microchemical studies and origin of the Agbaja Phanerozoic Ironstone Formation, Nupe Basin, Nigeria: a product of a ferruginized ooidal kaolin precursor not identical to the Minette-type. Mineralium Deposita 34: 284-296.
32. Fagel N, Andre L, Chamley H, Debrabant P, Jolivet L (1992) Clay sedimentation in the Japan Sea since the Early Miocene: influence of source-rock and hydrothermal activity. Sedimentary Geology 80: 27-40.

CHAPTER 17

Oxalate secretion by ectomycorrhizal *Paxillus involutus* is mineral-specific and controls calcium weathering from minerals

A. Schmalenberger[1,2,3], A. L. Duran[2], A. W. Bray[4], J. Bridge[1], S. Bonneville[4,+], L. G. Benning[4,5], M. E. Romero-Gonzalez[1], J. R. Leake[2] & S. A. Banwart[1]

ABSTRACT

Trees and their associated rhizosphere organisms play a major role in mineral weathering driving calcium fluxes from the continents to the oceans that ultimately control long-term atmospheric CO_2 and climate through the geochemical carbon cycle. Photosynthate allocation to tree roots and their mycorrhizal fungi is hypothesized to fuel the active secretion of protons and organic chelators that enhance calcium dissolution at fungal-mineral interfaces. This was tested using $^{14}CO_2$supplied to shoots of *Pinus sylvestris* ectomycorrhizal with the widespread fungus *Paxillus involutus* in monoxenic microcosms, revealing preferential allocation by the fungus of plant photoassimilate to weather grains of limestone and silicates each with a combined calcium and magnesium content of over 10 wt.%. Hyphae had acidic surfaces and linear accumulation of weathered calcium with secreted oxalate, increasing significantly in sequence: quartz, granite < basalt, olivine, limestone < gabbro. These findings confirmed the role of mineral-specific oxalate exudation in ectomycorrhizal weathering to dissolve calcium bearing minerals, thus contributing to the geochemical carbon cycle.

1. INTRODUCTION

The formation of soil, which provides the foundations of terrestrial ecosystems and agriculture, is based on the weathering of rock-forming minerals[1]. Calcium and magnesium are abundant in the Earth's crust[2] where calcium in particular precipitates with the bulk of the Earth's orthophosphate. Biological acquisition of growth-limiting mineral phosphorus is therefore expected to mobilise calcium in excess of biological requirements through biologically-driven co-dissolution from phosphorus-bearing minerals. Quantification of weathering activities, particularly for calcium- and magnesium-bearing rocks and minerals, is a key component in predicting rates of soil formation, calcium and magnesium fluxes from the continents to the oceans, and fluxes of biologically essential elements including phosphorus. The release of such key nutrients ultimately control the biomass and productivity in long-term ecosystem developments[3,4,5].

Until recently, continental-scale biological weathering has been largely attributed to plant root activities[6,7]. However, 80–90% of all plant species associate with mycorrhizal fungal partners[8,9]. Mycorrhizal mycelial networks are several times longer and much finer than plant roots, in the case of ectomycorrhiza often extending to 200 m in length g^{-1} soil[10]. These networks are supported by up to 30% of the net photosynthate of their plant hosts[10]. In return, the mycelial networks enhance plant suptake of growth-limiting nutrients, in particular phosphorus, from the soil environment thereby enhancing ecosystem productivity[11,12,13]. In typical ectomycorrhizal forests, which comprise over 35% of the global forest cover14, the transfer of photosynthate from tree-to-mycorrhizal fungi amounts to a biotic energy flux of 1000–6000 kJ m^{-2} a^{-1}, (based on solar energy capture of average insolation for temperate forests)15. This substantial energy input from plants into mycorrhiza is not only used for catabolic reactions (e.g. respiration) by the fungi, but also allocated to nutrient acquisition from minerals through, amongst other processes, the exudation of organic acids[8,16] that accelerate chemical dissolution of oxide and silicate minerals. Despite increasing awareness of the potential importance of these plant-fungal partnerships in mineral weathering, a paucity of measurements has quantified their contributions to element mass transfers. Likewise, the extent to which plant photosynthate is directed to fungal-mineral interfaces has received limited attention to date[17], but has recently been shown to be a major control on mycorrhizal-fungal driven mineral weathering rates[18].

A number of ectomycorrhizal fungi, including *Paxillus involutus*, can be grown in pure culture, where the fungi can synthesise and secrete up to 1 mmol oxalate g^{-1} of fungal biomass (DW)[19,20,21]. Both calcium and bicarbonate ion concentrations influence oxalic acid secretion by *P. involutus* grown with nitrate[19,20,21]. The secretion of oxalic acid is thought to play a major role in the dissolution of apatite, releasing phosphorus for uptake by ectomycorrhizal fungi and immobilizing the released calcium into calcium oxalate[22]. However, other elements may also influence fungal oxalate secretion. For example, *Pinus sylvestris* symbiotic with *Hebeloma crustuliniforme* has shown trends for oxalate exudation to be higher when exposed to rock grains containing different amounts of potassium[23]. This is consistent with studies of fungi in ectomycorrhizal symbiosis with trees acting as biosensors actively foraging for nutrients in rocks and minerals, distinguishing mineralogy such as apatite from quartz, and different grain sizes[17,24,25].

In the current literature, most mineral weathering data are derived from solution-based abiotic dissolution experiments[26,27,28]. There, weathering of rocks and minerals is controlled primarily by interactions in aqueous solutions[29] in which rates are dependent on the distance from equilibrium, pH, ionic strength, redox state, and temperature. Biological weathering processes are distinct in their spatial precision, often targeted at specific nutrient-rich minerals in soils, evident as preferential colonization of mycorrhizal hyphae on grains of certain rock types[24]. Some of the most biologically active forest soils that experience intense weathering from organic acid exudation by mycorrhizal fungi are freely-draining and seasonally dry[30]. Consequently, there is a need to study weathering in environments where soils are not water saturated for extended periods of time[31], e.g. in the absence of gravitational water[32]. Indeed, weathering of biotite and apatite in microcosms by *P. involutus* in symbiosis with host trees has revealed that significant chemical and physical alteration of minerals can take place in conditions where the minerals are not immersed in water[22,33,34,35].

The aim of the present study was to test the hypotheses that (i) ectomycorrhizal weathering is ultimately driven by fungal allocation of photoassimilate received from a host plant (in this case, *Pinus sylvestris*), (ii) the release of calcium from minerals, and its accumulation as a weathering product by a symbiotic fungal partner (in this case, *P. involutus*), is linked to mineral-specific rates of fungal secretion of oxalic acid and the formation of calcium oxalate. To address these hyphotheses, monoxenic microcosms were designed with mineral weathering arenas, containing grains of a range of important carbonate and silicate rocks and minerals (limestone, basalt, gabbro, granite, olivine, microcline) of various calcium and phosphorus concentrations and a control mineral (quartz, 0.01% CaO, 0.0006% P_2O_5). These rocks/minerals were accessible only to mycorrhizal fungal hyphae extending from the roots of the host plants[17] and were hydrologically isolated from the root/mycorrhiza compartments to mimic unsaturated soil environments in which ectomycorrhizal fungi are most active[9]. The allocation of recent photosynthate carbon to *P. involutus*, and its subsequent deployment in weathering arenas was quantified using $^{14}CO_2$ pulse-labelling and tracing. Quantification of calcium oxalate accumulation on symbiotic ectomycorrhizal fungal hyphae revealed the extent to which calcium mass transfer, resulting from plant-drived fungal weathering is linked to oxalate secretion.

2. RESULTS

2.1 Rock/mineral composition

Concentrations in the rocks and minerals were expressed as weight (wt. %) binary oxide (Table 1). Calcium oxide (CaO) ranged from 32% in the limestone, to 15% in gabbro, 11% in basalt, 4% in granite and less than 0.2% in olivine and microcline. For magnesium oxide, (MgO) the highest concentrations were found in olivine (45%), followed by limestone (14%), gabbro (10%) and basalt (8%). In contrast, the highest amounts of potassium oxide (K_2O) were identified in microcline (15%) and granite (4%). Phosphorus concentrations were low in all rocks/minerals (0.004–0.36% P_2O_5) but varied by almost two orders of magnitude (Table 1) around the average concentration of phosphorus in the earth's crust2. The bulk chemical compositions of the rock/mineral samples are reflected in the mineralogy, as determined by Rietveld refinement of XRD patterns (Table 2). Samples of granite contained anorthite (16%, $CaAl_2Si_2O_8$) as the only calcium bearing mineral. Basalt samples contained anorthite (59%), and augite (20%, $(Ca,Na)(Mg,Fe,Al)(Si,Al)_2O_6$), while samples of gabbro contained anorthite (56%) and diopside (13%, $CaMgSi_2O_6$) as the major calcium bearing minerals. Limestone was primarily comprised of dolomite (74%, $CaMg(CO_3)_2$).

Plant-to-fungal carbon allocation to weathering arenas. Amounts of ^{14}C fixed by *Pinus sylvestris* and allocated from roots to weathering arenas colonised by *P. involutus* were lowest in wells with quartz, granite and microcline with below 3300 disintegrations per minute (DPM) g^{-1} dry weight (DW) (Fig. 1). A higher ^{14}C allocation of over 5900 DPM g^{-1} DW was recorded in the wells with olivine although this difference was not significant ($P > 0.05$). ^{14}C allocations for colonised wells with basalt and gabbro were close to 8000 DPM g^{-1} DW and for basalt significantly higher than quartz and microcline ($P < 0.05$). The highest ^{14}C carbon allocations of over 22,000 DPM g^{-1} DW were detected in colonised wells with limestone ($P < 0.05$).

Table 1: Major elements (% wt.) are expressed as oxides of silicon (SiO_2), aluminium (Al_2O_3), magnesium (MgO), potassium (K_2O), sodium (Na_2O), calcium (CaO), iron (Fe_2O_3) and phosphorus (P_2O_5). Values of 0.001 to 0.01 incur an error rate of better than 10% wt; nd = values below 0.001 wt %.

	SiO_2	Al_2O_3	CaO	MgO	K_2O	Na_2O	Total base cation oxides	Fe_2O_3	P_2O_5
Quartz	99.90	0.05	0.01	nd	0.01	nd	0.02	0.02	0.006
Microcline	63.97	18.66	0.06	0.02	15.17	0.94	16.18	0.11	0.092
Olivine	43.77	0.90	0.17	45.12	0.01	0.11	45.41	8.99	0.004
Perlite*	75.94	12.75	0.90	0.16	3.28	3.61	7.95	0.83	0.047
Granite	64.88	16.39	3.58	2.26	4.19	3.23	13.25	4.10	0.361
Basalt	46.58	18.39	10.93	8.17	0.30	2.12	21.51	11.63	0.246
Gabbro	46.49	19.64	14.94	10.42	0.08	1.44	26.87	6.10	0.025
Limestone	16.05	2.07	31.75	13.88	0.32	0.30	46.25	1.48	0.111

*Perlite was added to the main rooting compartment of all mesocosms, but was not present in the wells to which the other rocks and minerals listed here were added.

2.2 Determination of pH of *P. involutus* hyphae in weathering arenas

The pH values of the mycorrhizal hyphae removed from the grains of granite, basalt and limestone were determined with the fluorescent molecular probe SNARF4F through confocal laser scanning microscopy (CLSM). While the probe signal on the hyphae taken from wells with limestone and basalt resulted in preferential emission of fluorescent light around 580 nm (green), fungal samples taken from the granite wells typically showed stronger emissions at 640 nm than at 580 nm. Calibration of the signals revealed a pH from 4.9 to below 4.6 (pH 4.6 was the lowest calibration point, Fig. 2a,b) for basalt and limestone derived samples. In contrast, hyphae taken from the granite wells had variations in the fluorescence signal in the range of pH 6.2 to 6.6 (Fig. 2c).

Table 2: Mineralogy of rocks/minerals as determined by Rietveld refinement of XRD patterns (%).

		Feldspars					Micas		Pyroxenes			
	Quartz	Orthoclase	Albite	Anorthite	Microcline	Sanidine	Biotite	Muscovite	Augite	Wollastonite	Diopside	Enstatite
Rock/ mineral	SiO_2	$KAlSi_3O_8$	$NaAlSi_3O_8$	$CaAl_2Si_2O_8$	$KAlSi_3O_8$	$(K,Na)(Si,Al)_4O_8$	$K(Mg,Fe)_3(AlSi_3O_{10})(F,OH)_2$	$KAl_2(AlSi_3O_{10})(F,OH)_2$	$(Ca,Na)(Mg,Fe,Al)(Si,Al)_2O_6$	$CaSiO_3$	$CaMgSi_2O_6$	$Mg_2Si_2O_6$
Granite	26	6	25	16	15	0	4	6	0	0	0	0
Basalt	0	5	13	59	0	0	0	0	20	0	0	0
Gabbro	0	0	7	56	0	0	0	0	2	3	13	8
Limestone	22	0	0	0	0	0	0	0	0	0	0	0
Quartz	100	0	0	0	0	0	0	0	0	0	0	0
Olivine	0	0	8	0	0	0	0	0	0	0	0	0
Microcline	0	2	0	0	95	3	0	0	0	0	0	0

	Olivines				Carbonates			
	Faylite	Forsterite	Hornblende	Kaolinite	Calcite	Dolomite	Aragonite	Calcite Magnesian
Rock/mineral	Fe_2SiO_4	Mg_2SiO_4	$Ca_2(Mg,Fe,Al)_5(Al,Si)_8O_{22}(OH)_2$	$Al_2Si_2O_5(OH)_4$	$CaCO_3$	$(CaMg)(CO_3)$	$CaCO_3$	$(Ca,Mg)CO_3$
Granite	0	0	0	0	0	0	0	0
Basalt	0	0	4	0	0	0	0	0
Gabbro	0	6	5	0	0	0	0	0
Limestone	0	0	0	0	2	74	1	1
Quartz	0	0	0	0	0	0	0	0
Olivine	2	85	0	5	0	0	0	0
Microcline	0	0	0	0	0	0	0	0

2.3 Qualitative analysis of the chemistry of symbiotic hyphae and attached secondary minerals from weathering arenas through micro Fourier transform infrared (μFT-IR) spectroscopic analysis

All sampled *P. involutus* hyphae removed from microcosms with grains of olivine, basalt, limestone and gabbro showed a characteristic peak at 1317–1324 wavenumber cm^{-1} in the μFT-IR spectrum, identical to the main peak for calcium oxalate monohydrate[36] (Fig. 3). The fungal amide peaks I and II were overlaid with a broader second calcium oxalate peak at approximately 1628 cm^{-1} with the result that the amide II peak (1550 cm^{-1})[37] was only detectable as an inflexion point (as a result of putative calcium oxalate accumulation) from the 1628 cm^{-1} maximum. A characteristic peak of calcium oxalate between 1317–1322 cm^{-1}, and a partial overlay of the amide peaks at 1628 cm^{-1} were detectable, associated with some hyphal samples removed from two of six wells with granite and microcline while all hyphal samples removed from wells with quartz grains showed no presence of calcium oxalate. Three additional characteristic peaks were detected in all fungal samples in the region of 1030–1160 cm^{-1} referred to here as carbon-hydroxyls, esters and carbonyl groups. This region has also been referred to as polysaccharides of fungal mycelia[37].

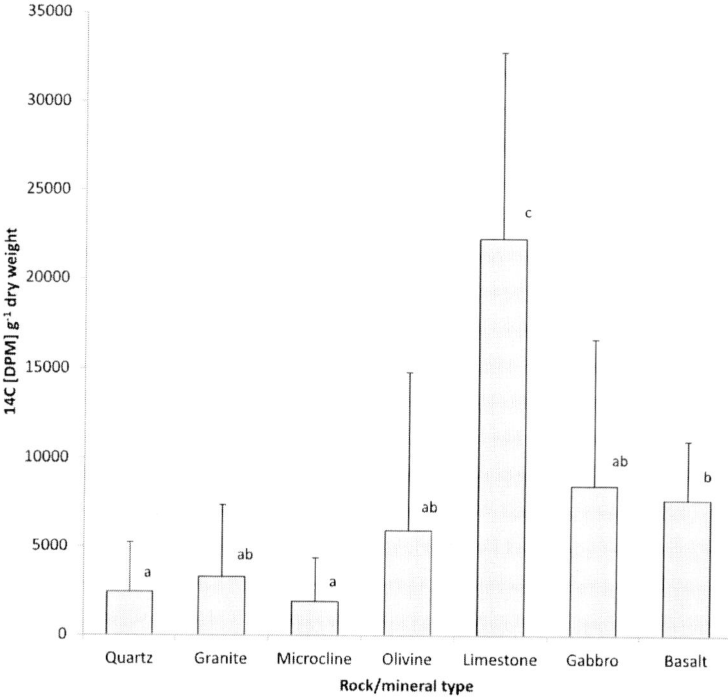

Figure 1: Quantification of ^{14}C pulse-labelled photosynthate allocated by *Pinus sylvestris* to *P. involutus*, and its subsequent deployment in biomass and exudates in fungal-colonised weathering arenas containing quartz, granite, microcline, basalt, olivine, limestone and gabbro. Error bars indicate standard deviation. Means sharing the same letter are not significantly different ($P > 0.05$, ANOVA, Games-Howell).

2.4 Qualitative analysis of crystals formed on hyphae employing scanning electron microscopy with energy dispersive X-ray spectroscopy (SEM-EDS)

P. involutus hyphae were taken from a well in representative microcosms with limestone, basalt and granite grains respectively and analysed using SEM-EDS. The electron backscatter images of all the fungal hyphae from wells with limestone and basalt showed bright crystalline structures, while only a few of these particles were found associated with hyphae removed from a microcosm with granite grains (Supplementary Fig. S1). EDS analysis of selected areas with those particles at the surface of mineral grains confirmed the presence of calcium. EDS from hyphal samples from basalt and limestone wells showed peaks at K_α (emission line of electron transition to K shell) of calcium that reached about 3000 and 6000 X-ray counts per minute, while comparatively smaller peaks from selected areas of hyphal samples from granite only reached levels of less than 800 X-ray counts per minute.

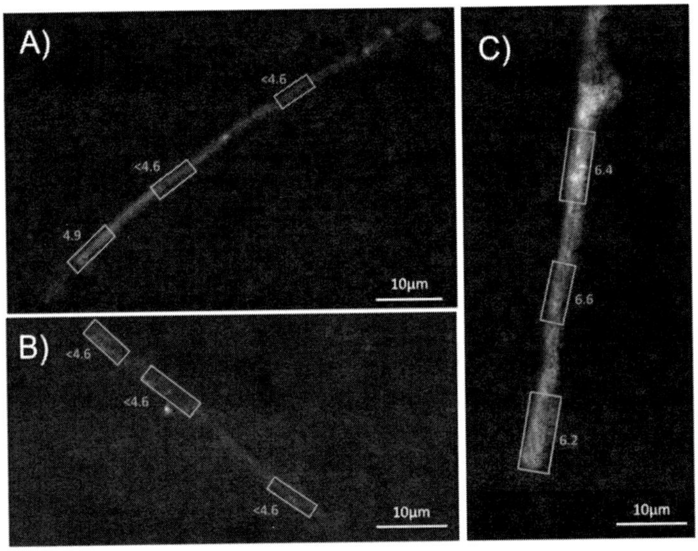

Figure 2: Confocal laser scanning micrococope images from symbiotic *P. involutus* hyphae stained with 5 μM SNARF4F (Invitrogen) in water and emmissions detected at 580 and 640 nm with a Zeiss Axioplan and a 510 meta detector, excited with a 488 nm argon laser (see materials for details). Fluorescence intensities revealed pH values for (A) *P. involutus* from basalt, pH < 4.6–4.9, (B) *P. involutus* from limestone, pH < 4.6, (C) *P. involutus* from granite, pH = 6.2–6.6.

2.5 Quantitative analysis of oxalate and calcium in and attached to the surface of symbiotic hyphae using ion chromatography and inductively coupled plasma mass spectrometry

Amounts of oxalate per unit biomass of hyphae increased in order from quartz, granite, microcline, basalt, olivine, limestone to gabbro (Table 3).

Hyphae of *P. involutus* sampled from wells with quartz and granite contained less than 10 mg oxalate per g of biomass (DW, including secondary minerals, Table 3). Significantly higher ($P < 0.05$) concentrations of oxalate were found in the presence of basalt, olivine and limestone in the range of 60–80 mg oxalate g^{-1} biomass (DW, including secondary minerals). Over 190 mg oxalate g^{-1} biomass (DW, including secondary minerals) was found in the presence of gabbro, which was higher ($P < 0.05$) than for all other minerals or rocks tested.

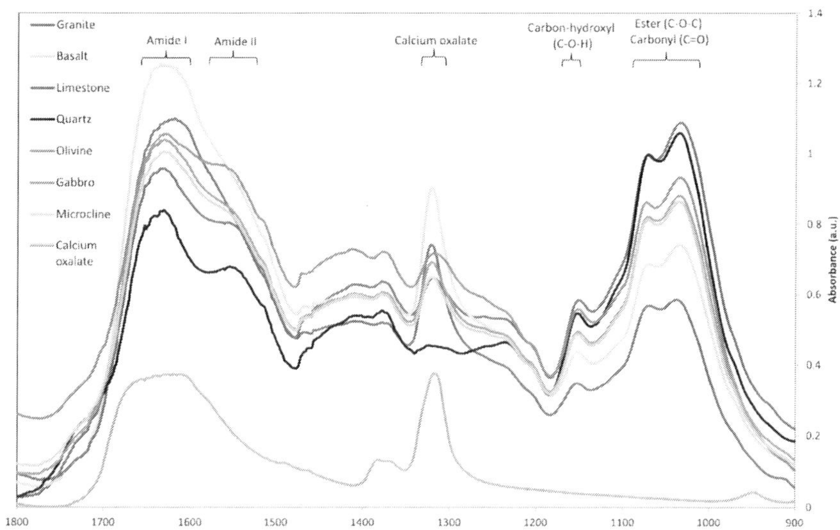

Figure 3: Micro Fourier transform infrared (μFT-IR) spectra from symbiotically grown hyphae of *P. involutus* with adhering secondary minerals taken from weathering arenas in microcosms with minerals/rocks of quartz (black), granite (blue), microcline (bright green), basalt (orange), olivine (dark red), limestone (bright red) and gabbro (dark green). Calcium oxalate was measured in comparison as a fine powder (grey). Values were obtained from combined readings from 6 different locations of each sample in triplicates at 100 repeats each (1800 repeats per spectrum in graph).

While very low concentrations of calcium were detected associated with fungal biomass growing with quartz and granite (below 4 mg g^{-1}biomass DW, including secondary minerals), higher values were found with hyphae growing on microcline and olivine (20–40 mg g^{-1} biomass DW including secondary minerals). Hyphae in the presence of basalt, limestone and gabbro (>10% CaO in parental rocks, Table 1) showed significantly higher

($P < 0.05$) accumulation of calcium (40–120 mg g^{-1}biomass DW including secondary minerals) with gabbro providing significantly higher values than all other tested minerals and rocks ($P < 0.05$; Table 3). In contrast, hyphae of *P. involutus* grown in monoculture on Modified Melin Norkran's agar contained only 1.4 mg g^{-1} biomass DW calcium. Accumulation of oxalate and calcium per unit length of hyphae per day showed close similarities to the total accumulation, with hyphae from weathering arenas of gabbro exceeding the accumulation of 2.1 and 2.7 10^{-10} mol m^{-1} d^{-1} oxalate and calcium, respectively (Table 4).

Table 3: Accumulation of oxalate and calcium (mg g^{-1}) biomass (DW) on *P. involutus* hyphae (incl. secondary minerals).

	Oxalate		Calcium	
Weathering arena Rock/mineral	**mean**	**±**	**mean**	**±**
Quartz	3.08[a]	3.17	2.69[a]	2.39
Granite	5.58[a]	5.76	3.07[a]	2.72
Microcline	35.04	7.33	22.37	2.84
Olivine	72.36[b]	19.21	32.99[ab]	17.75
Basalt	63.75[b]	32.4	43.64[b]	18.66
Limestone	77.30[b]	4.58	57.15[b]	8.28
Gabbro	197.91[c]	20.77	112.73[c]	11.53

Calcium and oxalate accumulation per g biomass (DW, Table 3) correlated closely (calcium:oxalate, $R^2 = 0.909$) at a slope of 1.37 (Fig. 4). For olivine and granite the mass ratio of calcium:oxalate was closer to 1.0 while for limestone and basalt it was closer to 1.4. The noble agar base in the wells was a potential source of calcium. However, concentrations were only in the region of 11.5 (+/−0.02) ng g^{-1} (wet weight, WW). Outside the wells, the base agar concentrations ranged around 16–19 ng g^{-1} WW for calcium that was separated from the fungus by a cellophane sheet. The perlite was a further potential source of calcium (Table 1).

Table 4: Oxalate and calcium accumulation rates (10^{-10} mol) per m hyphae and day ($m^{-1} d^{-1}$).

Weathering arena rock/mineral	Oxalate		Calcium	
	mean	±	mean	±
Quartz	0.033[a]	0.035	0.065[a]	0.613
Granite	0.034[a]	0.030	0.047[a]	0.047
Microcline	0.383	0.053	0.551	0.675
Olivine	0.610[b]	0.159	0.591[ab]	0.718
Basalt	0.732[b]	0.521	1.149[ab]	0.569
Limestone	0.845[b]	0.367	1.395[ab]	0.414
Gabbro	2.184[c]	0.250	2.795[b]	0.485

R-square value of linear regression fit is 0.909, at a slope of 1.37 (black line) in comparison to a slope of 1 (dashed line).

2.6 Mycorrhizal weathering rates compared to abiotic weathering rates

This study's fungal-driven calcium accumulation rates (moles Ca $m^{-2} s^{-1}$) were based on the maximum theoretical extent of hypha-mineral contact area as the mass transfer interface and represent lower bound values as; i) large proportions of the hyphae have been observed to be not in direct contact with rock/minearls, thus the maximal fungal-mineral contact area may be greatly overestimated and as a consequence the calculated calcium accumulation rate per square metre underestimated, possibly by a factor of 10 based on visual inspection of microcosms and optical microscopy micrographs; ii) duration of colonization of weathering arenas may have been up to two weeks shorter than reported due to biweekly colonization inspections; iii) fungal oxalate secretion may not have taken place at a constant rate during the colonization of the weathering arenas; iv) not all calcium and oxalate may have been transformed into secondary minerals, for example, some calcium may be transported to the plant host, and not included in this quantification.

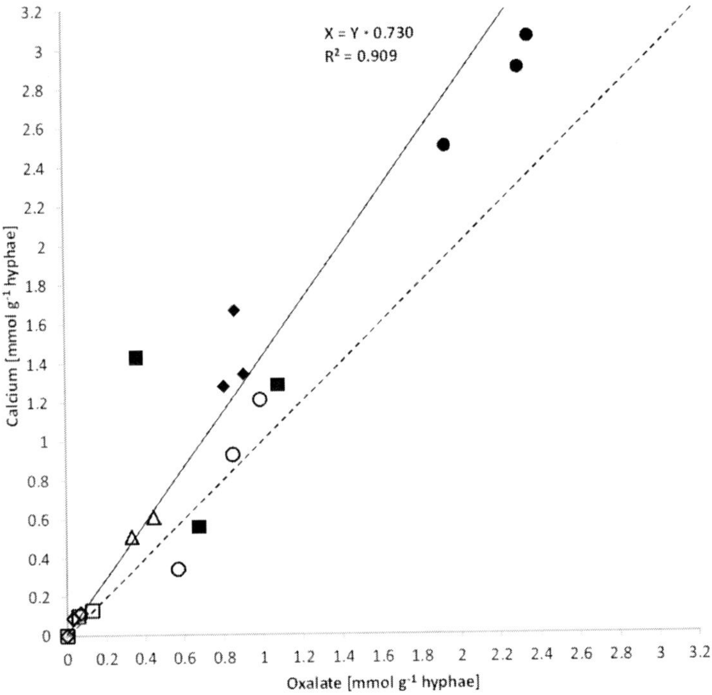

Figure 4: Correlation of oxalate and calcium accumulation rates in mmol g^{-1} DW from quartz (empty diamond), granite (empty square), microcline (empty triangle), olivine (empty circle), basalt (filled square), limestone (filled diamond) and gabbro (filled circle).

Accumulation rates of calcium through fungal-mineral interaction in monoxenic microcosms in this study were greater than reported abiotic stoichiometric weathering rates of pure calcium bearing mineral phases in laboratory experiments at the same temperature[26,28,38]. Based on the presence of anorthite, the fastest weathering calcium bearing mineral in basalt and gabbro, previously reported abiotic weathering rates at pH 4 were 10–100 times lower than the here quantified ectomycorrhiza weathering rates (even without considering the factors discussed above that are likely to have underestimated this biotic weathering). However, calcium accumulation rates from granite in this study were similar to the abiotic dissolution rates reported for anorthite. Furthermore, calcium accumulation from limestone weathering in this study was at least three orders of magnitude lower than reported abiotic dissolution of dolomite[38] (Table 5a,b).

Table 5: (A) Calcium accumulation rates based on theoretical maximum fungal-rock contact area[ł]. (B) Chemical weathering at 288 K in water* [mol m^{-2} s^{-1}]. łcontact area as calculated in methods. ± indicates standard deviation. *modified (Arrhenius correction to 288 K) from26,28,38.

	(A)			(B)		
Rock/mineral	**Major calcium bearing minerals [>1 mol%]**	**Calcium accumulation [mol m^{-2} s^{-1}]**				
		R_1	**±**		**pH4**	**pH7**
Granite	Anorthite	3.2×10^{-11}	3.2×10^{-12}	Anorthite	2×10^{-11}	6×10^{-12}
Basalt	Anorthite, augite, hornblende	5.6×10^{-10}	2.7×10^{-10}	Diopside	3×10^{-11}	1×10^{-13}
Limestone	Dolomite	6.6×10^{-10}	5.6×10^{-11}	Augite	2×10^{-12}	$>10^{-14}$
Gabbro	Anorthite, diopside, augite, hornblende	1.1×10^{-9}	5.7×10^{-10}	Hornblende	1×10^{-12}	2×10^{-13}
				Dolomite	1×10^{-6}	

3. DISCUSSION

The formation of secondary minerals on hyphae in this study was confirmed to be calcium oxalate using μFT-IR[39] and these findings were verified via identification of calcium as one main component of the crystals using SEM-EDS. The overlay of the calcium oxalate spectrum and the amide I and II peaks of the fungal biomass allowed a semiquantitative evaluation suggesting that exudation of higher quantities of oxalate were correlated to rock/mineral samples containing more than 10% wt of calcium and magnesium combined (olivine, basalt, gabbro, and limestone). No other types of oxalate crystals such as magnesium oxalate (glushinskite) were identified via μFT-IR. Likewise, SEM-EDS identfied no major cations other than calcium in the crystals. Minor variations in the IR peak position of calcium oxalate were most likely associated with crystaline water (in mono- and dihydrate forms)[40]. The lack of accumulation of other oxalate based crystals such as magnesium oxalate suggested either calcium-selective dissolution or that nutrients other than calcium such as magnesium may have dissolved and been taken up by the fungal hyphae and transported to the plant host.

In this study, the pH at the hyphal-surfaces was lowered from above 6 to below 5 in weathering arenas with basalt and limestone when compared to granite. By analogy with abiotic dissolution mechanisms, this drop in pH, putativley associated to the potential uptake of cations by the hyphae and the exudation of oxalate ($\sim 10^{-10}$ mol m^{-1} d^{-1}; Table 4), would accelerate calcium-bearing silicate and carbonate mineral dissolution that includes the release of calcium and production of calcium oxalate on the fungal hyphae as identified via μFT-IR. These findings are in accord with previous observations of symbiotic *P. involutus* hyphae on biotite, where pH decrease and organic acid exudation, amongst other processes, was identified as an important factor in fungal directed weathering[33]. The lower pH values for hypha in contact with basalt and limestone are associated with relatively greater calcium release, compared with higher pH values for hypha in contact with granite which exhibited far less calcium release (Fig. 2). Because the associated alkalinity production from calcium dissolution should increase interfacial pH, this indicates that the relatively lower interface pH in the presence of the greater calcium dissolution for basalt and limestone is controlled independently by biologicaly-induced mechanisms such as cation uptake and oxalate production resulting in calcium oxalate precipitation, rather than as a consequence of the abiotic calcium dissolution reaction. A biologically-driven regulation of interface pH is also supported by the correlating measurements of greater ^{14}C photosynthate allocation to basalt and limestone compared to granite (Fig. 1).

The fungal allocation of ^{14}C-labelled carbon received from tree photosynthesis differed between groups of rocks and minerals, with low allocations to granite, microcline and quartz, medium-high allocations to basalt, gabbro and olivine and significantly higher allocation to limestone grains (Fig. 1). The accumulation of calcium on the hyphae correlated with ^{14}C allocation ($R^2 = 0.688$ for silicate rocks and minerals; Supplementary Figure S2), with low amounts associated with hyphae in the presence of granite and quartz, and significantly higher amounts of calcium were associated with hyphae taken from arenas with basalt, limestone and gabbro (Table 3). Likewise, accumulation of oxalate was lowest in/on hyphae in the presence of granite and quartz, and significantly higher in/on hyphae taken from arenas with olivine, basalt, limestone and gabbro. This is consistent with a previous study22, employing almost identical microcosms with *P. sylvestris* symbiotic with *P. involutus*, in which smallest amounts of ^{14}C-labelled recent photosynthate were traced into fungal-colonised weathering arenas containing quartz compared to much higher rates of ^{14}C allocation into those containing apatite grains.

Consequently, the evidence suggests that the allocation of photosynthate in ectomycorrhizal *P. involutus* mycelial networks is intimately linked to factors that enhance the dissolution reactivity of calcium-bearing silicate and carbonate minerals. These include; active lowering of pH, increased exudation of oxalate as a ligand to accelerate element release from mineral surfaces, resulting in calcium oxalate precipitation. This reduces the dissolved calcium activity thereby maintaining high solid-solution concentration gradients favouring dissolution of calcium-bearing silicate and carbonate minerals. These factors all correlate with the observed biological effects of increased photoassimilate allocation, oxalate exudation and the release of elements from minerals. However, this link was weaker in weathering arenas with limestone, in which ^{14}C accumulation was higher than for other minerals but the accumulations of oxalate-bound calcium was significantly less than in the presence of gabbro. This may be due to some of the ^{14}C in the case of the limestone arising from abiotic precipitation / exchange of ^{14}C into the calcium carbonate mineral.

The stoichiometry of calcium oxalate predicts a one-to-one ratio for the crystalline weathering product, whereas in this study ratios (calcium:oxalate) across the arenas with different minerals and rocks were somewhat higher at 1.37 ($R^2 = 0.909$; Fig. 4, Table 3). Calcium abundance exceeding that of oxalate may have come in part from calcium in the fungal biomass which would not be part of the secondary minerals.

While only in the weathering arena with gabbro the fungal biomass was singificantly higher than in the quartz controls, recent investigations of rock grains buried under angiosperm trees also revealed significantly higher hyphal colonization rates (quantified as hyphal length) of basalt compared to granite and quartz24. Findings in the present study suggest that after initial exploratory colonization of weathering arenas, the presence of certain rocks/minerals have subsequently resulted in enhanced weathering activity, but not necessarily in significant increases in fungal biomass. However, photosynthate (^{14}C) allocation was elevated in weathering arenas of higher weathering activity (calcium oxalate accumulation), albeit not always significantly. In this study, ^{14}C allocation was measured at the end of the experiment over a period of 19 hours. The biological activity of the fungus in growth and secretion of oxalate during the 19 h of ^{14}C exposure is likely to have changed very much from the initial colonization to the later stages of mineral dissolution throughout the experimental period. Some of the high variability in ^{14}C allocation by the fungus among the measured replicates may reflect temporal differences in rates of colonization of the wells containing minerals.

The present study revealed that the accumulation of oxalate, not the accumulation of fungal biomass ($P = 0.217$), served as a proxy of fungal weathering activity by symbiotic *P. involutus* and was linked to the photoassimilate allocation at the weathering interface. It is unlikely that biological regulation of weathering activity was solely based on calcium release, since calcium requirements by the plant host (10 to 50 $mg\,g^{-1}$(DW) in plant tissue)41 are rather limited. The mobilization of other nutrients, essential for growth are likely to have played a key role. Only traces of essential nutrients such as phosphorus, magnesium or potassium are present in quartz, thus photosynthate allocation to quartz by the fungus would be uneconomical as opposed to more nutrient-rich rocks or minerals. Olivine, limestone, gabbro and basalt contained high levels of magnesium, but in different mineral forms (forsterite, dolomite, diopside and augite, respectively). Basalt and granite contained the highest amounts of phosphorus. However, XRD analysis was unable to determine its mineral form. Since about 95% of phosphorus in Earth's crust is associated with calcium and this element controls P solubility in most soils42, the calcium release by mycorrhiza may be a constitutive response to facilitate phosphorus uptake by plants since this is one of the primary functions of mycorrhiza[9,12].

Quirk and colleagues concluded in their work that mycorrhizal fungi concentrate their activity on weatherable nutrients[24]. Indeed, the abiotic kinetic reactivity of calcium bearing minerals alone could not explain the level of accumulated calcium oxalate in this study. The abiotic dissolution rate of calcium from dolomite, the dominating calcium bearing mineral in limestone, is about five orders of magnitude higher than from anorthite[26,28] (at pH 4), which is highly abundant in basalt and gabbro (Table 2). However, biotic calcium dissolution in this study was significantly higher from gabbro than limestone. These findings suggest that ectomycorrhizal weathering is highly targetted and enhanced by calcium-bearing minerals but the rate of oxalate release by *P. involutus* is not controlled soley by the calcium concentration in a mineral or rock. The accumulation of calcium oxalate indicates rates of calcium dissolution that exceed biological demand by the fungi and the plants, and the relatively low solubility of calcium oxalate. High levels of calcium may actually be toxic to fungi. Wood rotting fungi have been reported to form caclium oxalate crystals as a form of detoxification[43], thus calcium oxalate formation in this study may have reduced a possible negative effect of free calcium on fungal growth.

The weathering of calcium-bearing minerals is likely to have released other nutrient elements that may have been taken up by the fungal hyphae and transported to the plant host, and would therefore not be detected on the fungal hyphae. Indeed, Smits and colleagues found evidence of phosphorus translocation from apatite through *P. involutus* to the plant host *P. sylvestris*[22] and calcium accumulation most likely as calcium oxalate crystals have been reported in mesh bags with apatite buried in forest soils[44]. However, the presence of apatite in the present study could only be inferred from normative calculations. In olivine, calcium and phosphorus concentrations are very low. Nevertheless, oxalate accumulation on fungal hyphae in contact with olivine was found to be significanlty higher than with quartz or granite. This suggests that nutrients such as magnesium, abundant in olivine, basalt, gabbro and limestone may have been targetted by the fungus as well. The accumulation rates of oxalate in this study (Table 4) were of the same order of magnitude as in pine microcosms with *Hebeloma crustuliniforme* as fungal symbiont[23].

Calcium concentrations in the plant host are about an order of magnitude lower than its typical concentrations in the earth's crust2. The relatively low plant calcium demands means that the rate of calcium oxalate accumulation on the fungal hyphae in our study can be used as a proxy for estimating ectomycorrhizal weathering rates. These weathering rates were highest for gabbro calculated at 1.1×10^{-9} mol m^{-2} s^{-1} calcium and exceeded independently reported abiotic weathering rates of the calcium bearing minerals in gabbro (anorthite, diopside, augite, hornblende) by a factor of 100 and exceeded abiotic weathering rates by a factor of 10 in the case of calcium bearing minerals in basalt and olivine (Table 5). Unlike weathering rates from this study, abiotic weathering rates taken from the literature were carried out with individual mineral components only, thus abiotic weathering rates of minerals in rocks as used in this study may not be exactly the same. Since the calculations behind the ectomycorrhizal weathering rates assumed that all fungal hyphae were in contact with the rocks or minerals, the actual fungal weathering rate was potentially much higher as microscopic observations suggested that the vast majority of fungal hyphae were actually aerial. Considering the given uncertaincies from above, the weathering rates reported here are considered to be lower bound values.

The extent of calcium oxalate acumulation on the ectomycorrhizal hyphae alongside photosynthate translocation to the fungal hyphae clearly show that ectomycorrhizal weathering is indeed ultimately driven by the carbon supply provided by the plant. However, calcium abundance in the silicate rocks, minerals and limestone tested appears not to be the sole driver for targetted fungal weathering via exudation of oxalate to mobilize nutrients. This study provides indirect evidence that plant and fungal nutrient element contents in minerals, such as magnesium or phosphorus may contribute to the specific allocation of fungal exudates and biological weathering activity, but the main chemical sink for secreted oxalate is the formation of crystaline calcium oxalate, irrespective of the chemistry of the minerals.

Despite the limitations of deducing weathering rates from the accumulation of calcium in secondary minerals, calculations presented in this study clearly indicate that ectomycorrhial weathering has the potential to be several orders of magnitude faster than abiotic weathering rates. However, interaction of ectomycorrhizal hyphae in soils with other microbes such as bacteria may increase or decrease biological weathering. Oxalate may serve as a carbon source for a group of soil bacteria[45] while at the same time soil bacteria may enhance biological weathering through organic acid release[27]. Furthermore, ectomycorrhiza other than *P. involutus* may show different rates of weathering. Thus, further research is needed to quantify the effect of other microbiota in biological weathering in soils in the future. In conclusion, ectomycorrhizal weathering has the potential to dominate chemical dissolution rates in the vadose zone of temperate forests where low levels of gravitational water and high ectomycorrhizal activity occur. The evidence provided in this study will further support long-term carbon cycle models where ectomycorrhiza-weathering is increasingly recognized as a driver of land-to-ocean calcium fluxes, influencing the long-term concentration of CO_2 in the atmosphere[46,47]. This has been shown to occur via non-linear feedbacks, where as atmospheric CO_2 decreases through being sequestered into marine calcium and magnesium carbonates, rates of tree-mycorrhiza-driven weathering also decrease as a result of restricted photosynthate allocation with falling CO_2[18,48].

4. METHODS

4.1 Rock and mineral substrates used for fungal weathering experiments

Samples of rocks (granite, basalt, limestone, gabbro) and minerals (quartz, microcline and olivine) were individually fractured using a jaw-crusher and further broken in an agate mortar before sieving to obtain a 0.5–1.0 mm size fraction. These rock/mineral grains were then washed ultrasonically in deionised water to remove fine particles, dried at 80 °C and subsequently sterilised by autoclaving.

The chemical composition of the crushed rock/mineral samples was determined by X-ray Fluorescence (XRF) analysis using an XRF spectrometer (PANalytical Axios Sequential system; School of Earth, Atmospheric and Environmental Science, University of Manchester, UK). The mineralogy of each crushed sample was quantified by X-ray diffraction (XRD) using a Bruker D8 Advance Diffractometer with a scan range between 2 to 75° 2θ and at a 0.01° step size. The relative percentages of the mineral phases were determined by Rietveld refinement using the TOPAS software (version 4.2, Bruker AXS).

4.2 Monoxenic microcosms with plant host and symbiotic fungus

Pinus sylvestris seeds from the Forestry Commission, England were surface sterilized in two rounds of 30% hydrogen peroxide solution for 5 min each, rinsed in sterile distilled water and aseptically germinated over a period of 4 weeks on 1.5% plant agar as described elsewhere[49]. Germination took place in climate controlled plant growth chambers at 15 °C day and 10 °C night temperatures, with an 18 hour photoperiod at a photon flux density of 250 μmol m^{-2} s^{-1} and relative humidity of 60–75%. These growth conditions were maintained throughout the experimental stages of the study. The seedlings were transferred individually to sterile 10 × 10 cm square Petri dishes with roots draped on a cellophane sheet on top of 10% Modified Melin Norkran's medium[50] and inoculated with the ectomycorrhizal fungus *Paxillus involutus*. The shoots of *P. sylvestris* protruded outside the dish through a narrow hole that was sealed around the plant stem with sterile anhydrous lanolin (BDH, Prolabo) to exclude microbial contamination[49]. After 10 weeks, the plants had formed mycorrhizal roots and were transplanted into sterile experimental microcosms that were assembled in a laminar air-flow cabinet. The square Petri dish microcosms contained 100 mL modified Rorison's[51] nutrient solution (Supplementary Table S1) beneath a cellophane sheet and was solidified with 1.5% noble agar (Becton Dickinson, Oxford, UK). A single-grain-thickness bed of acid-washed and sieved perlite (volcanic glass) grains of 2.0–2.4 mm was set in a 20 mL layer of noble agar on top of the cellophane (solutions, perlite and cellophane were autoclaved). Six autoclaved plastic wells of 20 mm diameter were inserted as weathering arenas[22]. To these wells, 500 μL of 0.8% noble agar was added and overlaid with a disc of cellophane onto which 0.4 or 0.5 g of sterile mineral/rock samples was placed (one type of rock/mineral per microcosm).

The majority of the microcosms were harvested after 106–125 days of incubation. However, three microcosms with quartz, olivine and granite were harvested after 225–312 days due to late fungal colonization of the wells. All microcosms were watered with sterile water (perlite layer) in 90 day intervals (weathering arenas did not receive any added water). On harvesting, weathering arenas were removed from the microcosms, oven dried (80 °C) and stored in desiccators. Individual hyphae were then removed under 10–50 times magnification using fine forceps, and freed from adhering test mineral grains until all visible hyphae were removed from the wells, leaving the rock/mineral grains behind.

4.3 Quantification of plant-to-fungal carbon allocation to the weathering arenas

After 16–19 weeks of growth, a subset of microcosms (eight microcosms per rock/mineral with four wells each) were sealed in gas tight clear acrylic labelling chambers (675 cm^3) and exposed to 1.036 MBq of $^{14}CO_2$gas, in an illuminated fume hood (photon flux density of 250 µm m^{-2} s^{-1}) for pulse labelling as described earlier[49]. Microcosms were destructively harvested approx. 19 hours after labelling, coincident with the time of greatest ^{14}C allocation to the mycorrhizal hyphae determined in earlier trials[52]. The wells with rocks/mineral grains were divided into two equal subsamples and dried at 80 °C overnight. Dry weight measurements were recorded and one half of each well sample was oxidised with a Packard 307 sample oxidiser (Packard Instruments, Meriden, CT) to determine total ^{14}C concentration as described previously[49]. The oxidiser operated at >97% recovery rate with <0.08% between sample carry-over. Scintillation counting was carried out on a Tri-Carb 3100 TR Liquid Scintillation Analyser (Packard Instruments). The remaining well samples were in part used for the quantification of oxalate and calcium concentrations.

4.4 Fungal pH measurements

Using fine forceps, individual fungal hyphae were taken from intact mesocosms with live plants and fungi growing in symbiosis. Hyphae were removed from wells containing basalt, granite and limestone grains, from triplicate microcosms, and immediately subjected to confocal laser scanning microscopy (CLSM) to determine pH using the molecular probe SNARF4F (Invitrogen, Carlsbad, CA). CLSM was carried out using a Zeiss Axioscope with Meta 510 detector (Zeiss, Jena, Germany) and an argon multiphoton laser at 488 nm. The Meta 510 detector was set up to detect signals at distinct wavelengths of 544–587 nm and 619–661 nm. A 5 µM solution of the SNARF4F molecular probe was distributed directly onto individual fungal hyphae. At a final concentration of 5 µM the SNARF4F molecular probe emits at around 580 nm in acidic conditions and at around 640 nm in neutral conditions. The ratios of the fluorescence signals were used for pH calibration and calculation as described previously33.

4.5 Micro Fourier transform infrared spectroscopy (μFT-IR) of fungal hyphae

Measurements by μFT-IR were performed on a Perkin-Elmer Spotlight imaging system (Perkin-Elmer, Waltham, MA). μFT-IR spectra were collected over the range of 4000-700 cm^{-1} wavenumber at a resolution of 1 cm^{-1}, a beam diameter of 6.25 μm and an aperture of 50 × 50 µm in reflective mode on an aluminum block. Background removal from the aluminum block was carried out using the Spectrum Spotlight software (Perkin-Elmer). Each of our test rock/minerals (powdered material), and calcium oxalate monohydrate (powdered, purity 99%, Alpha Aesar, Karlsruhe, Germany) together with hyphae from a monoculture of *P. involutus* grown on cellophane over nutrient agar were scanned for reference purposes. *P. involutus* hyphae were collected from weathering arenas (fresh and dried samples without rock or mineral particles attached) as described above, and μFT-IR spectra obtained from samples derived from at least three different microcosms (of each rock/mineral type), from a total of six wells. Five spectra from different locations were acquired for each well at 100 scans each to increase the representativeness of the spectra per well and subsequently combined to a single spectrum. Scanning areas were selected that contained sufficient biomass to obtain a clear signal without adverse scattering effects.

4.6 Scanning electron microscopy with energy dispersive X-ray spectroscopy (SEM-EDS)

SEM-EDS analysis was carried out on sub samples of *P. involutus* hyphae from the wells in the microcosms analyzed through non-destructive μFT-IR (see above) with a CamScan MKII (Cambridge, UK). Fungal hyphae were prepared on double sided carbon tape and coated with carbon using a Speedivac 12E6/1598 carbon coater (Crawley, UK). Backscattering images were recorded to find elements with a high atomic number at an acceleration voltage of 15 kV. Selected crystalline structures were investigated for calcium content using EDS.

4.7 Quantification of calcium and oxalic acid via inductively coupled plasma mass spectrometry (ICP-MS) and ion chromatography

Quantification of calcium and oxalic acid was carried out on handpicked samples of *P. involutus* hyphae with attached secondary minerals. From triplicate microcosms with quartz, granite, olivine, basalt, limestone and gabbro hyphae were taken (microcline was sampled in duplicate due to the loss of a microcosm). *P. involutus* hyphae were weighed to a precision of 10^{-6} g using a Mettler Toledo MT5 micro balance (Mettler-Toledo Ltd., Greifensee, Switzerland) in acid-washed glass HPLC vials (1.0 mL). Samples were boiled for 30 minutes in 0.5 M sulfuric acid in volumes of 0.3 mL and the aqueous phase was subsequently divided: (i) 0.1 mL was subjected to direct ICP-MS analysis (after a 20 times dilution on an Agilent 7500CX, Agilent Technologies; limit of detection 3 µg calcium L^{-1}) and (ii) the remnant 0.2 mL was used for oxalic acid extraction.

The quantification of calcium in the hyphal samples was carried out by measuring Ca44 via ICP-MS (Agilent 7500cx, Santa Clara, CA). Isotope fractionation by the fungus was assumed to be negligible since previous studies reported isotopic discrimination of calcium of below 1 ppm in ectomycorrhizal trees53.

Oxalic acid quantification was adapted from Lapeyrie and colleagues[19]. Briefly, 0.2 mL of the aqueous phase (see above) was mixed with 0.3 mL of organic solvent tributylphosphate (Acros Organics, Antwerpen, Belgium) in 1.5 mL reaction tubes for 10 min at room temperature on an overhead shaker (RM-2, Elmi, Riga, Latvia) at 20 rpm. The organic phase containing the putative oxalate from the extraction was then pipetted to a new reaction tube and was mixed with 0.2 mL of 2 M NaOH and centrifuged at 16000 *g* for 10 min. 160 µL of the aqueous phase (with the putative oxalate) was neutralized with 160 µL 2 M HCl. The concentration of oxalate was quantified with a ICS1500 ion chromatography system using a 10 µL injection loop, an AS auto sampler, an AS23 column with AG23 guard (all Dionex, Sunnyvale, CA) and a sodium (bi)carbonate based eluent54.

Absolute quantifications of calcium and oxalate were calculated per gram dry weight of the fungal biomass including the mass of attached secondary minerals. Additionally, based on the observed timescale of weathering arena colonization (last time point of arenas without fungal colonization to harvest, based on photographic records), and the conversion of fungal biomass into hyphal length (8.64×10^4 m g^{-1})55 it was possible to calculate calcium and oxalate accumulation rates per metre of hyphae, per day.

4.8 Calcium accumulation rate calculation

Rate calculations were based on (i) the quantification of calcium within and attached to the sampled fungal hyphae (see above), ii) the fungal biomass present in the weathering arenas at the time of harvesting, and iii) the duration of fungal colonisation of the weathering arenas. The measured fungal biomass was used to calculate a weathering arena specific, maximum possible fungal-mineral contact area (***A***, m^2), following (Eq. 1):

$$\boldsymbol{A} = m \cdot l \cdot w \tag{1}$$

where m is the measured mass of fungi in the weathering arena (g), l represents the length of hyphae per gram of biomass (8.64×10^4 m g^{-1})55, and w an assumed hyphal width from previous observations (5×10^{-6} m)35,56. Fungal calcium (Ca) accumulation rates (R, $moles_{Ca}$ m^{-2} s^{-1}), driven by fungal mineral weathering, were calculated as follows (Eq. 2):

$$R = \frac{\mathrm{Ca}}{t \cdot \boldsymbol{A}} \tag{2}$$

where Ca is the number of moles of calcium measured by ICP-MS, t represents the time, in seconds, of the colonization of the wells, and ***A***the maximum hypha-mineral contact area.

Published chemical dissolution rate laws for minerals in aqueous laboratory solutions are based on 298 K[26,28]. In contrast, dissolution *in vivo* for temperate forests takes place under much cooler conditions. In this study, microcosms were incubated with 8 h night at 283 K and 16 h day at 288 K. Therefore, the published abiotic dissolution rates presented in this study were corrected from 298 K to 288 K using the Arrhenius equation as described previously[57]. Abiotic dissolution rates for individual minerals were, where needed, corrected for rate of calcium release from the stoichiometric dissolution rates reported.

5. STATISTICAL ANALYSIS

Analyis of variance (ANOVA) was tested for all analysed data using Tukey HSD post hoc test where the data meet the requirements of homoscedasticity, as confirmed by Leven's test. When the raw data showed distributions that were not not normally distributed they were transformed by taking the square root or the log to the base 10. Where the Levene's test indicated heteroscedasticity in both the raw and transformed data, the Games-Howell posthoc test was used, provided a Welch test indicated significant differences between the sample means. When several weathering arenas were analysed from a single microcosm, the values were used to create means to exclude any non-independent artefacts in the statistical analysis. Error ranges in all tables are based on standard deviation.

CONTRIBUTIONS

The corresponding author A.S. prepared the manuscript and all figures and tables within the manuscript and was the leading researcher of the analytical experiments presented in the manuscript. Co-authors A.B., S.B., L.B., J.L. and S.A.B. were deeply involved in the manuscript revision. Co-author A.D. contributed to the microcosm system setup and harvest and conducted the ^{14}C experiment (Figure 1). Co-author A.B. conducted the current version of the weathering rate calculation (Table 5). Co-author J.B. contributed to the general weathering rates analysis. Co-authors S.B., A.B. and L.B. contributed to the XRD analysis (Table 2). Co-author M.R.-G. contributed to the FTIR spectra interpretation (Figure 3).

ACKNOWLEDGEMENTS

We would like to thank Cathy Shields for her assistance in the SEM-EDS analysis and the Natural Environment Research Council NERC for funding this project (NE/C521001/1).

ADDITIONAL INFORMATION

Competing financial interests: The authors declare no competing financial interests.

REFERENCES

1. Anderson, I. C. & Cairney, J. W. G. *Diversity and ecology of soil fungal communities: Increased understanding through the application of molecular techniques.* Environ. Microbiol. **6**, 769–779 (2004).

2. Brantley, S. L. et al. *Twelve testable hypotheses on the geobiology of weathering.* Geobiology **9**, 140–165 (2011).
3. Turner, B. L. et al. *Soil microbial biomass and the fate of phosphorus during long-term ecosystem development.* Plant Soil**367**, 225–234 (2013).
4. Peltzer, D. A. et al. *Understanding ecosystem retrogression.* Ecol. Monogr. **80**, 509–529 (2010).
5. Turner, B. L., Condron, L. M., Wells, A. & Andersen, K. M. *Soil nutrient dynamics during podzol development under lowland temperate rain forest in New Zealand.* Catena **97**, 50–62 (2012).
6. Berner, E. K., Berner, R. A. & Moulton, K. L. in Treatise on Geochemistry (eds H. D. Holland & K. K. Turekian) **Ch. 5.06**, 169–188 (Pergamon, 2003).
7. Taylor, L. L. et al. *Biological weathering and the long-term carbon cycle: integrating mycorrhizal evolution and function into the current paradigm.* Geobiology **7**, 171–191 (2009).
8. Landeweert, R., Hoffland, E., Finlay, R. D., Kuyper, T. W. & Van Breemen, N. *Linking plants to rocks: Ectomycorrhizal fungi mobilize nutrients from minerals.* Trends Ecol. Evol. **16**, 248–254 (2001).
9. Smith, S. E. & Read, D. J. Mycorrhizal symbiosis. (Academic Press, Amsterdam, The Netherlands, 2008).
10. Leake, J. R. et al. *Networks of power and influence: the role of mycorrhizal mycelium in controlling plant communities and agroecosystem functioning.* Can. J. Bot. **82**, 1016–1045 (2004).
11. Jones, D. L. *Organic acids in the rhizosphere - a critical review.* Plant Soil **205**, 25–44 (1998).
12. Parniske, M. *Arbuscular mycorrhiza: the mother of plant root endosymbioses.* Nature Rev. Microbiol. **6**, 763–775 (2008).
13. Read, D. J., Leake, J. R. & Perez-Moreno, J. *Mycorrhizal fungi as drivers of ecosystem processes in heathland and boreal forest biomes.* Can. J. Bot. **82**, 1243–1263 (2004).
14. Bonan, G. B. *Forests and climate change: forcings, feedbacks, and the climate benefits of forests.* Science **320**, 1444–1449 (2008).
15. Klass, D. L. in Encyclopedia of Energy (ed C. J. Cleveland) 193–212 (Elsevier, 2004).
16. Hoffland, E. et al. *The role of fungi in weathering.* Front. Ecol. Environ. **2**, 258–264 (2004).
17. Leake, J. R. et al. *Biological weathering in soil: The role of symbiotic root-associated fungi biosensing minerals and directing photosynthate-energy into grain-scale mineral weathering.* Mineral. Mag. **72**, 85–89 (2008).
18. Quirk, J., Andrews, M. Y., Leake, J. R., Banwart, S. A. & Beerling, D. J. *Ectomycorrhizal fungi and past high CO_2 atmospheres enhance mineral weathering through increased below-ground carbon-energy fluxes.* Biol. Lett. **10** (2014).
19. Lapeyrie, F., Chilvers, G. A. & Bhem, C. A. *Oxalic acid synthesis by the mycorrhizal fungus Paxillus involutus (Batsch EX FR) FR.* New Phytol. **106**, 139–146 (1987).
20. Lapeyrie, F. *Oxalate synthesis from soil bicarbonate by the mycorrhizal fungus Paxillus involutus.* Plant Soil **110**, 3–8 (1988).
21. Lapeyrie, F., Ranger, J. & Vairelles, D. *Phosphate-solubilizing activity of ectomycorrhizal fungi in vitro.* Can. J. Bot. **69**, 342–346 (1991).
22. Smits, M. M., Bonneville, S., Benning, L. G., Banwart, S. A. & Leake, J. R. *Plant-driven weathering of apatite – the role of an ectomycorrhizal fungus.* Geobiology **10**, 445–456 (2012).
23. Van Hees, P. A. W. et al. *Oxalate and ferricrocin exudation by the extramatrical mycelium of an ectomycorrhizal fungus in symbiosis with Pinus sylvestris.* New Phytol. **169**, 367–378 (2006).
24. Quirk, J. et al. *Evolution of trees and mycorrhizal fungi intensifies silicate mineral weathering.* Biol. Lett. **8**, 1006–1011 (2012).
25. Wallander, H. *Uptake of P from apatite by Pinus sylvestris seedlings colonised by different ectomycorrhizal fungi.* Plant Soil**218**, 249–256 (2000).
26. Brantley, S. L. in Kinetics of water-rock interaction (eds S. L. Brantley, J. D. Kubicki & A. F. White) 151–210 (Springer, 2008).

27. Hausrath, E. M., Neaman, A. & Brantley, S. L. *Elemental release rates from dissolving basalt and granite with and without organic ligands*. Am. J. Sci. **309**, 633–660 (2009).
28. McAdam, A. C., Zolotov, M. Y., Sharp, T. G. & Leshin, L. A.*Preferential low-pH dissolution of pyroxene in plagioclase-pyroxene mixtures: implications for martian surface materials.* ICARUS **196**, 90–96 (2008).
29. Ullman, W. J., Kirchman, D. L., Welch, S. A. & Vandevivere, P.*Laboratory evidence for microbially mediated silicate mineral dissolution in nature*. Chem. Geol. **132**, 11–17 (1996).
30. Griffiths, R. P., Baham, J. E. & Caldwell, B. A. *Soil solution chemistry of ectomycorrhizal mats in forest soil*. Soil Biol. Biochem. **26**, 331–337 (1994).
31. Ilvesniemi, H. et al. *Water balance of a boreal Scots pine forest*. Boreal Environ. Res. **15**, 375–396 (2010).
32. Gabler, R. E., Petersen, J. F. & Trapasso, L. M. Essentials of Physical Geography. 8 edn, (Thompson Higher Education, 2006).
33. Bonneville, S. et al. *Tree-mycorrhiza symbiosis accelerate mineral weathering: evidences from nanometer-scale elemental fluxes at the hypha-mineral interface*. Geochim. Cosmochim. Ac. **75**, 6988–7005 (2011).
34. Bonneville, S. et al. *Plant-driven fungal weathering: Early stages of mineral alteration at the nanometer scale*. Geology **37**, 615–618 (2009).
35. Saccone, L. et al. *High resolution characterization of ectomycorrhizal fungal-mineral interactions in axenic microcosm experiments*. Biogeochemistry **111**, 411–425 (2012).
36. Hunt, J. M., Wisherd, M. P. & Bonham, L. C. *Infrared absorption spectra of minerals and other inorganic compounds*. Anal. Chem.**22**, 1478–1497 (1950).
37. Naumann, A. *A novel procedure for strain classification of fungal mycelium by cluster and artificial neural network analysis of Fourier transform infrared (FTIR) spectra*. Analyst **134**, 1215–1223 (2009).
38. Gautelier, M., Oelkers, E. H. & Schott, J. *An experimental study of dolomite dissolution rates as a function of pH from −0.5 to 5 and temperature from 25 to 80°C*. Chem. Geol. **157**, 13–26 (1999).
39. Tewari, J. P., Shinners, T. C. & Briggs, K. G. *Production of calcium oxalate crystals by two species of Cyathus in culture and infested plant debris*. Z. Naturforsch. C-a J. Biosci. **52**, 421–425 (1997).
40. Maurice-Estepa, L., Levillain, P., Lacour, B. & Daudon, M.*Advantage of zero-crossing-point first-derivative spectrophotometry for the quantification of calcium oxalate crystalline phases by infrared spectrophotometry*. Clin. Chim. Acta **298**, 1–11 (2000).
41. Marschner, H. Mineral nutrition of higher plants. 2nd edn, (Academic Press, 1995).
42. Jahnke, R. A. **Vol. 50** Global Biogeochemical Cycles (eds S. S. Butcher, R. J. Charlson, G. H. Orians & G. V. Wolfe) 301–315 (Elsevier, 1992).
43. Jarosz-Wilkolazka, A. & Gadd, G. M. *Oxalate production by wood-rotting fungi growing in toxic metal-amended medium*. Chemosphere **52**, 541–547 (2003).
44. Wallander, H., Johansson, L. & Pallon, J. *PIXE analysis to estimate the elemental composition of ectomycorrhizal rhizomorphs grown in contact with different minerals in forest soil*. FEMS Microbiol. Ecol. **39**, 147–156 (2002).
45. Sahin, N. *Oxalotrophic bacteria*. Res. Microbiol. **154**, 399–407 (2003).
46. Taylor, L. L., Banwart, S. A., Leake, J. R. & Beerling, D. J. *Modeling the evolutionary rise of ectomycorrhiza on sub-surface weathering environments and the geochemical carbon cycle*. Am. J. Sci. **311**, 369–403 (2011).
47. Taylor, L. L., Banwart, S. A., Valdes, P. J., Leake, J. R. & Beerling, D. J. *Evaluating the effects of terrestrial ecosystems, climate and carbon dioxide on weathering over geological time: a global-scale process-based approach*. Philos. T. R. Soc. B **367**, 565–582 (2012).
48. Quirk, J., Leake, J. R., Banwart, S. A., Taylor, L. L. & Beerling, D. J.*Weathering by tree-root-associating fungi diminishes under simulated Cenozoic atmospheric CO_2 decline*. Biogeosciences **11**, 321–331 (2014).
49. Leake, J. R., Donnelly, D. P., Saunders, E. M., Boddy, L. & Read, D. J.*Rates and quantities of carbon flux to ectomycorrhizal mycelium following ^{14}C pulse labeling of Pinus sylvestris seedlings: effects of litter patches and interaction a wood-decomposer fungus*. Tree Physiol. **21**, 71–82 (2001).

50. Marx, D. H. *Influence of ectotrophic mycorrhizal fungi on resistance of pine roots to pathogenic infections. I. Antagonism of mycorrhizal fungi to root pathogenic fungi and soil bacteria.* Phytopathology **59**, 153–163 (1969).
51. Hewitt, E. J. Sand and Water Culture Methods used in the Study of Plant Nutrition. 2nd edn, (Commonwealth Agric. Bureaux Tech. Comm. No. 22, 1966).
52. Duran, A. L. The effect of mineral particle grain size and mineralogy on ectomycorrhizal-mineral interactions PhD thesis, Univeristy of Sheffield, (2014).
53. Hoff, C. J., Bryce, J. G., Hobbie, E. A., Colpaert, J. V. & Bullen, T. D. in American Geophysical Union, Fall Meeting. B53B-05.
54. Schmalenberger, A., Telford, A. & Kertesz, M. A. *Sulfate treatment affects desulfonating bacterial community structures in Agrostis rhizospheres as revealed by functional gene analysis based on asfA.* Eur. J. Soil Biol. **46**, 248–254 (2010).
55. Wallander, H., Göransson, H. & Rosengren, U. *Production, standing biomass and natural abundance of ^{15}N and ^{13}C in ectomycorrhizal mycelia collected at different soil depths in two forest types.* Oecologia **139**, 89–97 (2004).
56. Gazzè, S. A. et al. *Nanoscale observations of extracellular polymeric substances deposition on phyllosilicates by an ectomycorrhizal fungus.* Geomicrobiol. J. **30**, 721–730 (2013).
57. Rosling, A., Lindahl, B. D., Taylor, A. F. S. & Finlay, R. D. *Mycelial growth and substrate acidification of ectomycorrhizal fungi in response to different minerals.* FEMS Microbiol. Ecol. **47**, 31–37 (2004).

CHAPTER 18

Factors affecting the Nd^{3+} (REE^{3+}) luminescence of minerals

Christoph Lenz[1], Dominik Talla, Katja Ruschel, Radek Škoda[2], Jens Götze[3], Lutz Nasdala[1]

Institut für Mineralogie und Kristallographie, Universität Wien, Althanstraße 14, 1090 Wien, Austria

Institute of Geological Sciences, Masaryk University, Kotlářská 2, 61137 Brno, Czech Republic

Institut für Mineralogie, TU Bergakademie Freiberg, Brennhausgasse 14, 09596 Freiberg, Germany

ABSTRACT

In this paper, possibilities and limits of the application of REE^{3+} luminescence (especially the $Nd^{3+}\,{}^4F_{3/2} \rightarrow {}^4I_{9/2}$ emission) as structural probe are evaluated. Important factors controlling the Nd^{3+} luminescence signal are discussed, including effects of the crystal-field, crystal orientation, structural state, and temperature. Particular attention was paid to the study of the accessory minerals zircon ($ZrSiO_4$), xenotime–(Y) (YPO_4), monazite–(Ce) ($CePO_4$) and their synthetic analogues. Based on these examples we review in short that (1) REE^{3+} luminescence can be used as non-destructive phase identification method, (2) the intensities of certain luminescence bands are strongly influenced by crystal orientation effects, and (3) increased widths of REE^{3+}-related emission bands are a strong indicator for structural disorder. We discuss the potential of luminescence spectroscopy, complementary to Raman spectroscopy, for the quantitative estimation of chemical (and potentially also radiation-induced) disorder. For the latter, emissions of Nd^{3+}-related centres are found to be promising candidates.

1. INTRODUCTION

The luminescence emission of rare-earth elements (REEs) in general is well-studied, stimulated by their importance in various modern technological applications such as lighting, colour-television screens, solid-state lasers, phosphors, and chromophores in different host materials (e.g., Belsky and Krupa 1999; Kenyon 2002; Bünzli and Piguet 2005; Liu and Jacquier 2005). In the Earth sciences, one challenging task is the detection of REEs in different host minerals using luminescence methods; a large variety of mineral systems have been studied already (Tarashchan 1978; Ohnenstetter et al. 1991; Habermann et al. 1996; Gaft et al. 1999; Götze et al. 1999; Blanc et al. 2000; Waychunas 2002; Nasdala et al. 2004; Gaft et al. 2005; Czaja et al. 2008). Synthetic minerals individually- or multi-doped with REEs have also been studied routinely, aiming at a better understanding of the luminescence in their natural analogues. For the example of zircon and synthetic $ZrSiO_4$, such investigations have been done by Cesbron et al. (1993, 1995), Blanc et al. (2000), Hanchar et al. (2001), and Friis et al. (2009). Time-resolved luminescence techniques have improved the detection of various luminescence centres utilizing their different decay times (Reisfeld et al. 1996; Gaft et al. 1999; Gaft et al. 2001). Furthermore, unravelling the internal zoning of the REE distribution within crystals, especially of accessory minerals, may provide valuable information on their primary formation and post-growth history. Such patterns are visualised easily using luminescence techniques (Hanchar and Miller 1993; Hanchar and Rudnick 1995; Rakovan and Reeder 1996; Götze 2000, 2002; see also the review chapter by Götze et al. in this special issue).

In addition to traditional direct imaging, the luminescence of crystals and other geological samples is studied increasingly using (hyperspectral) mapping techniques (see the review chapter by MacRae et al. in this special issue). The need four sound spectral interpretation in reducing such hyperspectral data sets has increased the interest in a detailed understanding of REE luminescence features (Nasdala et al. 2004; MacRae et al. 2005). For a number of minerals, luminescence-intensity based quantifications of trace REEs have been proposed (Barbarand and Pagel 2001; Habermann 2002; Richter et al. 2003; MacRae et al. 2005; 2012). Moreover, traces of REE^{3+} in minerals and mineral-based ceramics are used as structural probes studied with respect to order–disorder phenomena, in particular the accumulation of structural damage as caused by irradiation. For instance, spectral parameters of Eu^{3+} emissions have been used for the characterisation of REE-substituted cation-sites in glasses and crystalline materials in process of evaluation of potential nuclear waste forms (Ollier et al. 2003; Reisfeld et al. 2004; Reisfeld 2005; Ternane et al. 2005). Seydoux-Guillaume et al. (2002) and Panczer et al. (2012) proposed that the REE-dominated luminescence emission of monazite–(Ce) may bear quantitative information on the accumulated radiation-damage in this mineral.

In accordance to the finding of the latter authors, we consider the luminescence signal of Nd^{3+}as particularly promising structural probe. The $^4F_{3/2} \rightarrow {}^4I_{9/2}$ electronic transition of Nd^{3+}causes luminescence emission in the near-infrared (NIR) spectral range between 10,600 and 11,800 cm^{-1} (830–940 nm). This emission is a promising candidate because : (1) the luminescence signal can be collected with standard Si-based detectors; (2) it is effectively excited by standard lasers, e.g., Ar^+ (488 nm, 514 nm) as well as via accelerated electrons (cathodoluminescence, CL); (3) other luminescence emissions in the respective wavenumber range are rare; and (4) it has been identified in many mineral hosts, especially in REE-bearing accessory minerals. The application of Nd^{3+} (REE^{3+}) luminescence as a structural probe requires a detailed understanding of the influencing factors and their calibration, respectively. Here we discuss the basic effects of (1) the crystal field environment; (2) crystal orientation; (3) structural disorder; and (4) temperature on the Nd^{3+} luminescence as representative example. Similar to Nd^{3+}, other appropriate REE^{3+} can be used, e.g., Pr^{3+}, Sm^{3+}, Eu^{3+}, but are not discussed in this work. In this study, the Nd^{3+} hosts zircon, xenotime–(Y) and monazite–(Ce) and their synthetic analogues have been studied in detail.

2. SAMPLES & METHODS

2.1 Samples and preparation

Luminescence investigations were performed on a large variety of specimens including monazite–(Ce), Zircon and xenotime–(Y) and/or their synthetic analogues (see Table 1). An overview of the chemically homogeneous, natural monazite–(Ce) samples measured here, including their origins and ages, is given in Ruschel et al. (2012). Only samples of these authors that were annealed in air to reconstitute their crystalline state were chosen for the study of chemical-induced disorder, to avoid biases of the results due to effects of self-irradiation damage on the luminescence parameters obtained.

Table 1 Overview of samples studied in detail

Samples		Substitution type	Reference
Monazite-(Ce)	natural	huttonite + cheralite	all homogeneous, dry-annealed monazite–(Ce) samples from Ruschel et al. (2012)
	synthetic	$Ce_{1-x}REE_x[PO_4]$ substitution	Nd-, Gd- and/or La-containing homogeneous samples from Ruschel et al. (2012)
	synthetic	cheralite + "Na-Th substitution" $Ce_{1-x-y}\,Ca_{0.5x}Na_{y/3}Th_{0.5+2y/3}[PO_4]$	this study; see in "Details on synthesis procedures"
Zircon	synthetic	almost pure crystals doped with trace Nd^{3+} without charge compensation	this study; see in "Details on synthesis procedures"
	synthetic	almost pure crystals doped with trace Nd^{3+} incorporated via xenotime substitution, $Zr_{1-x}Nd_x[Si_{1-x}P_xO_4]$	this study; see in "Details on synthesis procedures"
Xenotime-(Y)	synthetic	crystals doped with Nd^{3+} incorporated via $Y_{1-x}Nd_x[PO_4]$ substitution	Talla et al. (2011)

In addition, a range of flux-grown, synthetic samples were studied, including $Ce_{1-x}LREE_xPO_4$ (with LREE = La, Nd, Gd) of Ruschel et al. (2012) and $CePO_4$ samples with predominant *cheralite* (2 $REE^{3+} \leftrightarrow Ca^{2+} + Th^{4+}$) or *huttonite* ($P^{5+} + REE^{3+} \leftrightarrow Si^{4+} + Th^{4+}$) substitution (synthesised in this study; see below).

The orientation-dependence of the luminescence was studied for Nd-doped YPO_4 samples synthesised by Talla et al. (2011), and Nd-doped $ZrSiO_4$ crystals produced in this study (see sub-chapter "Details on synthesis procedures" below). An overview of the samples investigated is given in Table 1.

Doubly polished thin sections (thicknesses ~30 μm) attached to a glass slide were prepared with respect to the crystallographic orientation (xenotime and zircon samples), or as randomly cut slabs (monazite samples from Ruschel et al. 2012). For electron microprobe and cathodoluminescence analysis the sections were coated with carbon.

2.2 Details on synthesis procedures

Synthetic monazite–(Ce) crystals with predominant cheralite or huttonite substitution were produced by a flux method using the slow-cooling technique. Sodium polyphosphate was preferentially used as flux material because of its good solubility in water, making the extraction of the resulting product straightforward (and fast in comparison with $Pb_2P_2O_7$, commonly used for REE phosphate syntheses; Boatner 2002; Talla et al. 2011). Platinum crucibles with a volume of 100 cm^3 were used as reaction vessels. For each batch, 10 grams of $NaPO_3$ were used as flux material and mixed with $CeCl_3.7H_2O$, setting the molar ratio Ce/(Ce+Na) in the melt to 0.12 (Talla et al. 2011). A total of five samples of Th-doped monazite–(Ce) were produced using $Th(NO_3)_4.5H_2O$. In the first three, Th was added in varying amounts, the weight ratio Th/(Th+Ce) in the melt being 400, 2,000 and 10,000 ppm. Charge balance was provided using the *cheralite substitution* mechanism (2 $REE^{3+} \leftrightarrow Ca^{2+} + Th^{4+}$) with $CaCO_3$ (analytical grade) as source of Ca. A fourth, un-doped sample was also produced for comparison (blank). In the fifth sample, the *huttonite substitution* ($P^{5+} + REE^{3+} \leftrightarrow Si^{4+} + Th^{4+}$) was used as the mechanism for introducing Th into monazite–(Ce) with silica as source of Si. The weight fraction Si/(Si+Ce+Th) in the melt was adjusted to 0.02. All reagents were of analytical grade except for $NaPO_3$ and the cerium chloride. Preliminary tests showed that pollutant trace Nd present in the commercial $CeCl_3.7H_2O$ results in minute concentrations of Nd in the $CePO_4$ samples grown, which, however, yield easily measurable Nd^{3+}-luminescence emissions. A separate Nd source was therefore not added to the crucible.

Table 2 Chemical formulae of synthetic Th-doped monazite-(Ce) samples in atomic percent per formula unit (apfu), calculated from average electron-microprobe data of different chemically heterogeneous zones. The distortional behaviour of all substituents (Δ) and heterovalent substituents (Δ_{hetero}) are approximated with a weighted ionic radii difference between the predominant cation/anion [Ce/P in monazite–(Ce)] and each substituent (see Eq. 1)

		Chemical formula (apfu)									
Sample	Point	Ce	Nd	Na	Ca	Th	U	P	Si	Δ	Δ_{hetero}
C1 *(blank)*	1	0.999	0.001	bdl	bdl	bdl	bdl	0.999	bdl	0.00012	0.00009
C2	2	0.998	0.001	bdl	0.001	bdl	bdl	0.999	bdl	0.00011	0.00008
C3	3	0.997	0.002	bdl	bdl	bdl	bdl	0.999	bdl	0.00015	0.00008
	4	0.939	0.001	0.010	0.013	0.037	0.001	0.997	bdl	0.00405	0.00023
	5	0.941	0.001	0.009	0.014	0.036	0.001	0.999	bdl	0.00377	0.00374
	6	0.915	0.001	0.015	0.018	0.051	0.001	0.999	bdl	0.00522	0.00520
C4	7	0.864	0.001	0.023	0.029	0.083	0.001	0.999	bdl	0.00850	0.00847
	8	0.931	0.001	0.012	0.015	0.040	0.001	1.000	bdl	0.00410	0.00407
	9	0.997	0.001	0.001	0.001	bdl	bdl	0.999	bdl	0.00010	0.00004
CSi	10	0.987	0.012	bdl	bdl	0.001	bdl	1.000	bdl	0.00044	0.00002
	11	0.960	bdl	0.010	0.001	0.001	bdl	0.999	bdl	0.00266	0.00264
	12	0.867	bdl	0.040	0.002	0.089	0.001	0.999	bdl	0.00805	0.00803
	13	0.909	bdl	0.027	0.002	0.061	bdl	0.999	0.001	0.00556	0.00554

After being filled, the Pt crucibles were capped by a loose lid. Crucibles were first heated to a peak temperature of 1,170 °C, at a rate of 230 K per hour. This temperature was kept for 2 hours, and reduced afterwards to 900 °C, at a rate of −1.35 K per hour. The furnace was then switched off, and it was not opened until samples had cooled down slowly to room temperature. The crystals produced were about 0.3–1.0 mm in size. Their colours range from light green (low-Th syntheses) to colourless (elevated Th content). The total amount of crystals produced was much smaller when higher concentrations of heterovalent substituents were present. A detailed micro-chemical characterisation of these synthetic samples is given in Table 2.

Synthesis of zircon crystals (up to 1.5 mm) was accomplished by a flux technique similar to that described by Cesbron et al. (1993, 1995) and Hanchar et al. (2001) with an adjusted synthesis route combining the evaporation of the highly volatile MoO_3 with a rapid cooling rate. This allowed us to use an alkali-free mixture of 0.53 g ZrO_2 with 0.26 g SiO_2 and 10 g MoO_3, preventing the incorporation of Li^+ into the zircon crystals (Hanchar et al. 2001). Preliminary test runs showed that the resulting ratios of 0.07 mol% (Zr+Si)/(Zr+Si+Mo) and 1:1 Zr/Si resulted in the growth of larger, well-developed crystals. To produce Nd-doped $ZrSiO_4$, a minor amount of Nd was added. This was done in two different ways, first by adding Nd_2O_3 without any additional charge-compensating species, and second by adding $NdPO_4$ ensuring charge balance via the *xenotime substitution* ($Zr^{4+} + Si^{4+} \leftrightarrow REE^{3+} + P^{5+}$). The molar ratio Nd/(Nd+Zr) in the melt was adjusted to 0.6 mol%. The temperature path consisted of heating the loosely capped Pt crucibles to 1,100 °C, at a rate of 216 K per hour. Without a soaking interval, the temperature was then reduced to 700 °C, at a rate of −26.5 K per hour. Afterwards, samples were left in the furnace to cool slowly. Crystals were extracted manually (i.e., without the need to use any chemical reagents) from the crucible without problems. It was found that the size of the crystals grown was influenced strongly by the evaporation surface of the melt during the experiment. A small evaporation surface prevents rapid vaporisation of Mo-flux, the solvent of all reagents. Therefore, thinner crucibles result in the formation of larger single-crystals.

2.3 Analytical methods

Electron microprobe analyses were performed using a Cameca SX100 electron microprobe with the following measurement conditions: accelerating voltage 15 kV, beam current 20 nA, and 8 μm beam diameter. A defocused beam was in order to minimize the loss of Na during analysis. The peak counting times were 20 s for major elements and 40 to 60 s for minor and trace elements; counting times for the background (measured on both, the high- and low-energy side) were set to half of the respective peak counting-time. The following natural and synthetic standards were used: $Si_{K\alpha}$-natural sanidine, Eifel; $Na_{K\alpha}$-natural albite, Amelia; $P_{K\alpha}$-$LaPO_4$; $Ca_{K\alpha}$,$Th_{M\alpha}$-synthetic $CaTh(PO_4)_2$; $Ce_{L\alpha}$-synthetic $CePO_4$; $Nd_{L\beta}$-synthetic $NdPO_4$; $U_{M\beta}$-synthetic U. The raw data were corrected using the PAP routine (Pouchou and Pichoir 1991). Elemental X-ray maps were acquired at following conditions: accelerating voltage 15 kV, beam current 60 nA, and <1 µm beam diameter. Signals of $Na_{K\alpha}$, $P_{K\alpha}$, $Ce_{L\alpha}$, $Nd_{L\beta}$, and $Th_{M\alpha}$, and backgrounds for each element were collected. The data were acquired in stage moving mode, with 1 µm step width and 50 ms dwell time. The collected peak intensities were corrected for the background intensities and converted to absolute concentrations. Finally colour-coded element distribution maps were produced to visualize internal distribution patterns for certain chemical components.

Cathodoluminescence (CL) spectroscopy was performed using a hot cathode luminescence microscope (HC1-LM) with a high-vacuum chamber ($< 10^{-6}$ bar). Spectral acquisition was done with an Acton Spectra Pro 2300i spectrometer with a charge-coupled device (CCD) detector, which is attached to the CL microscope by a silica-glass fibre-guide. Measurements were taken with a 100 µm slit, a 150 s line grid, an accelerating voltage of 14 kV and a current of 0.2 mA with a current density of approximately 10 $\mu A\ mm^{-2}$. Calibration was done with the mercury vapour discharge emission. The optical aperture produces a spot size of 30 µm. Cathodoluminescence examinations were carried out on polished thin sections that are coated with carbon to prevent any build-up of electrical charge during CL operation (further details in Neuser et al. 1995).

Room-temperature laser-induced photoluminescence measurements were carried out by means of a Horiba Jobin Yvon LabRam-HR (high resolution) system equipped with an Olympus BX41 optical microscope, a grating with 600 grooves per millimetre, and a Si-based, Peltier-cooled CCD detector. Luminescence spectra were excited using the continuous 532 nm emission of a frequency-doubled Nd:YAG laser (34 mW on the sample surface). With the system operated in confocal mode and an Olympus 100× objective (numerical aperture = 0.9), the lateral resolution was better than 1.5 µm, and the depth resolution (with the laser beam focused at the sample surface) was approximately 3 µm. The spectral resolution in the NIR was determined to be ~2 cm^{-1}. Wavenumber calibration was done using the Rayleigh line and Ne lamp emissions; the wavenumber accuracy was better than 0.5 cm^{-1}. Temperature-dependent measurements were done adding a Linkam FTIR 600 liquid-nitrogen cooling stage. The temperature accuracy was better than 2 K. Photoluminescence (hyperspectral) maps were obtained using a software-controlled x–y stage, with a step width of 3 µm.

Luminescence bands were fitted assuming Lorentzian-Gaussian (pseudo-Voigt) band shapes. Background correction turned out to be unnecessary.

2.4 Fingerprints of Nd^{3+} luminescence emission: generalities

Luminescence emissions are generated through the (radiative) release of energy during electronic transitions. Emission spectra of REE^{3+} in crystalline hosts are characterised by very sharp bands in contrast to the luminescence of $3d$ or $4d$ elements (Blasse and Grabmaier 1994). This unique characteristic of REE^{3+} luminescence is determined by their particular electron configuration. Trivalent rare earth ions (Ce^{3+} to Yb^{3+}) have an incompletely filled $4f$ shell which is shielded by outer filled $5s^2$ and $5p^6$ orbitals. In consequence, the influence from ligands in the host matrix is small (but of crucial importance). The narrow luminescence bands originate from intra-configurational $4f^n$ electronic transitions. The distinct $4f^n$ energy levels result from different electronic interactions. The four kinds of electronic interactions are illustrated in Fig. 1 for the example of Nd^{3+} with $4f^3$ electron configuration. The energetic state of $4f^n$electrons in free REE^{3+} ions splits into spectroscopic *terms* due to the repulsion of unpaired $4f$ electrons (*electron–electron interaction*). In REE^{3+} ions, this term-splitting is on the order of ca. 10^4 cm^{-1} (Marfunin 1979). While the possible influence of negative charges from ligands is low, these terms further split into spectroscopic *levels* due to their *spin-orbit coupling* (on the order of ca. 10^3 cm^{-1}). Note that the separation of energy levels by these interactions applies to free REE^{3+} ions.

An important result is that REE^{3+} optical transitions appear at similar energy ranges in very different host materials. A complete diagram of possible energy levels for each REE^{3+} is given in Carnall et al. (1989). If REE^{3+} ions are further surrounded by ligands e.g., in molecules (*ligand field*) or solids (*crystal field*), the electronic levels split into *sublevels* (*Stark level* splitting on the order of ca. 10^2 cm^{-1}; cf. Fig. 1). According to the *crystal field theory* (cf. Burns 1993) ligand-directed orbitals experience an energy destabilization due to repulsion with the negative charge of the ligand. Non-directed orbitals are stabilized, as their potential energy drops. The crystal field splitting depends on a variety of parameters connected to the site of the REE^{3+} within the crystal structure, e.g., site symmetry, the interatomic distance to the ligands, charge of ligands etc. Given the different characteristics of cation sites among minerals, REE^{3+}spectra may provide site-specific information on the short-range order.

Figure 2 exemplifies the Nd^{3+} photoluminescence of five different hosts: the accessory minerals xenotime–(Y), monazite–(Ce), titanite, and synthetic cubic zirconia (YCZ) and Yttrium-aluminium garnet (YAG). In all hosts the Nd^{3+}-related ($^4F_{3/2} \rightarrow {}^4I_{9/2}$) luminescence emission can be detected within the spectral range between 10,600 and 11,800 cm^{-1} (940–840 nm). Yttrium-stabilized zirconia (fluorite structure type, *Fm3m*) and YAG (garnet structure type, *Ia3d*) are well known synthetic gemstones, both with cubic symmetry. Xenotime–(Y) crystallizes in the tetragonal space group $I4_1/amd$. Monazite ($P2_1/n$) and titanite ($P2_1/a$) have monoclinic crystal structures. The completely different crystal field around the Nd^{3+}-centre is reflected by strongly differing photoluminescence spectra (Fig. 2). Note that in many minerals probably more than one substitutional REE-site exists (e.g., apatite, Czaja et al. 2009).

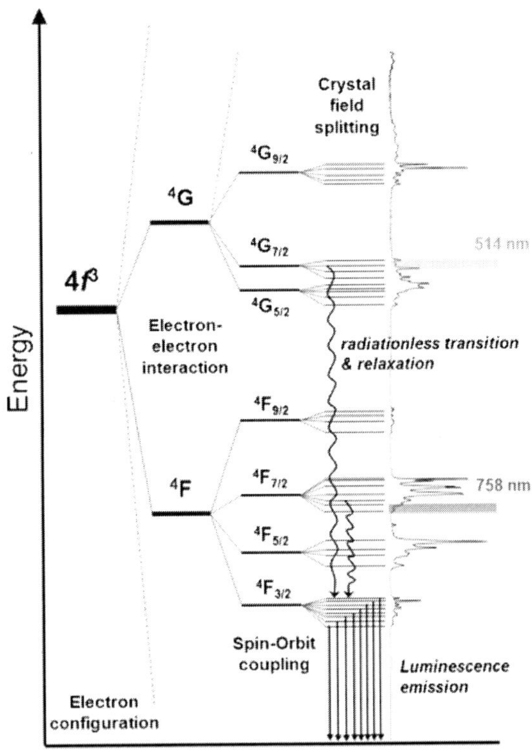

Figure. 1 Schematic illustration of the generation of electronic states for the example of Nd^{3+} with $4f^3$ electronic configuration. As examples for typical photoluminescence excitations, energies of photons of 514 and 758 nm laser light are marked on the right. After various excitation and subsequent relaxation processes, electrons eventually reach the lowest excited level ($^4F_{3/2}$). The transition from this level to the ground state is typically a radiative process that results in the emission of light, observed in the near-infrared range of the electromagnetic spectrum

In these cases standard steady-state luminescence techniques obtain the luminescence signal from all potential sites possibly differing from sample to sample. Time-resolved luminescence spectroscopy or site-selective spectroscopy may offer differentiation possibilities (Dexpert-Ghys et al. 1984; Dexpert-Ghys et al. 1996; Gaft et al. 1997; Piriou et al. 2001).

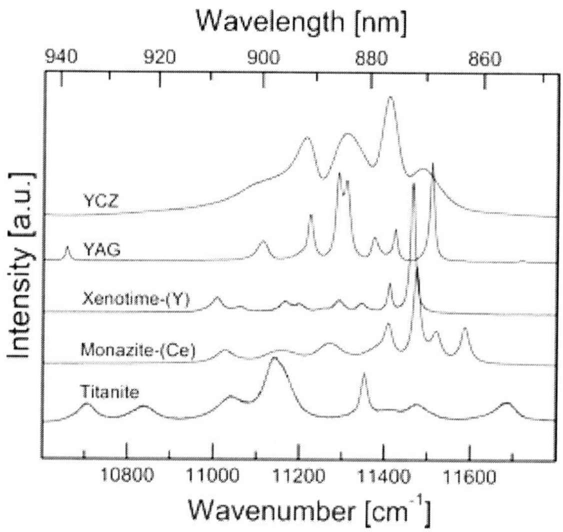

Figure. 2 Photoluminescence spectra (532 nm excitation) of Nd^{3+} ($^4F_{3/2} \rightarrow {}^4I_{9/2}$) in different host minerals: Yttrium-stabilized cubic zirconia (YCZ), yttrium-aluminium garnet (YAG), xenotime–(Y), monazite–(Ce), and titanite. Different cationic environments of the substituted Nd^{3+} cause dissimilar crystal field splittings and hence fingerprint-like luminescence patterns

Comparing the isostructural minerals monazite–(Ce) ($CePO_4$) and gasparite–(Ce) ($CeAsO_4$), both with C_1 cation site symmetry, reveals a comparatively close Nd^{3+}-luminescence pattern with only minor band shifts (Fig. 3a). Note that the band width is strongly influenced by chemical heterogeneity and intensity ratios by the crystal orientation (in more detail discussed below).

The minerals xenotime–(Y) (YPO_4) and chernovite–(Y) ($YAsO_4$) are isostructural with zircon ($ZrSiO_4$). All crystallize in space group $I4_1/amd$ with D_{2d} cation site symmetry. Whereas xenotime–(Y) and chernovite–(Y) show similarities in the Nd^{3+}-luminescence spectra, zircon differs remarkably, despite having the same site symmetry (Fig. 3b). Synthetic Nd^{3+}:$ZrSiO_4$charge-balanced with P^{5+} and without, as well as reference zircons from various locations were compared; their Nd^{3+} luminescence shows a similar pattern to that of synthetic $Nd^{3+}:ZrSiO_4$shown in Fig. 3b. This example demonstrates the impact of a modified *electronic* structure on the luminescence pattern: xenotime-(Y) and chernovite-(Y) have an isotypic crystal structure with chemically related elements; both have trivalent Y^{3+} on the cationic and $P^{5+}/As^{5+}O_4$ on the anionic position. Zircon is different, having tetravalent Zr^{4+} at the cation and $Si^{4+}O_4$ at the anion sites, respectively. A different charge distribution at the substitutional REE^{3+} site remarkably affects the crystal field splitting, and therefore the REE^{3+} luminescence signal.

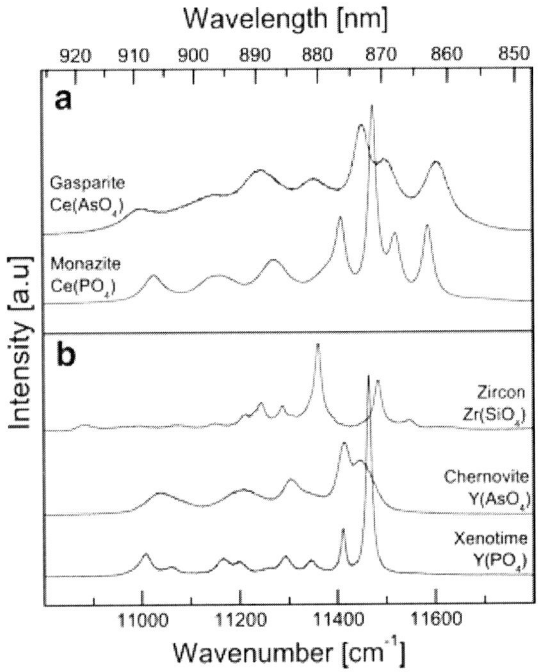

Figure. 3 Photoluminescence spectra (532 nm excitation) of Nd^{3+} ($^4F_{3/2} \rightarrow {}^4I_{9/2}$) in two groups of isostructural minerals. **A** Gasparite-(Ce) and monazite-(Ce) with cation site-symmetry C_1. **B** Zircon, chernovite-(Y), and xenotime-(Y) with cation site-symmetry D_{2d}

Luminescence studies on synthetic mineral analogues provide both: (1) a better identification of a REE in a specific host since the luminescence signal of many REE^{3+} overlap and a discrimination is complicated in natural specimens (e.g., Sm^{3+}–Pr^{3+}); and (2) a fast mineral identification when the individual sublevel splitting is known, potentially applied to igneous/metamorphic or sedimentary petrology (Richter et al. 2008).

2.5 Orientation-dependence of REE^{3+} luminescence

As mentioned above, the crystal-field splitting depends on the site symmetry of the REE^{3+}-substituted cation site. A direct consequence is that the luminescence intensity of sublevels depends on the orientation of the measuring direction to certain symmetry elements of the crystal structure. Even though being a well-known luminescence phenomenon (Owen et al. 1998; Barbarand and Pagel 2001; Finch et al. 2003), the orientation-dependence of REE^{3+}emissions is often neglected, which may result in biased conclusions. To give two examples: Czaja et al. (2009) proposed to use the intensity ratio of emissions related to Pr^{3+} and Sm^{3+} to determine Pr/Sm ratios in apatite. Habermann (2002) applied REE-cathodoluminescence intensity for a semi-quantitative estimation of REE trace element concentration in calcite. In those papers, however, authors failed to consider possible luminescence intensity differences due to the orientation-dependence of the emissions analysed. For this reason, the orientation dependence of REE^{3+} luminescence on the example of Nd^{3+} is exemplified here for the sake of completeness (cf. Cesbron et al. 1995, Gaft et al. 2005).

Depending on the luminescence technique applied, different aspects of orientation effects can be discerned. In laser-induced photoluminescence studies the polarisation of the incident laser has to be considered, because most standard apparatus provide a strongly linear-polarised beam due to polarising effects of optical components (e.g., resonator, mirrors). The coupled interaction of the laser-beam polarisation and the orientation of the crystal leads to four independent variables: (1) the direction of the laser beam relative to the crystal [e.g., *x*(*yy*)*y*, cf. Porto notation in Porto and Krishnan (1967)]; (2) the polarisation of the laser beam relative to the crystal [*y*(*xy*)*y*)]; (3) the polarisation of the luminescence emission relative to the crystal [*y*(*yx*)*y*], and (4) the direction of luminescence emission detected relative to the crystal [*y*(*yy*)*x*]. Most spectrometer systems used in geoscientific research are coupled to microscopes, in which the laser beam and the emission detected proceed in opposite direction through the observing objective lens (x(yy)*x*, quasi-backscattering set-up). Hence, the latter variable cannot be analysed separately using such a system set-up. Figure 4 shows the other three aspects with the example of laser-induced photoluminescence spectra of Nd^{3+} ($^4F_{3/2} \rightarrow {}^4I_{9/2}$) in synthetic tetragonal Nd^{3+}:YPO_4 (xenotime structure type, $I4_1/amd$). Please note that the measurement conditions, e.g., laser power, measurement point on the sample, accumulation time etc., were kept strictly identical. Figure 4a demonstrates the effect of a polarised laser beam with the electric field vector (*E*) aligned along different crystallographic directions; here the laser beam is directed along the crystallographic *x* axes; the polarisation of the electric field vector is aligned in *y* and z, respectively. The polarisation of the emission is neglected (no polarisation filters), all intensity of the randomly polarised signal in plane $z - y$ is detected. Changing the polarisation direction of the laser beam causes intensity ratios of the luminescence sublevels to vary significantly. The main sublevel band at 11,467 cm^{-1} is most strongly affected.

A polarisation filter placed in the emission path-way reveals that the luminescence emission itself is polarised (Fig. 4b). Analysing the luminescence emission with polarisation filters in different directions leads to significant variations of intensity ratios. Turning the sample perpendicular to its tetragonal main axis (z), with the laser polarization direction and the polarization direction of the analysed emission held constant, results in intensity increases in all other sublevel bands relative to the main at 11,467 cm^{-1} (Fig. 4c).

In contrast to laser-induced photoluminescence, the effect of beam polarisation is without significance using an unpolarised electron or ion beam with cathode-/ionoluminescence techniques. The polarisation of the luminescence emission is of importance especially if the signal is analysed with polarisation filters, which is commonly not done in CL measurements. In Fig. 5 the cathodoluminescence spectra of synthetic xenotime–(Y) and zircon crystals are presented. Although the luminescence was excited with an unpolarized electron beam there are significant effects of different analysing directions relative to the crystal. A potential semi-quantitative estimation of trace elements via cathodoluminescence intensity, as for instance applied to apatite, therefore needs a broad range of carefully calibrated conditions, e.g., standard material, system stability, system response (Cesbron et al. 1993; Barbarand and Pagel 2001). Moreover, other effects, e.g., concentration quenching, quenching/sensitizing by other impurities (Kempe and Götze 2002) and, in particular, the analysing direction relative to the crystal, have to be considered (cf. Barbarand and Pagel 2001).

2.6 Effects of the real structure on Nd^{3+} luminescence: the example of monazite–(Ce)

Real structure is defined as the entity of deviations from the ideal chemical composition and structure. A common feature is the presence of defects, including point (e.g., substitutions, vacancies, interstitials), line, or three-dimensional defects (e.g., holes, surface defects, grain boundaries). Typical effects on light-spectroscopic signals are band broadening and shifting.

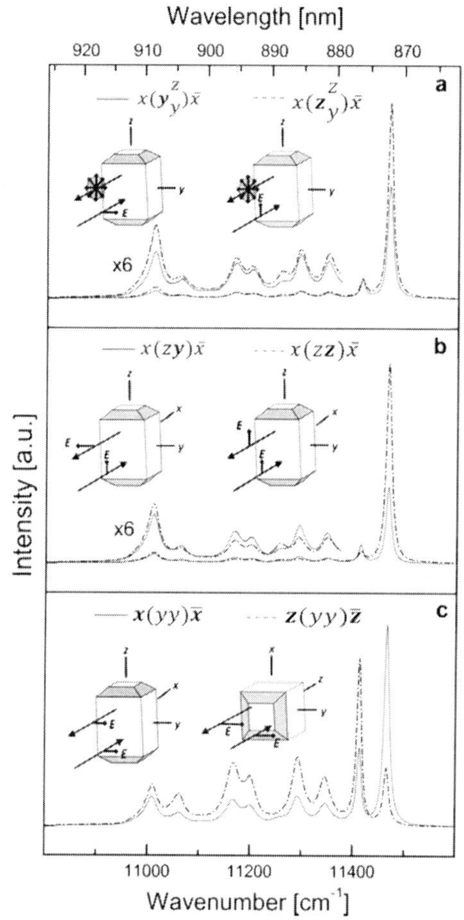

Figure. 4 Orientation-dependence of the laser-induced PL (532 nm excitation) of Nd^{3+} ($^4F_{3/2}$ → $^4I_{9/2}$) in synthetic tetragonal Nd^{3+}:YPO_4 (xenotime; $I4_1/amd$). Note that all spectra have the same vertical scaling, intensities are therefore directly comparable. **A** Effect of the polarization of the incident laser beam. **B** Effect of the polarization of the luminescence emission. **C** Effect of the direction (relative to the crystal) of the excitation/emission analysed. Small schematic sketches are inserted for better illustration of the Porto notation quoted above

In optical spectroscopy this phenomenon is commonly termed *inhomogeneous broadening* (cf. Macfarlane 1990; cf. Skinner and Moerner 1996). The sharpness of optical transitions, observed either in absorption or emission processes, depends theoretically on the lifetime of the excited electronic state. This is due to the quantum-mechanical uncertainty and includes both radiative and non-radiative processes (Blasse and Grabmaier 1994). For excited states of the f^n configuration, radiative rates are typically in the millisecond and sub-millisecond time scale. This is because these transitions are parity-forbidden and often spin-forbidden as well (Blasse and Grabmaier 1994; Gaft et al. 2001; Liu and Jacquier 2005). At hypothetical ideal conditions, absorption and emission bands of REE have very narrow Lorentzian shapes. These conditions include: REE substituents of the same type with no cross-interaction, a perfect crystal without defects/strain, and without any lattice vibrations (phonons) achieved at very low temperature (ideally 0 K with hypothetically no vibronic coupling). The observed absorption/emission band is considered as superposition of all electronic transitions of a certain number of REE substituents in the area analysed (*homogeneous broadening*, Fig. 6a). However, effects of the real structure (Fig. 6b) perturb the local environment of the REE centre. As a consequence, the crystal field around the REE cations is randomly distorted to a certain amount and electronic transition energies shift slightly. The observed spectral profiles can then be considered as a broad envelope over much narrower homogeneous lines introducing a Gaussian character to the band shape (schematically illustrated in Fig. 6b).

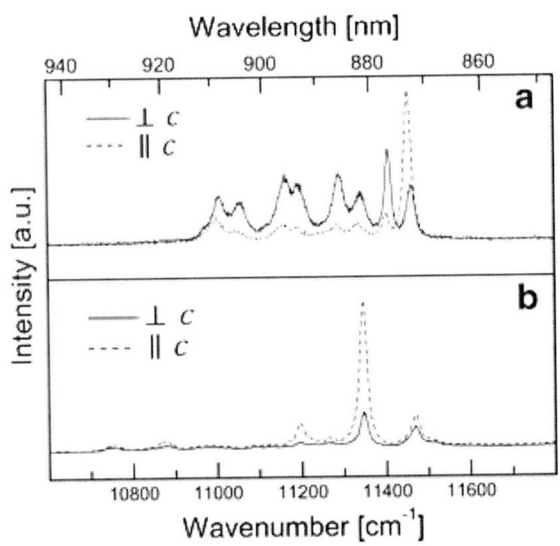

Figure. 5 Orientation dependence of the CL emission of Nd^{3+} ($^4F_{3/2} \rightarrow {}^4I_{9/2}$). **A** Synthetic Nd^{3+}:YPO_4 (xenotime). **B** Synthetic Nd^{3+}:$ZrSiO_4$ (zircon). Note that all spectra have the same vertical scaling, intensities are therefore directly comparable. The emitted CL light was obtained in directions perpendicular and parallel to the *c*-axis of the respective crystal. Note that intensity ratios among sublevel bands vary appreciably

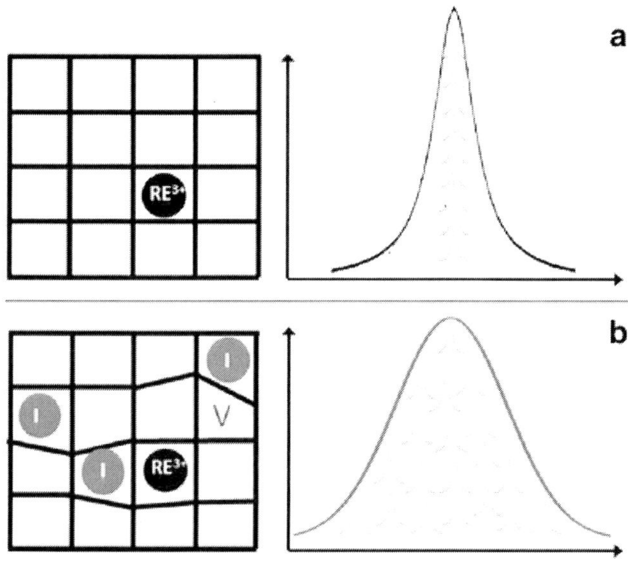

Figure. 6 Schematic illustration (simplified) of the band broadening related to optical transitions as affected by the sample's real structure (modified after Skinner and Moerner 1996). **A** Nearly "ideal" crystal (merely doped with low levels of an individual REE), assuming the absence of notable amounts of defects and strain. The observed absorption or emission band, respectively, is considered as superposition of all electronic transitions of a certain number of REE substituents in the area analysed (*homogenous broadening*). **B** "Real" crystal with notable amounts of defects such as impurities (I), vacancies (V), and non-formula constituents, resulting in notable lattice strain. Here, the crystal field around the REE site is perturbed randomly. The observed spectral profile is a broad envelope over much narrower homogenous lines (*inhomogeneous broadening*)

To illustrate the effect of the real structure on REE luminescence, we use the example of chemically induced disorder in monazite–(Ce) for three reasons: First of all, almost all natural monazite-group minerals contain Nd^{3+} and show Nd-luminescence. Second of all, the monazite structure has an extremely broad range of chemical compositions, so that a variety of substitution mechanisms can be studied. Third, Ruschel et al. (2012) have recently studied the structural disorder of monazite–(Ce) using Raman spectroscopy (band broadening of the symmetric stretching vibration of PO_4 tetrahedrons; A_{1g} mode). This was done with the basic objective to investigate, and distinguish among, disturbing effects of the chemical composition and/or structural damage as caused by self-irradiation due to the incorporation of actinide elements. Panczer et al. (2012) showed that the incorporation of non-formula elements affects the sharp sublevel luminescence bands of Nd^{3+} as well. In the present paper we address the question whether Nd^{3+}-luminescence spectroscopy can be used for a quantitative estimation of the chemically induced structural disorder.

Monazites are monoclinic orthophosphates with the general formula LREE[PO_4]. These minerals have an extremely broad range of chemical compositions with Ce being the predominant cation in naturally occurring species. Solid solutions with other minerals of the monazite group allow extensive substitutions of Ce by other LREEs (especially La, Nd, Sm). In addition to the more limited incorporation of heavy REEs, significant amounts of the actinide elements Th and U, and to much lesser extent radiogenic Pb, may also be present (Williams et al. 2007). The incorporation of non-REE ions in the wt% range is commonly explained by the two coupled substitutions: (1) $2REE^{3+} \leftrightarrow Ca^{2+} + Th^{4+}$ (*cheralite substitution*) and (2) $P^{5+} + REE^{3+} \leftrightarrow Si^{4+} + Th^{4+}$ (*huttonite substitution*). The continuity in the monazite-huttonite series extends to a maximum of 30 at% huttonite in natural monazites (Della Ventura et al. 1996; Kucha 1980; Förster and Harlov 1999). The substitution mechanisms for Th^{4+} can be applied to U^{4+}, also. For details on the crystal chemistry of the monazite structure the reader is referred to reviews by Kolitsch and Holstam (2004) and Clavier et al. (2011).

Here we present Nd^{3+} luminescence data for three different sample sets reflecting different substitution mechanisms: (1) a series of homogeneous synthetic $Ce_{1-x}LREE_xPO_4$ crystals doped with large amounts of other REEs, forming a solid solution with LREE = La, Nd, Gd (for detailed characterisation see Ruschel et al. 2012); (2) synthetic samples doped with Th^{4+}compensated with Ca^{2+} and/or Na^+ (see details in the sample description above); and (3) homogeneous natural monazite–(Ce) samples reflecting combined substitution mechanisms in nature including the huttonite substitution with charge compensation by Si^{4+} on anion-sites (for detailed characterisation see Ruschel et al. 2012).

Figure 7 shows that the exclusive substitution of homovalent REE^{3+} for Ce^{3+} on the cation-site causes only minor band broadening. Monazites–(Ce) with heterovalent substitutions, such as Th^{4+}, Ca^{2+} on cation and Si^{4+} on the anion site, show stronger band broadening and shifting (Fig. 7). In this case, the FWHM (full width at half maximum) of the sublevel bands lowers the probability of a precise deconvolution/fitting due to an effective band overlap. Hence, the less affected out of the eight distinctive sublevel bands at ca. 11,030 abs. cm^{-1} is chosen for the interpretation of the FWHM hereafter.

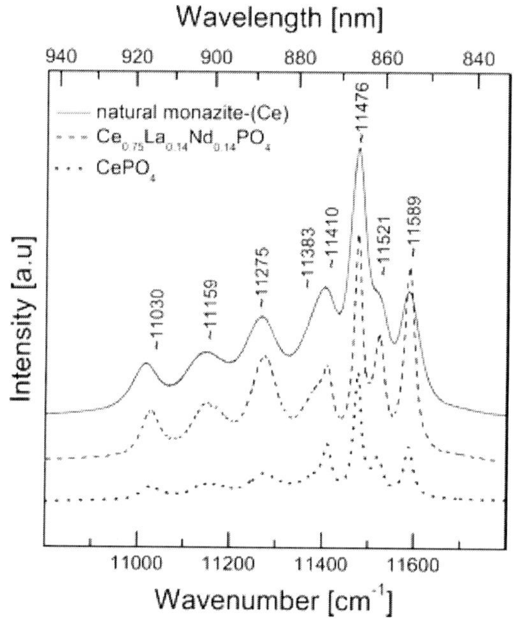

Figure. 7 Photoluminescence spectra (532 nm excitation) of Nd^{3+} ($^4F_{3/2} \rightarrow {}^4I_{9/2}$) of annealed, Th-rich natural monazite-(Ce) from Central Australia (sample ECA6 of Ruschel et al. 2012) in comparison with La,Nd-doped and un-doped $CePO_4$. Band positions are quoted for un-doped $CePO_4$

To describe chemically induced disorder we introduce a *distortion parameter* (Δ) which reflects the distortional behaviour of every substituent on regular cation and anion sites. This is approximated with a summation of the weighted ionic radii difference between the predominant cation/anion (Ce/P in monazite–Ce) and each substituent:

$$\Delta = \sum_{cs} c_{cs} * |r_c - r_{cs}| + \sum_{as} c_{as} * |r_a - r_{as}| \tag{1}$$

where c is the concentration of cation (c_{cs}) or anion substituents (c_{as}) in atomic percent per formula unit (apfu), and r the ionic radius of the predominant cation (r_c) or anion (r_a), and of the cation (r_{cs}) or anion substituent (r_{as}), respectively. For monazite–(Ce), the cationic radii in 9-fold and anionic radii of P and Si in 4-fold coordination after Shannon (1976) are used. This formalism takes into account that large ionic radii differences between predominant and substitutional ions should have much more distortional impact than smaller. Additionally, this impact is weighted with the concentration of each ion providing an effective distortional impact which is summed over all substitutional ions present.

Figure 8 shows the effects of chemical substitution on spectral parameters of the Nd^{3+} ($^4F_{3/2} \rightarrow {}^4I_{9/2}$) luminescence emission. An increase of the distortion parameter (Δ), calculated from chemical data of all investigated monazite-(Ce) samples (i.e., Table 2), causes a band shift to lower wavenumbers. In this case, the band shift is linear to the total sum of weighted radii differences of all substituents (Δ). This indicates that the substitution of smaller ions than Ce causes a reduced cell volume. Clavier et al. (2011) summarized data suggesting a complete solid solution in the synthetic system $LaPO_4$–$ThSiO_4$ with linearly decreasing cell parameters with increasing Th content. Alike, unit cell parameters decrease linearly with the substitution ratio in the $La_xGd_{1-x}PO_4$ monazite solid solution (Clavier et al. 2011).

Interestingly, the FWHM of the Nd^{3+}-luminescence bands is mainly controlled by heterovalent substituents (Figs. 7 and 8b). The Δ_{hetero} parameter in Fig. 8b was calculated applying Eq. 1 only to heterovalent substituents present in the samples analysed. The comparison of spectral parameters of the Ce_{1-x} $REE_x[PO_4]$ solid solution (filled triangle, Fig. 8) gives an illustrative example: The extensive substitution of homovalent REE^{3+} causes large band shifts (Fig. 8a), but have no effect on the FWHM (Fig. 8b). Heterovalent substitutions appear to have a much higher impact on the electronic structure and therefore on the perturbation of electronic states than homovalent ones (see the discussion on the crystal field above). Hence, the FWHM of luminescence bands is rather a measure of the perturbed electronic structure which is in this case connected to chemically induced disorder via substitution of heterovalent ions.

The estimation of structural disorder by radii differences applied above is based upon a rigid hard-sphere model of ions and cannot take into account lattice distortion by other real-structure defects, e.g., vacancies. Nevertheless, the good fit to a linear trend suggests that the inferred disorder is mainly controlled by the different chemical substitution mechanisms in the synthetic and natural monazites analysed in this study. Panczer et al. (2012) correlate the FWHM of Nd-luminescence bands of natural monazites with the thorium content (wt%) without regarding other heterovalent substitutions. For comparison, we show a similar correlation of our samples in Fig. 8c. The variance of FWHM among the natural samples (open dots) is greater than for the correlation in Fig. 8b. The FWHMs of synthetic multi-doped monazite samples (cross, Fig 8) show an individual correlation with the ThO_2 content. Therefore, the correlation of the FWHM – as potential estimation of structural disorder – with the ThO_2 content alone is imprecise if variable substitution mechanisms determine the chemical composition.

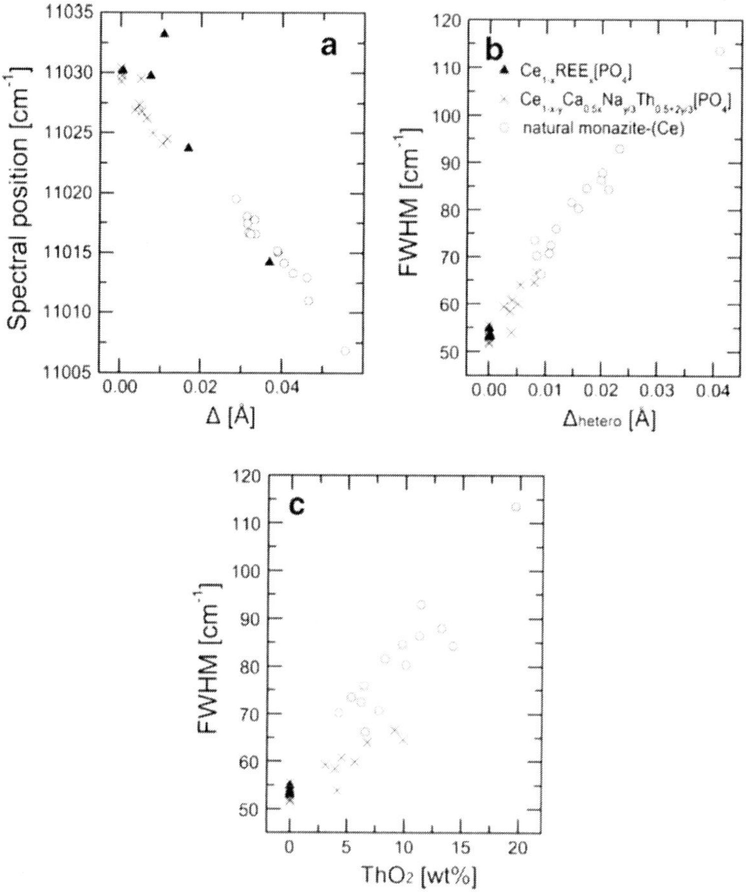

Figure. 8 Effects of chemical substitutions in various natural and synthetic monazite samples on spectral parameters of the Nd^{3+} ($^4F_{3/2} \rightarrow {}^4I_{9/2}$) luminescence emission, shown for the ~11,030 cm^{-1} sublevel band (see Fig. 7). **A** The increase of the distortion parameter Δ (see text and Eq. 1) causes a band shift to lower wavenumbers. **B** The increase of the heterovalent distortion parameter Δ_{hetero} (see text) causes a linear increase of the band FWHM. **C** Band FWHM and ThO_2 content show a less pronounced correlation, making potential estimates of the chemically induced structural disorder from the ThO_2 content alone rather imprecise

In this study, potential effects of corpuscular self-irradiation on the structural disorder are excluded by using synthetic or annealed samples. The estimation of the chemical impact on the REE-luminescence band widths is of major importance for the discrimination and quantification of irradiation-induced disorder by REE-luminescence spectroscopy (see the review chapter of Nasdala et al. in this special issue). The irradiation-induced disorder accumulated in geologic timescales can be quantified if the chemical-dependent contribution is known. Hence, REE-luminescence spectroscopy may be a complementary technique to Raman spectroscopy, especially for minerals where distinctive Raman bands are absent (e.g., titanite or pyrochlore).

Furthermore, information on the chemically induced structural disorder derived from REE-luminescence data may help to study the dynamics in solid solutions and the interpretation of zoning patterns. This is illustrated in Fig. 9, which presents element distribution (Fig. 9a) and hyperspectral Nd-luminescence maps (Fig. 9b) of a synthetic Th-doped monazite–(Ce). The single crystal shown is characterised by two major zones of significantly different chemical composition. The inner area consists of $CePO_4$ with extremely low amounts of trace elements below the EPMA detection limits, and represents an early stage of crystal growth in the flux. The outer zone (ca. 30–70 μm in width) contains elevated amounts of Th, Ca, and Na and is hence depleted in Ce (average chemical formula $Ce_{0.854}Th_{0.084}Ca_{0.039}Na_{0.023}PO_4$). The rim of the crystal (i.e. the outermost ca. 5–10 μm of the outer zone) is strongly enriched in Th. Note that the Th-doped synthetic samples contain well-detectable amounts of Na (Table 2), which we assign to their growth in a sodium-polyphosphate flux (cf. description of synthesis details above). Table 2 presents element quantities (quoted in apfu) of several measurement points on the synthetic Th-doped monazite–(Ce) samples. It was observed for all individual analyses that after subtraction of the respective Th amount related to the cheralite component ($2REE^{3+} \leftrightarrow Ca^{2+} + Th^{4+}$; Ca:Th = 1:1), the ratio of the remaining Na–to–Th is close to 1:2. This suggests a substitution mechanism according to

$$Na^{+} + 2Th^{4+} \leftrightarrow 3Ce^{3+} \tag{2}$$

To the best of our knowledge, this Na–Th substitution has not been described for natural monazite–(Ce) thus far.

The Nd-luminescence parameters are very sensitive to the structural disorder as introduced by the presence of non-formula elements (cf. again Fig. 9b). The FWHM of the ~11,030 cm^{-1}sublevel band broadens from 53 cm^{-1} in the case of nearly pure $CePO_4$ (inner area) up to 65 cm^{-1} near the Th-rich rim of the crystal, accompanied by a shift of the band position from 11,031 cm^{-1} to 11,022 cm^{-1}. In all interior regions and zones, the concentration of Nd was below the EPMA detection sensitivity. We speculate that Nd may be slightly enriched in the outer zone, coupled to the other substituents; this is concluded from the generally higher Nd-luminescence intensity in this area. Other potential causes of lateral differences in the Nd-emission intensity – such as quenching and/or sensitizing (e.g., Marfunin 1979) or effects of structural disorder (e.g., Nasdala et al. 2006) – can be excluded in our case of synthetic, mildly Nd-contaminated specimens.

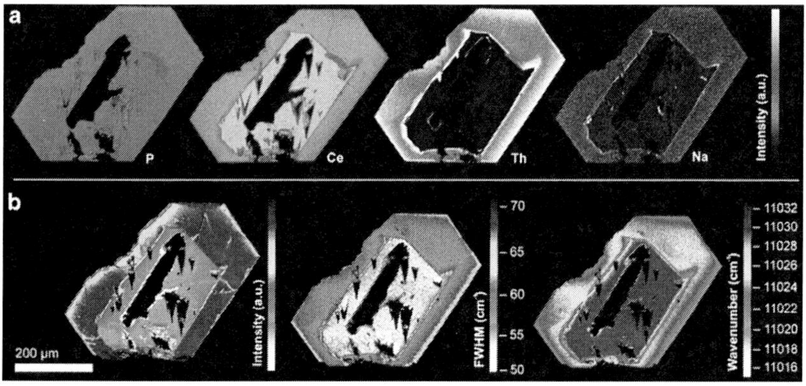

Figure. 9 Element distribution maps (**A**) and hyperspectral Nd^{3+} maps (**B**) of a synthetic Th-doped monazite single crystal. The latter shows distributions of spectral parameters of the ~11,030 cm^{-1} sub-band of the $^4F_{3/2} \rightarrow {}^4I_{9/2}$ electronic transition of Nd^{3+}. Irregularities in the inner zone (i.e. black triangular areas) are flux inclusions and outbreaks from polishing. Note that changes in Nd-luminescence parameters (especially the FWHM) correlate closely with changes in the incorporation of non-formula elements

2.7 Temperature dependence

An additional important factor affecting the REE-luminescence band width (and marginally the band position) is temperature. Increasing temperature enhances the extent of lattice vibrations (phonons). Lattice vibrations induce an oscillation of interatomic distances at the luminescence centre within the crystal structure. In consequence, the potential energy of electronic states and thus the transition energy of corresponding optical phenomena are distributed around an equilibrium value. The observed band shape is again a sum of all constituent transitions over a specific time interval in the analysed area. Luminescence measurements are commonly done at very low temperatures to avoid this additional perturbation (vibronic coupling). In general, lowering the temperature allows an improved separation of the obtained luminescence features, but also enables transitions from excited levels, which have previously been depopulated via phonons (non-radiative transitions). The temperature effect on REE band widths is comparatively small compared with *d-d* or *d-f* transitions, where absorption and luminescence band widths are much larger. This is because *f*-electrons and their inter-transitions are well shielded by the outermost filled $5s^2 5p^6$ orbitals, so that the interaction with the ligands is small.

Figure 10 illustrates the effect of temperature on the Nd^{3+} luminescence with the monazite–(Ce) example discussed above: The FWHMs of all the eight observed sublevel bands of the $Nd^{3+}(^4F_{3/2} \rightarrow {}^4I_{9/2})$ emission increase with increasing temperature. The FWHM of broad sublevel bands increases much more than those of the thinner ones (Fig. 10). Similarly, broad bands show the strongest positional shift. However, the shift of the broadest band at 11,159 cm^{-1} does not exceed 9 cm^{-1} (0.7 nm) from 80 K to 300 K. The low sensitivity of the band position to higher temperature is a useful feature for high temperature experiments with diamond anvil cells (DAC), where optical-luminescence pressure-gauges are used successfully. Datchi et al. (1997; 2007) promote the distinct luminescence emission of Sm^{2+} (intra $4f^6$; $^5D_0 \rightarrow {}^7F_0$) doped SrB_4O_7 as viable temperature-insensitive alternative to the Cr^{3+} (R-lines) luminescence of ruby (Cr^{3+}:Al_2O_3).

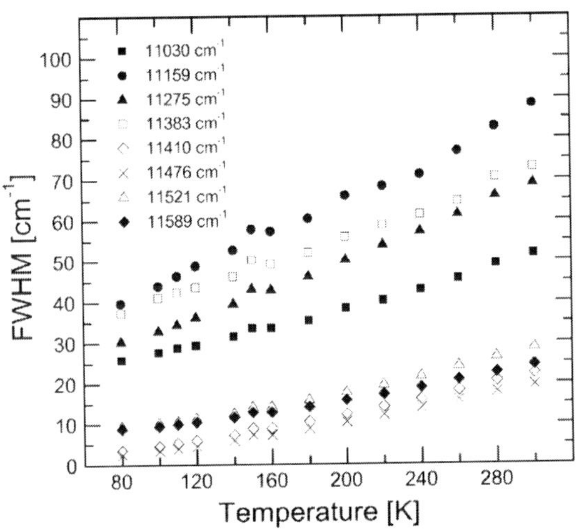

Figure. 10 Effect of temperature on the FWHM of eight distinctive luminescence sublevel bands (Nd^{3+}, $^4F_{3/2} \rightarrow {}^4I_{9/2}$) of synthetic $CePO_4$. Temperature increase causes general broadening of sublevel bands, with broader bands showing stronger absolute increase of their FWHM, compared to narrower bands

The prediction of FWHMs and shifts of REE bands cannot be deduced from first principles. The temperature not only affects the electronic interaction within the luminescent REE^{3+} ion, but also the surrounding ligands of the crystal field. An anisotropic thermal expansion of the crystal structure stresses the cation site (Mogilevsky et al. 2007; Jardin et al. 2008; Li et al. 2009) and results in a anisotropic crystal field splitting, which is further influenced by local defects. This consideration is of crucial importance for the evaluation of band widths for the quantification of chemically induced (see above) or radiation-induced disorder (see the review chapter of Nasdala et al. in this special issue). Figure 11 shows the FWHM of the 11,030 cm^{-1} Nd-sublevel band of different monazite-(Ce) samples with various chemical compositions. Chemically induced disorder not only causes higher FWHMs but also influences the impact of the temperature effect. High chemical disorder inhibits the effect of temperature broadening, possibly due to structural relaxation induced by the presence of non-formula elements. Measuring the FWHMs for the quantification of e.g., chemically induced disorder gives different absolute values at different temperatures, whereby the relative significance is not changed drastically. Hence, an absolute FWHM correlation with effects connected to structural disorder should be done at constant temperature, which is done throughout this study.

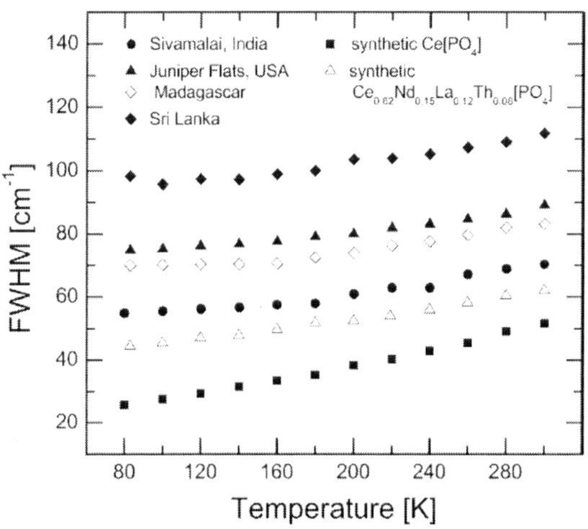

Figure. 11 Influence of the chemical composition on the FWHM-temperature dependence, for the example of the ~11,030 cm^{-1} Nd^{3+} sublevel. Chemically induced disorder not only causes FWHM increases (cf. Fig. 8a) but also decreases the effect of temperature on the band broadening, possibly due to structural relaxation by the substituents

All luminescence spectra presented in the sections above are recorded at room temperature (298 K), because low temperature measurements using cooling stages or other liquid-nitrogen set-ups entail considerable effort; especially for measurements with high lateral resolution for hyperspectral luminescence maps (cf. Fig. 9).

CONCLUDING REMARKS

First of all, the luminescence emission of REE^{3+} in minerals depends on the type of incorporated REEs. The energetic positions of their spectroscopic levels determine the luminescence emission pattern (a chart of spectroscopic levels of all REE^{3+} was given by Carnall et al. 1989). Commonly, natural samples contain more than one type of REE^{3+}, often resulting in extensively overlapping luminescence patterns. Due to electronic shielding of the outermost $5s^2$ and $5p^6$ orbitals, the influence on the $4f$ transitions of the crystal field is weak and a specific optical transition of a REE^{3+} appears with comparable spectral energies in very different mineral hosts (Fig. 2). Nevertheless, the crystal field splits spectroscopic levels into sublevels depending on the distribution of the electron density at the substituted lattice site (Fig. 2). Luminescence investigations of REE^{3+} in minerals hence provide first the opportunity to identify REEs in a specific host mineral. Second, they enable fast mineral identification, provided the individual sublevel splitting of a specific REE^{3+} in the mineral under investigation is known. With respect to the latter, the luminescence emission of Nd^{3+} was found to be a promising candidate because (1) the luminescence signal can be traced with standard Si-based detectors, (2) it is effectively excited by standard lasers as well as via cathodoluminescence, (3) other luminescence emissions in the respective spectral energy range are rare, and (4) a low number of sublevels simplifies the data reduction (i.e., fitting/deconvolution).

Even minute amounts of REE^{3+} cause detectable luminescence signals and can potentially be used for the qualitative identification of trace REEs. In contrast to optical absorption spectroscopy, the quantitative estimation of REEs concentrations via luminescence intensity is difficult for several reasons (see detailed discussions above). In particular, this study shows once more that the relative luminescence intensity of certain sublevels may depend strongly on the crystal orientation (Fig. 4). Furthermore, both the polarisation of the incident laser (in the case of laser-induced photoluminescence) and the polarisation of the luminescence signal are of importance when comparing relative luminescence intensities.

The FWHM of REE^{3+} luminescence sublevel bands contains information on the crystal-structural state and provides great opportunities to study different types of structural disorder with luminescence techniques (Fig. 9). In this study, the effect of chemically induced disorder on the luminescence signal of monazite-(Ce) has been described with a simple model based on ionic radii. The discrimination between chemically induced (Fig. 8) and irradiation-induced structural disorder is possible, however, when the influence of measurement temperature (Fig. 10 and 11) is considered.

ACKNOWLEDGEMENTS

We are indebted to U. Kolitsch and G. Giester for providing standard material for comparison purposes. Sample preparation was done by A. Wagner. The lead author is most grateful to M. Wildner for fruitful discussions. Constructive comments by D. Ohnenstetter, M. Gaft and an anonymous reviewer are gratefully acknowledged. Funding of this research was provided by the Austrian Science Fund (FWF) through grant P24448-N19 to LN, and the European Commission through project CZ.1.07/2.3.00/20.0052 "Research group for radioactive waste repository and nuclear safety" to DT and RS.

REFERENCES

1. Barbarand J, Pagel M (2001) Cathodoluminescence study of apatite crystals. Am Mineral 86:473–484
2. Belsky A, Krupa J (1999) Luminescence excitation mechanisms of rare earth doped phosphors in the VUV range. Displays 19:185–196
3. Blanc P, Baumer A, Cesbron F, Ohnenstetter D, Panczer G, Rémond G (2000) Systematic cathodoluminescence spectral analysis of synthetic doped minerals: anhydrite, apatite, calcite, fluorite, scheelite and zircon. In: Pagel M, Barbin V, Blanc P, Ohnstetter D (eds) Cathodoluminescence in geosciences. Springer, Berlin, pp 127–160
4. Blasse G, Grabmaier B (1994) Luminescence materials. Springer, Berlin, 248 p
5. Boatner LA (2002) Synthesis, structure, and properties of monazite, pretulite, and xenotime. Rev Mineral Geochem 48:87–121
6. Bünzli JCG, Piguet C (2005) Taking advantage of luminescent lanthanide ions. Chem Soc Rev 34:1048–1077
7. Burns RG (1993) Mineralogical applications of crystal field theory, Cambridge University Press
8. Carnall W, Goodman G, Rajnak K, Rana R (1989) A systematic analysis of the spectra of the lanthanides doped into single crystal LaF. J Chem Phys 90:3443–3457
9. Cesbron F, Ohnenstetter D, Blanc P, Rouer O, Sichere MC (1993) In corporation des terres rares dans des zircons de synthèse: etude par cathodoluminescence. C R Acad Sci II 316:1231–1238
10. Cesbron F, Blanc P, Ohnenstetter D, Rémond G (1995) Cathodoluminescence of rare earth doped zircons. I. Their possible use as reference materials. Scanning Microscopy Suppl 9:35–56

11. Clavier N, Podor R, Dacheux N (2011) Crystal chemistry of the monazite structure. J Eur Ceram Soc 31:941–976
12. Czaja M, Bodył S, Głuchowski P, Mazurak Z, Strek W (2008) Luminescence properties of rare earth ions in fluorite, apatite and scheelite minerals. J Alloy Compd 451:290–292
13. Czaja M, Bodył S, Lisiecki R, Mazurak Z (2009) Luminescence properties of Pr3+ and Sm3+ ions in natural apatites. Phys Chem Miner 37:425–433
14. Datchi F, LeToullec R, Loubeyre P (1997) Improved calibration of the SrB4O7:Sm2+ optical pressure gauge: advantages at very high pressures and high temperatures. J Appl Phys 81:3333–3339
15. Datchi F, Dewaele A, Loubeyre P, Letoullec R, Le Godec Y, Canny B (2007) Optical pressure sensors for high-pressure–high-temperature studies in a diamond anvil cell. High Press Res 27:447–463
16. Della Ventura G, Mottana A, Parodi GC, Raudsepp M, Bellatreccia F, Caprilli E, Rossi P, Fiori S (1996) Monazite-huttonite solidsolutions from the vico volcanic complex, Latium, Italy. Mineral Mag 60:751–758
17. Dexpert-Ghys J, Faucher M, Caro P (1984) Site selective spectroscopy and structural analysis of yttria-doped zirconias. J Solid State Chem 54:179–192
18. Dexpert-Ghys J, Mauricot R, Faucher M (1996) Spectroscopy of Eu3+ ions in monazite type lanthanide orthophosphates LnPO4, Ln= La or Eu. J Lumin 69:203–215
19. Finch A, Hole D, Townsend P (2003) Orientation dependence of luminescence in plagioclase. Phys Chem Miner 30:373–381
20. Förster H, Harlov D (1999) Monazite–(Ce)-huttonite solid solutions in granulite-facies Metabasites from the Ivrea-Verbano Zone, Italy. Mineral Mag 63:587–594
21. Friis H, Finch AA,Williams CT, Hanchar JM(2009) Photoluminescence of zircon (ZrSiO4) doped with REE3+ (REE = Pr, Sm, Eu, Gd, Dy, Ho, Er). Phys Chem Miner 37:333–342
22. Gaft M, Reisfeld R, Panczer G, Shoval S, Champagnon B, Boulon G (1997) Eu3+ Luminescence in high-symmetry sites of natural apatite. J Lumin 72:572–574
23. Gaft M, Reisfeld R, Panczer G, Uspensky E, Varrel B, Boulon G (1999) Luminescence of Pr3+ in minerals. Opt Mater 13:71–79
24. Gaft M, Panczer G, Reisfeld R, Uspensky E (2001) Laser-induced time-resolved luminescence as a tool for rare-earth element identification in minerals. Phys Chem Miner 28:347–363
25. Gaft M, Reisfeld R, Panczer G (2005) Luminescence spectroscopy of minerals and materials. Springer, Berlin, 372 p
26. Götze J (2000) Cathodoluminescence microscopy and spectroscopy in applied mineralogy. Freiberger Forschungsheft C 485, TU Bergakademie Freiberg, 128 p
27. Götze J (2002) Potential of cathodoluminescence (CL) microscopy and spectroscopy for the analysis of minerals and materials. Anal Bioanal Chem 374:703–708
28. Götze J, Habermann D, Neuser RD, Richter DK (1999) Highresolution spectrometric analysis of rare earth elements-activated cathodoluminescence in feldspar minerals. Chem Geol 153:81–91
29. Habermann D (2002) Quantitative cathodoluminescence (CL) spectroscopy of minerals: possibilities and limitations. Miner Petrol 76:247–259
30. Habermann D, Neuser RD, Richter DK (1996) REE-activated cathodoluminescence of calcite and dolomite: high-resolution spectrometric analysis of CL emission (HRS-CL). Sediment Geol 101:1–7
31. Hanchar J, Miller C (1993) Zircon zonation patterns as revealed by cathodoluminescence and backscattered electron images: implications for interpretation of complex crustal histories. Chem Geol 110:1–13
32. Hanchar J, Rudnick R (1995) Revealing hidden structures: the application of cathodoluminescence and back-scattered electron imaging to dating zircons from lower crustal xenoliths. Lithos 36:289–303
33. Hanchar JM, Finch RJ, Hoskin PWO, Watson EB, Cherniak DJ, Mariano AN (2001) Rare earth elements in synthetic zircon: part 1. Synthesis, and rare earth element and phosphorus doping. Am Mineral 86:667–680

34. Jardin R, Pavel CC, Raison PE, BouëxièreD, Santa-CruzH, Konings RJM, Popa K (2008) The high-temperature behaviour of PuPO4 monazite and some other related compounds. J Nucl Mater 378:167–171
35. Kempe U, Götze J (2002) Cathodoluminescence (CL) behaviour and crystal chemistry of apatite from rare-metal deposits. Mineral Mag 66:151–172
36. Kenyon A (2002) Recent developments in rare-earth doped materials for optoelectronics. Prog Quant Electron 26:225–284
37. Kolitsch U, Holtstam D (2004) Crystal chemistry of REEXO4 compounds (X= P, as, V). II. Review of REEXO4 compounds and their stability fields. Eur J Mineral 16:117–126
38. Kucha H (1980) Continuity in the monazite-huttonite series. Mineral Mag 43:1031–1034
39. Li H, Zhang S, Zhou S, Cao X (2009) Bonding characteristics, thermal expansibility, and compressibility of RXO4 (R = rare earths, X = P, as) within monazite and zircon structures. Inorg Chem 48:4542–4548
40. Liu G, Jacquier B (2005) Spectroscopic properties of rare earths in optical materials. Springer, Berlin, 550 p Factors affecting the Nd3+ (REE3+) luminescence of minerals 427
41. Macfarlane R (1990) Inhomogeneous broadening of spectral lines in doped insulators. J Lumin 45:1–5
42. MacRae CM, Wilson NC, Johnson SA, Phillips PL, Otsuki M (2005) Hyperspectral mapping –combining cathodoluminescence and Xray collection in an electron microprobe. Microsc Res Techniq 67:271–277
43. MacRea CM, Wilson NC, Torpy A, Davidson CJ (2012) Hyperspectral cathodoluminescence imaging and analysis extending from ultraviolet to near infrared. Microsc Microanal 18:1239–1245
44. Marfunin AS (1979) Spectroscopy, Luminescence, and Radiation Centers in Minerals. Springer, Berlin Heidelberg, New York, p 352
45. Mogilevsky P, Boakye E, Hay R (2007) Solid solubility and thermal expansion in a LaPO4-YPO4 system. J Am Ceram Soc 90:1899–1907
46. Nasdala L, Götze J, Hanchar J, Gaft M, Krbetschek M (2004) Luminescence techniques in earth sciences. In: Beran A, Libowitzky E (eds) Spectroscopic methods in mineralogy, vol 6, Eur mineral union notes in mineralogy. Eötvös Univ Press, Budapest, pp 43–91
47. Nasdala L, Kronz A, Hanchar JM, Tichomirova M, Davis DW, Hofmeister W (2006) Effects of natural radiation damage on back-scattered electron images of single crystals of minerals. Am Mineral 91:1739–1746
48. Neuser R, Bruhn F, Götze J, Habermann D, Richter D (1995) Kathodolumineszenz: methodik und anwendung. Zbl Geo Pal I 1995(1/2):287–306
49. Ohnenstetter D, Cesbron F, Rémond G, Caruba R, Claude JM (1991) Emission de cathodoluminescence de deux populations de zircons naturels: tentative d'interprétation. C R Acad Sci II 313:641–647
50. Ollier N, Concas G, Panczer G, Champagnon B, Charpentier T (2003) Structural features of a Eu3+ doped nuclear glass and gels obtained from glass leaching. J Non-Cryst Solids 328:207–214
51. Owen JJ, Cheetham AK, McFarlane RA (1998) Orientation-dependent fluorescence studies and spectroscopic analysis of doped barium yttrium fluoride upconversion laser crystals (BaY2-x-yYbxTmyF8). J Opt Soc Am B 15:684–693
52. Panczer G, De Ligny D, Mendoza C, Gaft M, Sexdoux-Guillaume A, Wang X (2012) Raman and fluorescence. In: Dubessy J, Caumon M-C, Rull F (eds) Applications of Raman spectroscopy to Earth sciences and cultural heritage. Eur Mineral Union Notes in Mineralogy, vol 12. Eur Mineral Union and Mineral Soc of Great Britain and Ireland, pp 61–82
53. Piriou B, Elfakir A, Quarton M (2001) Site-selective spectroscopy of Eu3+-doped sodium lead phosphate apatite. J Lumin 93:17–26
54. Porto S, Krishnan R (1967) Raman effect of corundum. J Chem Phys 47:1009–1013
55. Pouchou J-L, Pichoir F (1991) Quantitative analysis of homogeneous or stratified microvolumes applying the method "PAP". In:
56. Heinrich KFJ, Newbury DE (eds) Electron probe quantitation. Plenum Press, New York, pp 31–75

57. Rakovan J, Reeder RJ (1996) Intracrystalline rare earth element distributions in apatite: surface structural influences on incorporation during growth. Geochim Cosmochim Ac 60:4435–4445
58. Reisfeld R (2005) Spectroscopy of rare earth ions. In: Vaseashta A, Dimova-Malinovska D, Marshall JM (eds) Nanostructured and advanced materials for applications in sensor, optoelectronic and photovoltaic technology, Nato Sci Ser II math, vol 204. Springer, Berlin, pp 77–100
59. Reisfeld R, Gaft M, Boulon G, Panczer C, Jørgensen C (1996) Laserinduced luminescence of rare-earth elements in natural fluorapatites. J Lumin 69:343–353
60. Reisfeld R, Zigansky E, Gaft M (2004) Europium probe for estimation of site symmetry in glass films, glasses and crystals. Mol Phys 102:1319–1330
61. Richter D, Götte T, Götze J, Neuser R (2003) Progress in application of cathodoluminescence (CL) in sedimentary petrology. Miner Petrol 79:127–166
62. Richter D, Gorgen P, Götte T (2008) Monazite cathodoluminescence – a new tool for heavy mineral analysis of siliciclastic sedimentary rocks. Sediment Geol 209:36–41
63. Ruschel K, Nasdala L, Kronz A, Hanchar JM, Többens DM, Škoda R, Finger F, Möller A (2012) A Raman spectroscopic study on the structural disorder of monazite–(Ce). Miner Petrol 105:41–55
64. Seydoux-Guillaume AM, Wirth R, Nasdala L, Gottschalk M, Montel JM, Heinrich W (2002) An XRD, TEM and Raman study of experimentally annealed natural monazite. Phys Chem Miner 29:240–253
65. Shannon R (1976) Revised effective ionic radii and systematic studies of interatomic distances in halides and chalcogenides. Acta Crystallogr A 32:751–767
66. Skinner J, Moerner W (1996) Structure and dynamics in solids as probed by optical spectroscopy. J Phys Chem 100:13251–13262
67. Talla D, Beran A, Škoda R, Losos Z (2011) On the presence of OH defects in the zircon-type phosphate mineral xenotime, (Y, REE)PO4. Am Mineral 96:1799–1808
68. Tarashchan AN (1978) Luminescence of minerals. Naukova Dumka, Kiev, 296 p (in Russian)
69. Ternane R, Ferid M, Panczer G, Trabelsi-Ayadi M, Boulon G (2005) Site-selective spectroscopy of Eu3+-doped orthorhombic lanthanum and monoclinic yttrium polyphosphates. Opt Mater 27:1832–1838
70. Waychunas GA (2002) Apatite luminescence. In: Kohn ML, Rakovan J and Hughes JM (eds) Rev Mineral Geochem, vol 48. Mineral Soc Am, Washington DC, 701–742
71. Williams ML, Jercinovic MJ, Hetherington CJ (2007) Microprobe monazite geochronology: understanding geologic processes by integrating composition and chronology. Annu Rev Earth Planet Sci 35:137–175

INDEX